Hot Quarks 2016: Workshop for Young Scientists on the Physics of Ultrarelativistic Nucleus-Nucleus Collisions

Journal of Physics: Conference Series Volume 832

South Padre Island, Texas, USA
12 – 17 September 2016

ISBN: 978-1-5108-3986-1
ISSN: 1742-6588

Printed from e-media with permission by:

Curran Associates, Inc.
57 Morehouse Lane
Red Hook, NY 12571

Some format issues inherent in the e-media version may also appear in this print version.

Printed by Curran Associates, Inc. (2017)

For permission requests, please contact the Institute of Physics
at the address below.

Institute of Physics
Dirac House, Temple Back
Bristol BS1 6BE UK

Phone: 44 1 17 929 7481
Fax: 44 1 17 920 0979

techtracking@iop.org

Additional copies of this publication are available from:

Curran Associates, Inc.
57 Morehouse Lane
Red Hook, NY 12571 USA
Phone: 845-758-0400
Fax: 845-758-2633
Email: curran@proceedings.com
Web: www.proceedings.com

Table of contents

Volume 832

Hot Quarks 2016: Workshop for young scientists on the physics of ultrarelativistic nucleus-nucleus collisions

12–17 September 2016, South Padre Island, Texas, USA

Accepted papers received: 23 March 2017
Published online: 25 April 2017

Preface

Papers

Jets in the vacuum and in the medium

Heavy flavor, dileptons and photons

Correlations and Fluctuations

Relativistic hydrodynamics and collective phenomena

Baryons and Strangeness

IOP Conf. Series: Journal of Physics: Conf. Series **832** (2017) 011001 doi:10.1088/1742-6596/832/1/011001

Preface

The 7th edition of the Workshop for Young Scientists on the Physics of Ultra-relativistic Nucleus-Nucleus Collisions (**Hot Quarks 2016**) was held on South Padre Island, Texas, United States from September 12-17, 2016. Following the traditions of the conference, the meeting gathered almost 70 participants in the first years of their scientific careers. The present issue contains the proceedings of this workshop.

As in the past, the Hot Quarks workshop offered a unique atmosphere for lively discussions and interpretation of the current measurements from high-energy nuclear collisions. Dedicated time at the end of each session for questions, including anonymous questions from the "box", are crucial for this workshop. Recent results and upgrades at CERN's Large Hadron Collider (LHC) and Brookhaven's Relativistic Heavy Ion Collider (RHIC) were presented. Recent theoretical developments were also extensively discussed as well as the perspectives for future facilities such as the Facility for Antiproton and Ion Research (FAIR) at Darmstadt and the Electron-Ion Collider at Brookhaven. The conference's goal to provide a platform for young researchers to learn and foster their interactions was successfully met.

We wish to thank the sponsors of the Hot Quarks 2016 Conference, who supported the authors of this volume: European Laboratory for Particle Physics CERN (Switzerland), Cyclotron Institute at Texas A&M University (USA), ExtreMe Matter Institute EMMI (Germany), Helmholtz Association and GSI under grant VH-NG-822 (Germany), Helmholtz International Center for FAIR (Germany), National Science Foundation (USA), Netherlands Organization for Scientific Research (Netherlands), Nuclear Physics Institute of the CAS (Czech Republic), the Ministry of Education, Youth and Sport (Czech Republic) and 3 sponsors who wish to remain anonymous.

Javier López Albacete, Universidad de Granada (Spain)
Jana Bielcikova, Nuclear Physics Inst. of the Czech Academy of Sciences (Czech Republic)
Rainer J. Fries, Texas A&M University (USA)
Jiangyong Jia, Stony Brook University and Brookhaven National Laboratory (USA)
André Mischke, Utrecht University and Nikhef Amsterdam (The Netherlands)
Hannah Petersen, Goethe University, FIAS and GSI (Germany)
Lijuan Ruan, Brookhaven National Laboratory (USA)
Sevil Salur, Rutgers University, (USA)

IOP Conf. Series: Journal of Physics: Conf. Series **832** (2017) 011001 doi:10.1088/1742-6596/832/1/011001

IOP Conf. Series: Journal of Physics: Conf. Series **832** (2017) 011002 doi:10.1088/1742-6596/832/1/011002

Peer review statement

All papers published in this volume of *Journal of Physics: Conference Series* have been peer reviewed through processes administered by the proceedings Editors. Reviews were conducted by expert referees to the professional and scientific standards expected of a proceedings journal published by IOP Publishing.

IOP Conf. Series: Journal of Physics: Conf. Series **832** (2017) 012001 doi:10.1088/1742-6596/832/1/012001

Jets in heavy ion collisions

Christine Nattrass

Assistant Professor, Physics Department, University of Tennessee, Knoxville, TN, USA-37996

E-mail: christine.nattrass@utk.edu

Abstract. High energy collisions of heavy nuclei permit the study of nuclear matter at temperatures and energy densities so high that the fundamental theory for strong interactions, QCD, predicts a phase transition to a plasma of quarks and gluons. This matter, called a Quark Gluon Plasma (QGP), has been studied experimentally for the last decade and has been observed to be a strongly interacting liquid with a low viscosity. High energy partons created early in the collision interact with the QGP and provide unique probes of its properties. Hard partons fragment into collimated sprays of particles called jets and have been studied through measurements of single particles, correlations between particles, and measurements of fully reconstructed jets. These measurements demonstrate partonic energy loss in the QGP and constrain the QGP's properties. Measurements of the jet structure give insight into the mechanism of this energy loss. The information we have learned from studies of jets and challenges for the field will be reviewed.

1. Introduction

Jets are formed when a parton in a nucleon or nucleus scatters off of a parton in another nucleon or nucleus. After the scattering, the parton forms a parton shower which then hadronizes, leading to a collimated spray of hadrons. Jet production from collisions with a sufficiently high momentum transfer can be calculated perturbatively. The majority of hard scatterings are $2\rightarrow2$, meaning that a jet has a partner jet approximately 180° away in azimuth. Partons interacting with the Quark Gluon Plasma (QGP) formed in high energy nucleus-nucleus collisions may lose energy before the formation of the final hadrons, a process called jet quenching. Jet quenching is generally understood to occur dominantly through medium-induced gluon bremsstrahlung, with some contribution from collisional energy loss, leading to a broadening and softening of the distribution of particles in the jet. Quenching can be studied by measuring jets and the hadrons they comprise and this can be used to determine the properties of the QGP.

Jet quenching can be measured in heavy ion collisions using single high momentum hadrons, correlations between high momentum hadrons, fully reconstructed jets, and correlations between fully reconstructed jets and either other jets or hadrons. These measurements indicate that jets in heavy ion collisions are broadened and their fragmentation softened. However, a greater understanding of the bias in the jet population imposed by our measurement techniques and of cold nuclear matter effects is required to fully understand partonic energy loss in the QGP.

IOP Conf. Series: Journal of Physics: Conf. Series **832** (2017) 012001 doi:10.1088/1742-6596/832/1/012001

2. Single particle nuclear modification factor R_{AA}

Jet suppression can be quantified using measurements of the nuclear modification factor, R_{AA}

$$R_{AA} = \frac{dN^{AA}/dp_T}{T_{AA} d\sigma^{p+p}/dp_T} \tag{1}$$

where dN^{A+A}/dp_T is the spectrum of particles in A+A, $d\sigma^{pp}/dp_T$ is the cross section in p+p collisions, respectively, and $T_{AA} = N_{coll}/\sigma^{pp}_{inel}$ is nuclear overlap integral, the number of binary nucleon-nucleon collisions in an A+A collision divided by the inelastic cross section in p+p collisions. An $R_{AA} < 1$ indicates suppression and an $R_{AA} > 1$ indicates enhancement. Colored probes are expected to lose energy in the medium, but the QGP should be transparent to electromagnetic probes. At RHIC the direct photon R_{AA} is consistent with 1, indicating no energy loss in the medium [1]. All hadrons (unidentified hadrons [2–4], π^0 [5], η [6], ϕ [7], protons [8], J/ψ [9], ω [10], and $K^{\pm}$ [8]) and electrons from heavy flavor decays [11] have an $R_{AA} < 1$ for $p_T > 4$ GeV/c and approach a common value of $R_{AA} \approx 0.2$ at high momenta, consistent with substantial jet suppression. At the LHC single hadrons are also observed to be suppressed [12–14] while electromagnetic probes including direct photons [15], the W [16], and the Z [17] have an R_{AA} consistent with one.

Energy loss of a parton in the medium is defined theoretically by $\hat{q} = Q^2/L$ where Q is the momentum transfer from the parton to the medium and L is the path length traversed by the parton [18]. The single hadron R_{AA} was used by the JET collaboration to constrain the value of $\hat{q}$ to 1.2 ± 0.3 GeV^2/fm in Au+Au at $\sqrt{s_{\mathrm{NN}}} = 200$ GeV and 1.9±0.7 GeV^2/fm in Pb+Pb collisions at $\sqrt{s_{\mathrm{NN}}} = 2.76$ TeV for a 10 GeV/c quark at $\tau_0 = 0.6$ fm/c [19].

3. Measurements of reconstructed jets

All experiments primarily use the anti-k_T algorithm [20] to identify jet candidates, but there are several approaches to address the large combinatorial background from unrelated processes, called the underlying event, in heavy ion collisions. Without any background suppression and subtraction techniques, jet candidates may either be distorted by the inclusion of particles from the underlying event or be comprised entirely of particles from the underlying event. For R_{AA} measurements ALICE suppresses the background by requiring a hadron with $p_T > 5$ GeV/c to be included in the jet, measuring the average background by drawing random cones in the event, and subtracting the average background from the reconstructed jet energy [21]. ATLAS [21] and CMS [22] both have iterative procedures, first finding jet candidates, then calculating the estimated background by omitting high energy jet candidates from the background subtraction. For R_{AA} measurements ATLAS additionally requires calorimeter jets to be matched to a track jet comprised of tracks above 4 GeV/c. When the R_{AA} from each of these measurements is compared, the ALICE and CMS measurements are comparable and all measurements are in agreement at 100 GeV/c, but the ATLAS R_{AA} is roughly twice as high as the other measurements at 60 GeV/c. This is likely due to the different biases imposed by the methods.

The di-jet asymmetry is defined as the difference between the momenta of the leading and subleading jets divided by the sum of the momenta:

$$A_j = \frac{p_T^{leading} - p_T^{subleading}}{p_T^{leading} + p_T^{subleading}}. \tag{2}$$

An alternate definition uses the energies of the jets instead of the momenta. Even in p+p collisions, dijets' energies will not be perfectly balanced all of the time and there will be some spread of the distribution of A_J about zero. A_J in A+A collisions is therefore compared to that in p+p collisions. In A+A collisions, the leading and subleading jets' energies are even more

IOP Conf. Series: Journal of Physics: Conf. Series **832** (2017) 012001 doi:10.1088/1742-6596/832/1/012001

imbalanced, with the distribution of A_J no longer even centered about zero, demonstrating that jets are quenched in the QGP [23, 24]. When lower momentum particles are included in the reconstructed jets, the energy balance is restored [25]. This is one of the indications that the fragmentation function is softened in heavy ion collisions.

Fragmentation functions are the distribution of particles in a jet, usually measured as a function of $z = p_T^{hadron}/E_{jet}$ or $\xi = ln(1/z)$, and can be measured directly usiung fully reconstructed jets. These measurements also indicate that the fragmentation function is softened [26–28].

4. Di-hadron correlations

Jets can also be studied by looking at the correlations between two sufficiently high mometum hadrons. A high momentum trigger particle is used to define the coordinate system and the distribution of particles relative to that trigger particle is measured. There is a near-side peak arising from the same jet as that which created the trigger particle and an away-side peak centered around 180° away arising from the partner jet. These measurements allow studies of lower momentum jets but have a fairly low signal to background. Early studies were useful for demonstrating jet suppression at RHIC [29–33,33,34], before full jet reconstruction was possible. These measurements have a large combinatorial background which is correlated with the signal, quantified by the Fourier coefficients of the decomposition of the distribution of particles relative to the reaction plane, v_n. At low momenta, the non-zero v_n are understood to arise largely due to hydrodynamical flow of the medium [35–37]. A large fraction of published studies did not take odd v_n into account for the background subtraction, which does not qualitatively change the interpretation at high momenta (where the signal to background ratio is high) but substantially modifies the interpretation at lower momenta. Lower momenta data [38, 39] have been reanalyzed using a different background subtraction technique [40]. The reanalyzed data demonstrate that the away-side peak is strongly suppressed in Au+Au collisions and somewhat broader than that in d+Au collisions [41]. This is in contrast to the complete suppression seen at intermediate momenta [30] and the “Mach cone” structure seen at lower momenta [33, 33, 38, 42, 43], but consistent with other measurements demonstrating a softening and broadening of the fragmentation function.

5. Issues with the definition and bias

In order to compare to theory, even in elementary collisions, it is necessary for both theory and experiment to use the same jet finding algorithm [44]. The parton shower and the process of hadronization can move particles in or out of the jet cone. Even in p+p collisions, hadronization must be taken into account in perturbative calculations of jet cross sections in order to describe the data [45]. A measurement of a jet therefore should not be thought of as a direct measurement of the parton.

In addition, the measurement techniques introduced to suppress and subtract background in heavy ion collisions, introducing a bias in the surviving population of jets. This bias should be considered part of the definition of the measurement. Most measurements focus on narrow cone radii ($R < 0.3$) and either explicitly or implicitly require hard fragments. Since the background scales with the area of the jet, jets with narrow cone radii have less background and fewer fluctuations in the background. Selection of harder particles also reduces the background. However, these approaches bias the jet population towards quark jets rather than gluon jets. Measurements of jets in e^+e^- collisions indicate that gluon jets are broader and fragment into more, softer particles than quark jets [46, 47]. Even though there are more gluon jets produced, the bias towards quark jets may be significant in some kinematic regions. The selection of jets which fragment harder and into narrow cones also imposes a survivor bias, where jets that are retained in the population are less likely to have interacted significantly with the medium. By looking only at jets which do not look like the background, we are likely missing the most severely modified jets. These biases and their impact on the jet population needs to be understood better.

6. Cold nuclear matter effects

The goal of measurements of jets in heavy ion collisions is to look for the impact of hot nuclear matter on the parton, so any possible modifications of jet production in cold nuclear matter need to be understood for proper interpretation of measurements in $A+A$ collisions. However, there are a number of indications of cold nuclear matter effects, including indications of collective effects [48–50], suppression of the J/ψ [51] at midrapidity, and suppression of particles from the decays of hadrons containing heavy quarks at forward rapidities [52, 53]. The minimum bias nuclear modification factors in p+Pb and d+Au collisions, R_{pPb} and R_{dAu}, are consistent with one at midrapidity for single charged hadrons [54–57] and fully reconstructed jets [58–60], indicating no cold nuclear matter effects at midrapidity. ATLAS studies at forward rapidities indicate a centrality dependence that cannot be fully explained by a bias in the centrality selection [58, 59]. This indicates that cold nuclear matter effects are not fully understood and measurements need to be interpreted with care, particularly at forward rapidities.

7. Conclusions

Jet quenching is observed through measurements of the single particle R_{AA}, measurements of di-hadron correlations, and measurements of fully reconstructed jets. These measurements indicate that jets are broadened and their fragments are softened in heavy ion collisions relative to jets in $p+p$ collisions, as expected if partons lose energy through gluon bremsstrahlung or collisional energy loss. Our understanding of partonic energy loss has improved dramatically with the wealth of measurements and theoretical calculations, particularly through the quantitative constraint of $\hat{q}$ by the JET collaboration.

However, the biases in jet definitions imposed by the measurement techniques make it difficult to directly compare results for different experiments and to interpret the results. Nearly all measurement techniques bias the jet population towards jets which have had fewer interactions with the medium and towards quark jets. These effects need to be understood better quantitatively. This requires a significant effort from both theorists and experimentalists to understand the impact of the measurement technique. Furthermore, the observation of cold nuclear matter effects warrants further study in order to properly interpret results from hot nuclear matter.

Acknowledgments

I am grateful to Sevil Salur, Megan Connors, and Rosi Reed for productive conversations, which informed much of the discussion above, and to Adam Matyja and Redmer Bertens for comments on the manuscript. This work was supported in part by funding from the Division of Nuclear Physics of the U.S. Department of Energy under Grant No. DE-FG02-96ER40982.

References

[1] Afanasiev S *et al.* (PHENIX) 2012 *Phys. Rev. Lett.* **109** 152302 (*Preprint* `1205.5759`)
[2] Adams J *et al.* (STAR) 2003 *Phys.Rev.Lett.* **91** 172302 (*Preprint* `nucl-ex/0305015`)
[3] Adler S *et al.* (PHENIX) 2003 *Phys.Rev.Lett.* **91** 072301 (*Preprint* `nucl-ex/0304022`)
[4] Back B *et al.* (PHOBOS) 2004 *Phys.Rev.* **C70** 061901 (*Preprint* `nucl-ex/0406017`)
[5] Adare A *et al.* (PHENIX) 2008 *Phys. Rev. Lett.* **101** 232301 (*Preprint* `0801.4020`)
[6] Adare A *et al.* (PHENIX) 2010 *Phys. Rev.* **C82** 011902 (*Preprint* `1005.4916`)
[7] Adare A *et al.* (PHENIX) 2016 *Phys. Rev.* **C93** 024904 (*Preprint* `1509.06337`)
[8] Adare A *et al.* (PHENIX) 2013 *Phys. Rev.* **C88** 024906 (*Preprint* `1304.3410`)
[9] Adare A *et al.* (PHENIX) 2007 *Phys. Rev. Lett.* **98** 232301 (*Preprint* `nucl-ex/0611020`)
[10] Adare A *et al.* (PHENIX) 2011 *Phys. Rev.* **C84** 044902 (*Preprint* `1105.3467`)

IOP Conf. Series: Journal of Physics: Conf. Series **832** (2017) 012001 doi:10.1088/1742-6596/832/1/012001

[11] Adare A *et al.* (PHENIX Collaboration) 2011 *Phys.Rev.* **C84** 044905 (*Preprint* 1005.1627)
[12] Aamodt K *et al.* (ALICE) 2011 *Phys.Lett.* **B696** 30–39 (*Preprint* 1012.1004)
[13] Chatrchyan S *et al.* (CMS) 2012 *Eur.Phys.J.* **C72** 1945 (*Preprint* 1202.2554)
[14] 2016 Measurement of the charged particle nuclear modification factor in PbPb collisions at sqrt(sNN) = 5.02 TeV Tech. Rep. CMS-PAS-HIN-15-015 CERN Geneva URL cds.cern.ch/record/2155301
[15] Chatrchyan S *et al.* (CMS) 2012 *Phys. Lett.* **B710** 256–277 (*Preprint* 1201.3093)
[16] Chatrchyan S *et al.* (CMS) 2012 *Phys. Lett.* **B715** 66–87 (*Preprint* 1205.6334)
[17] Chatrchyan S *et al.* (CMS) 2011 *Phys. Rev. Lett.* **106** 212301 (*Preprint* 1102.5435)
[18] Salgado C A and Wiedemann U A 2003 *Nucl. Phys.* **A715** 783–786 (*Preprint* hep-ph/0209025)
[19] Burke K M *et al.* (JET) 2014 *Phys.Rev.* **C90** 014909 (*Preprint* 1312.5003)
[20] Cacciari M, Salam G P and Soyez G 2008 *JHEP* **04** 063 (*Preprint* 0802.1189)
[21] Adam J *et al.* (ALICE) 2015 *Phys.Lett.* **B746** 1–14 (*Preprint* 1502.01689)
[22] Khachatryan V *et al.* (CMS) 2016 (*Preprint* 1609.05383)
[23] Aad G *et al.* (ATLAS) 2010 *Phys.Rev.Lett.* **105** 252303 (*Preprint* 1011.6182)
[24] Chatrchyan S *et al.* (CMS) 2011 *Phys. Rev.* **C84** 024906 (*Preprint* 1102.1957)
[25] Adamczyk L *et al.* (STAR) 2016 (*Preprint* 1609.03878)
[26] Aad G *et al.* (ATLAS) 2014 *Phys.Lett.* **B739** 320–342 (*Preprint* 1406.2979)
[27] Chatrchyan S *et al.* (CMS) 2014 *Phys.Rev.* **C90** 024908 (*Preprint* 1406.0932)
[28] Chatrchyan S *et al.* (CMS) 2012 *JHEP* **1210** 087 (*Preprint* 1205.5872)
[29] Adler C *et al.* (STAR) 2003 *Phys.Rev.Lett.* **90** 082302 (*Preprint* nucl-ex/0210033)
[30] Adams J *et al.* (STAR) 2003 *Phys. Rev. Lett.* **91** 072304 (*Preprint* nucl-ex/0306024)
[31] Adams J *et al.* (STAR) 2004 *Phys.Rev.Lett.* **93** 252301 (*Preprint* nucl-ex/0407007)
[32] Adams J *et al.* (STAR) 2006 *Phys.Rev.Lett.* **97** 162301 (*Preprint* nucl-ex/0604018)
[33] Adare A *et al.* (PHENIX) 2007 *Phys.Rev.Lett.* **98** 232302 (*Preprint* nucl-ex/0611019)
[34] Adler S *et al.* (PHENIX) 2006 *Phys.Rev.* **D74** 072002 (*Preprint* hep-ex/0605039)
[35] Abelev B *et al.* (STAR) 2009 *Phys. Rev.* **C80** 064912
[36] Agakishiev G *et al.* (STAR) 2012 *Phys. Rev.* **C85** 014903
[37] Abelev B *et al.* (STAR) 2010 *Phys. Lett.* **B683** 123–128
[38] Agakishiev H *et al.* (STAR) 2010 (*Preprint* 1010.0690)
[39] Agakishiev H *et al.* (STAR) 2014 *Phys. Rev.* **C89** 041901
[40] Sharma N, Mazer J, Stuart M and Nattrass C 2016 *Phys. Rev.* **C93** 044915 (*Preprint* 1509.04732)
[41] Nattrass C, Sharma N, Mazer J, Stuart M and Bejnood A 2016 *Phys. Rev.* **C94** 011901 (*Preprint* 1606.00677)
[42] Adare A *et al.* (PHENIX) 2008 *Phys.Rev.* **C77** 011901 (*Preprint* 0705.3238)
[43] Abelev B *et al.* (STAR) 2010 *Phys.Rev.Lett.* **105** 022301 (*Preprint* 0912.3977)
[44] Huth J E *et al.* 1990 *1990 DPF Summer Study on High-energy Physics: Research Directions for the Decade (Snowmass 90) Snowmass, Colorado, June 25-July 13, 1990* pp 0134–136 URL http://lss.fnal.gov/cgi-bin/find_paper.pl?conf-90-249
[45] Abelev B *et al.* (ALICE) 2013 *Phys.Lett.* **B722** 262–272 (*Preprint* 1301.3475)
[46] Akers R *et al.* (OPAL) 1995 *Z. Phys.* **C68** 179–202
[47] Abreu P *et al.* (DELPHI) 1996 *Z. Phys.* **C70** 179–196
[48] Adam J *et al.* (ALICE) 2016 *Phys. Lett.* **B753** 126–139 (*Preprint* 1506.08032)
[49] Khachatryan V *et al.* (CMS) 2015 *Phys. Rev. Lett.* **115** 012301 (*Preprint* 1502.05382)
[50] Aad G *et al.* (ATLAS) 2014 *Phys. Rev.* **C90** 044906 (*Preprint* 1409.1792)
[51] Adare A *et al.* (PHENIX) 2013 *Phys. Rev. Lett.* **111** 202301 (*Preprint* 1305.5516)
[52] Adare A *et al.* (PHENIX) 2012 *Phys. Rev. Lett.* **109** 242301 (*Preprint* 1208.1293)
[53] Adare A *et al.* (PHENIX) 2014 *Phys.Rev.* **C89** 034915 (*Preprint* 1311.1427)
[54] Adler S S *et al.* (PHENIX) 2007 *Phys. Rev. Lett.* **98** 172302 (*Preprint* nucl-ex/0610036)
[55] Abelev B I *et al.* (STAR) 2010 *Phys. Rev.* **C81** 064904 (*Preprint* 0912.3838)
[56] Abelev B *et al.* (ALICE) 2013 *Phys. Rev. Lett.* **110** 082302 (*Preprint* 1210.4520)
[57] Aad G *et al.* (ATLAS) 2016 (*Preprint* 1605.06436)
[58] Adare A *et al.* (PHENIX) 2016 *Phys. Rev. Lett.* **116** 122301 (*Preprint* 1509.04657)
[59] Aad G *et al.* (ATLAS) 2014 (*Preprint* 1412.4092)
[60] Adam J *et al.* (ALICE) 2016 *Eur. Phys. J.* **C76** 271 (*Preprint* 1603.03402)

IOP Conf. Series: Journal of Physics: Conf. Series **832** (2017) 012002 doi:10.1088/1742-6596/832/1/012002

Towards a quantitative understanding of high p_T flow harmonics

Jorge Noronha

Instituto de Física, Universidade de São Paulo, C.P. 66318, 05315-970 São Paulo, SP, Brazil

E-mail: noronha@if.usp.br

Abstract. In this proceedings I briefly review the recent progress achieved on the calculation of v_n at high p_T via the coupling of a jet energy loss model with full event-by-event viscous hydrodynamics. It is shown that this framework can simultaneously describe experimental data for R_{AA}, v_2, and v_3 at high p_T. High p_T v_2 is found to be approximately linearly correlated with the soft v_2 on an event-by-event basis, which opens up a new way to correlate soft and hard observables in heavy ion collisions.

1. Introduction

The Quark-Gluon Plasma (QGP) formed in heavy ion collisions is the smallest (and the hottest) most perfect fluid ever made. Arguably, evidence for the formation of the QGP comes from three fronts: (i) the equation of state (EOS) computed using *lattice QCD* [1] shows that at temperatures $T > 200$ MeV (easily achievable in these collisions) quarks and gluons are not confined into hadrons, (ii) the large *anisotropic flow* of low p_T hadrons, computed using event-by-event viscous hydrodynamics simulations (for reviews see [2, 3]), requires the formation of a nearly inviscid medium that is not consistent with purely hadronic expectations [4–6], (iii) the large suppression of high p_T hadrons in AA collisions with respect to elementary pp collisions, a simple consequence of in medium *jet quenching* [7, 8]. The three items above are not disconnected. For instance, the EOS is used in the hydrodynamic modeling of the evolving QGP from which anisotropic flow coefficients are computed and this hydrodynamically expanding fluid serves as a background for the passage of jets.

While the QGP equation of state at zero baryon chemical potential is under control [9] and current event-by-event hydrodynamic simulations have achieved unprecedented levels of success [10] (and predictive power [11, 12]), the mechanism involving both the soft and hard phenomena responsible for the generation of anisotropic flow at high p_T remained somewhat elusive (see the discussion in [13]) and only now some of its features have become clearer. For instance, it was shown in [14, 15] that realistic event-by-event hydrodynamic modeling plays an important role in solving the high p_T $R_{AA} \otimes v_2$ puzzle (see [13] and refs. therein) in heavy ion collisions.

2. Results from the event-by-event viscous hydrodynamics + jet energy loss model

We use the v-USPhydro code [16–18] to model the expanding QGP fluid produced in PbPb collisions at $\sqrt{s_{NN}} = 2.76$ TeV and solve the viscous hydrodynamic equations event-by-event. The details about the parameters of the hydrodynamic calculations can be found in [14]. In order to vary the size of the initial energy density eccentricities present in the initial conditions

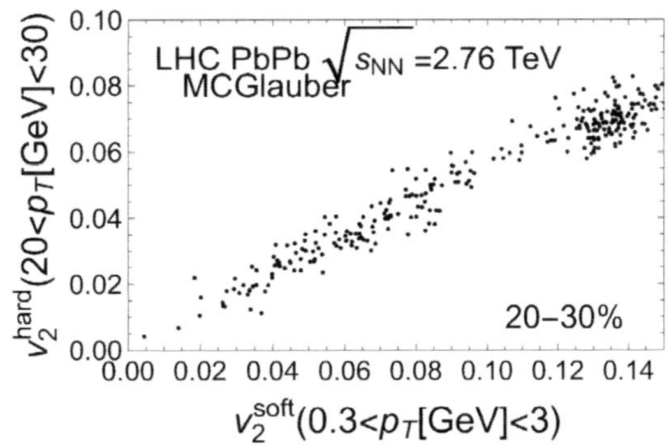

Figure 1. Correlation between the soft and the hard elliptic flow v_2 event-by-event.

for hydrodynamics, we used MCGlauber and MCKLN initial conditions [19] for the mid-central $20-30\%$ centrality class at the LHC. Our results for the low p_T soft v_2 and v_3 provide a good description of the data [20] (the shear viscosity over entropy density ratio is $\eta/s = 0.08$ in MCGlauber and $\eta/s = 0.11$ in MCKLN), as one can see in Fig. 1 of [14]. Once the hydrodynamic background is fixed and the low p_T azimuthal anisotropies are reproduced, one can use the spacetime profile of the hydrodynamic fields, computed event-by-event, in energy loss calculations.

Here the azimuthally averaged nuclear modification factor, $R_{AA}(p_T)$, and the high p_T azimuthal anisotropies (defined via a Fourier expansion of $R_{AA}(p_T,\phi)$) are investigated using the BBMG jet-energy loss [13, 21, 22]. In this model, the parton energy loss per unit length, dE/dL, is modeled as $\frac{dE}{dL} = -\kappa\, E^a(L)\, L^z\, T^c\, \zeta_q\, \Gamma_{\rm flow}$, where κ is the jet-medium coupling for quarks and gluons [13], T is the local temperature field along the jet path in the medium (with $c = 2+z-a$), ζ_q describes energy loss fluctuations [13], and the flow factor $\Gamma_{\rm flow}$ takes into account the boost from the local rest frame of the fluid. The parameters of the energy loss rate correspond to the pQCD-case defined in [13] ($a = 0, z = 1, c = 3, q = 0$), which gives a linear path length dependence for the energy loss, $dE/dL \sim L$. Effects from a quadratic path length dependence were investigated in [15].

High p_T azimuthal coefficients are defined via a correlation between soft and hard particles over many events, which necessarily implies that the geometrical fluctuations present in the soft sector are carried over to the hard sector. In fact, while the initial state eccentricities drive the soft flow harmonics in hydrodynamics through pressure and flow gradients, azimuthal momentum anisotropies at high p_T carry information about the initial state due to differences in the path length. Thus, one expects to find an approximate linear correlation between the soft and the hard elliptic flows event-by-event, since both are generated by the fluctuating initial spatial eccentricity ε_2. Such a linear correlation can be clearly seen in our model calculations displayed in Fig. 1 where each point corresponds to a hydro event (which contains many jets). This behavior motivated the more detailed study involving 4-particle cumulants of high p_T elliptic flow, which involves one hard particle correlated with 3 soft ones, performed in [15].

A comparison between our results for the π^0 $R_{AA}(p_T)$, (b) $v_2^{exp}(p_T)$, (c) $v_3^{exp}(p_T)$ in mid-central $\sqrt{s_{NN}} = 2.76$ TeV PbPb collisions at the LHC and the experimental data [23–27] can be found in Fig. 2. In these plots, v_2^{exp} and v_3^{exp} are computed via a soft-hard 2-particle correlator [3, 15], which correctly takes into account the way the high p_T measurement is performed. MCKLN initial conditions are solid red curves while the dotted-dashed black line stands for MCGlauber calculations. The black dotted line $\langle$MCGlauber$\rangle$ corresponds to an event averaged smoothed initial Glauber geometry, shown here for comparison. One can see that the results computed using MCKLN initial conditions provide a good description of the data. In fact, given

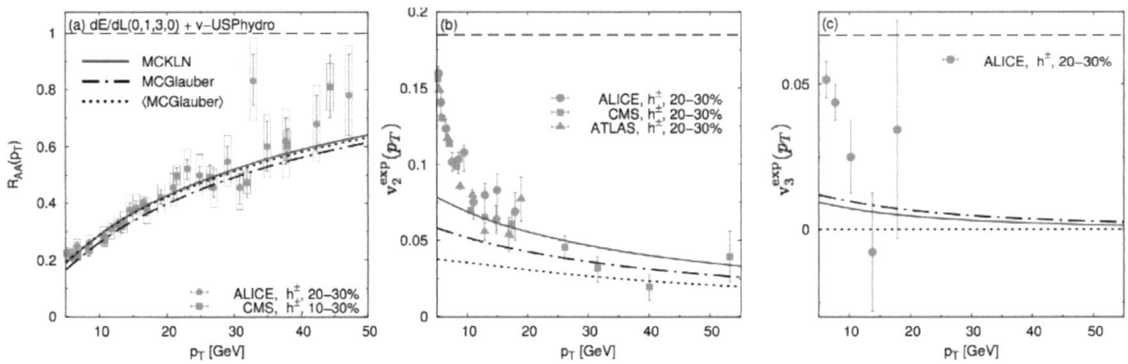

Figure 2. (Color online) Comparison between our results for the π^0 $R_{AA}(p_T)$, (b) $v_2^{exp}(p_T)$, (c) $v_3^{exp}(p_T)$ in mid-central $\sqrt{s_{NN}} = 2.76$ TeV PbPb collisions at the LHC to data [23–27].

that the high p_T flow harmonics are also determined by the initial eccentricities, initial conditions with an ε_2 larger than that found in MCGlauber, such as MCKLN (or IP-Glasma [10]), should provide a good description of v_2 at high p_T. On the other hand, the initial triangularity ε_3 is generally anti-correlated with ε_2 and this is why in Fig. 2 v_3 in MCGlauber is a bit larger than that found using MCKLN initial conditions.

3. Outlook

With the advent of realistic event-by-event viscous hydrodynamics + energy loss calculations [14, 15], it has become possible for the first time to simultaneously describe R_{AA}, v_2, and v_3 at high p_T in heavy ion collisions. This validates the idea [28] that jet energy loss determines the azimuthal anisotropy of the QGP at high p_T.

Many obvious improvements in our current model are needed. On the hydro side, other initial conditions such as IP-Glasma [10] and Trento [29] could be implemented and one needs to investigate the effects of a T-dependent η/s and bulk viscosity [30, 31] as well. In this regard, we remark that the effect of higher order transport coefficients [32] in the dynamical evolution of the non-conformal strongly coupled QGP remains largely unexplored. Furthermore, full 3+1 hydrodynamic evolution would be needed to compute the rapidity dependence of high p_T flow harmonics, in contrast to the boost-invariant scenario implemented here.

Regarding the energy loss model, even though our current implementation is extremely simplistic, it does seem to possess the necessary features to describe high p_T flow harmonics. With the new LHC data at $\sqrt{s_{NN}} = 5.02$ TeV one may be able to distinguish energy loss models with linear path length dependence from models where $dE/dL \sim L^2$, as shown in [15]. A necessary next step involves using more realistic energy loss models which contain the expected weak coupling QCD features as well non-perturbative effects from various sources (see [33]). It would be interesting to see how realistic event-by-event viscous hydrodynamic modeling affects the flow harmonics in the heavy flavor sector [34, 35].

By properly taking into account the effect of initial state fluctuations in the hydrodynamic medium, new observables [15] involving the correlation between soft and hard flow harmonics can now be investigated, in contrast to all the previous calculations that employed unrealistic, event-averaged hydrodynamic backgrounds. It would be interesting to see if the overall distribution of high p_T flow harmonics is similar to the one obtained at low p_T. At the moment, only calculations of the 4-particle cumulant $v_2\{4\}$ at high p_T [15] have been performed and higher order cumulants would be needed to assess the information contained in the event-by-event distributions. Better theoretical control of the fluctuations of flow harmonics at high p_T can be useful to distinguish between different energy loss models.

The type of analysis performed here must also be done in different collisions systems and different energies (see [15] for the case of PbPb at $\sqrt{s_{NN}} = 5.02$ TeV). A challenging feat would be to perform realistic jet energy loss + event-by-event hydrodynamic calculations that can be used to simultaneously investigate the soft and the hard flow harmonics in small systems, such as pA collisions. Finally, concerning the complete understanding of flow harmonics in heavy ion collisions, one may now say that we have a good (quantitative) understanding of the underlying mechanisms responsible for the observed azimuthal anisotropies at low $p_T < 3$ GeV and high $p_T > 10$ GeV. The hardest problem of quantitatively describing the non-monotonic behavior of flow harmonics at intermediate p_T, which I in jest have called the "uncanny valley", requires a novel self-consistent way to couple jets with hydrodynamics on an event-by-event basis that goes way beyond the (modest) attempt pursued in [36]. The solution to this problem remains, to the best of my knowledge, unknown. Perhaps some of the brave young minds that have contributed to make Hot Quarks 2016 a wonderful experience will lead the way towards solving this problem.

Acknowledgements

J.N. thanks the University of Houston for its hospitality and FAPESP and CNPq for support.

References

[1] Borsanyi S, Fodor Z, Hoelbling C, Katz S D, Krieg S and Szabo K K 2014 *Phys. Lett.* **B730** 99–104
[2] Heinz U and Snellings R 2013 *Ann. Rev. Nucl. Part. Sci.* **63** 123–151 (*Preprint* 1301.2826)
[3] Luzum M and Petersen H 2014 *J. Phys.* **G41** 063102 (*Preprint* 1312.5503)
[4] Noronha-Hostler J, Noronha J and Greiner C 2009 *Phys. Rev. Lett.* **103** 172302 (*Preprint* 0811.1571)
[5] Noronha-Hostler J, Noronha J and Greiner C 2012 *Phys. Rev.* **C86** 024913 (*Preprint* 1206.5138)
[6] Denicol G S, Gale C, Jeon S and Noronha J 2013 *Phys. Rev.* **C88** 064901 (*Preprint* 1308.1923)
[7] Gyulassy M and Plumer M 1990 *Phys. Lett.* **B243** 432–438
[8] Wang X N and Gyulassy M 1992 *Phys. Rev. Lett.* **68** 1480–1483
[9] Bazavov A *et al.* (HotQCD) 2014 *Phys. Rev.* **D90** 094503 (*Preprint* 1407.6387)
[10] Gale C, Jeon S, Schenke B, Tribedy P and Venugopalan R 2013 *Phys. Rev. Lett.* **110** 012302
[11] Noronha-Hostler J, Luzum M and Ollitrault J Y 2016 *Phys. Rev.* **C93** 034912 (*Preprint* 1511.06289)
[12] Niemi H, Eskola K J, Paatelainen R and Tuominen K 2016 *Phys. Rev.* **C93** 014912 (*Preprint* 1511.04296)
[13] Betz B and Gyulassy M 2014 *JHEP* **08** 090 [Erratum: JHEP10,043(2014)] (*Preprint* 1404.6378)
[14] Noronha-Hostler J, Betz B, Noronha J and Gyulassy M 2016 *Phys. Rev. Lett.* **116** 252301
[15] Noronha-Hostler J e a 2016 (*Preprint* 1609.05171)
[16] Noronha-Hostler J, Denicol G S, Noronha J, Andrade R P G and Grassi F 2013 *Phys. Rev.* **C88** 044916
[17] Noronha-Hostler J, Noronha J and Grassi F 2014 *Phys. Rev.* **C90** 034907 (*Preprint* 1406.3333)
[18] Noronha-Hostler J, Noronha J and Gyulassy M 2016 *Phys. Rev.* **C93** 024909
[19] Drescher H J and Nara Y 2007 *Phys. Rev.* **C75** 034905 (*Preprint* nucl-th/0611017)
[20] Chatrchyan S *et al.* (CMS) 2014 *Phys. Rev.* **C89** 044906 (*Preprint* 1310.8651)
[21] Betz B, Gyulassy M and Torrieri G 2011 *Phys. Rev.* **C84** 024913 (*Preprint* 1102.5416)
[22] Betz B and Gyulassy M 2012 *Phys. Rev.* **C86** 024903 (*Preprint* 1201.0281)
[23] Abelev B *et al.* (ALICE) 2013 *Phys. Lett.* **B720** 52–62 (*Preprint* 1208.2711)
[24] Chatrchyan S *et al.* (CMS) 2012 *Eur. Phys. J.* **C72** 1945 (*Preprint* 1202.2554)
[25] Abelev B *et al.* (ALICE) 2013 *Phys. Lett.* **B719** 18–28 (*Preprint* 1205.5761)
[26] Chatrchyan S *et al.* (CMS) 2012 *Phys. Rev. Lett.* **109** 022301 (*Preprint* 1204.1850)
[27] Aad G *et al.* (ATLAS) 2012 *Phys. Lett.* **B707** 330–348 (*Preprint* 1108.6018)
[28] Gyulassy M, Vitev I and Wang X N 2001 *Phys. Rev. Lett.* **86** 2537–2540 (*Preprint* nucl-th/0012092)
[29] Moreland J S, Bernhard J E and Bass S A 2015 *Phys. Rev.* **C92** 011901 (*Preprint* 1412.4708)
[30] Ryu S e a 2015 *Phys. Rev. Lett.* **115** 132301 (*Preprint* 1502.01675)
[31] Bernhard J E, Moreland J S, Bass S A, Liu J and Heinz U 2016 *Phys. Rev.* **C94** 024907 (*Preprint* 1605.03954)
[32] Finazzo S I, Rougemont R, Marrochio H and Noronha J 2015 *JHEP* **02** 051 (*Preprint* 1412.2968)
[33] Xu J, Liao J and Gyulassy M 2016 *JHEP* **02** 169 (*Preprint* 1508.00552)
[34] Prado C A G *et al.* 2016 (*Preprint* 1609.06093)
[35] Prado C A G *et al.* 2016 (*Preprint* 1611.02965)
[36] Andrade R P G, Noronha J and Denicol G S 2014 *Phys. Rev.* **C90** 024914 (*Preprint* 1403.1789)

IOP Conf. Series: Journal of Physics: Conf. Series **832** (2017) 012003 doi:10.1088/1742-6596/832/1/012003

Direct photon-hadron correlation measurement and a way towards photon-triggered jets at RHIC

Nihar Ranjan Sahoo (For the STAR Collaboration)
Cyclotron Institute, Texas A&M University, USA
E-mail: nihar@rcf.rhic.bnl.gov

Abstract. We report the results of $\gamma_{\rm dir}$- and π^0–hadron azimuthal correlations as a measure of the away-side jet-like correlated yields in central Au+Au and p+p collisions at $\sqrt{s_{NN}}$ =200 GeV in the STAR experiment. The charged-hadron per-trigger yields at mid-rapidity with respect to high-p_T $\gamma_{\rm dir}$ and π^0 in central Au+Au collisions are compared with p+p collisions. Within uncertainties, the same z_T($p_T^{\rm assoc}/p_T^{\rm trig}$) dependence of the suppression is observed for $\gamma_{\rm dir}$- and π^0- triggers. The results are compared with energy-loss model predictions. The $\gamma-$jet measurements can provide further understanding on the redistribution of in-medium energy loss. Ongoing $\gamma-$jet studies in the STAR experiment are also discussed.

1. Introduction

The azimuthal correlation of charged hadrons with respect to a direct-photon ($\gamma_{\rm dir}$) trigger was proposed as a promising probe to study the mechanisms of parton energy loss [1]. Since a $\gamma_{\rm dir}$-trigger escapes without interacting with the medium, it approximates the initial energy of the recoil parton, which is subject to medium modifications. The recoil parton of a $\gamma_{\rm dir}$ trigger is a quark in leading-order QCD processes, whereas that of a high-p_T π^0 trigger can be a quark or a gluon. In addition, a coincidence measurement of π^0 is biased to have been produced near the surface of the medium, while that of a $\gamma_{\rm dir}$ does not suffer from the same bias [2, 3]. Hence the comparison between the suppression of per-trigger away-side associated yields of $\gamma_{\rm dir}$ to those of π^0 triggers should exhibit differences due to both the color-factor dependence and the path-length dependence of energy loss.

2. STAR detectors and experimental techniques

The data were taken by the Solenoidal Tracker at RHIC (STAR) experiment in 2011 and 2009 for Au+Au and $p+p$ collisions at $\sqrt{s_{NN}}$ = 200 GeV, respectively. The Time Projection Chamber (TPC) is the main charged-particle tracking detector providing track information for the charged hadrons with $|\eta| < 1.0$ [4]. Events having a cluster with transverse energy $E_T > 8$ GeV, with $|\eta| \leq 0.9$, are selected for this analysis in the Barrel Electromagnetic Calorimeter (BEMC) [5]. The associated charged particles are selected in range 1.2 GeV/$c < p_T^{\rm assoc}$, whereas π^0 and $\gamma_{\rm dir}$ are triggered within $12 < p_T^{\rm trig} <$ 20 GeV/c. In order to distinguish a π^0, which at high p_T predominately decays to two photons with a small opening angle, from a single-photon cluster, a transverse shower-shape analysis is performed. A detailed discussion about the transverse shower profile (TSP) method and experimental techniques used in this analysis can be found in Ref. [6, 7].

IOP Conf. Series: Journal of Physics: Conf. Series **832** (2017) 012003 doi:10.1088/1742-6596/832/1/012003

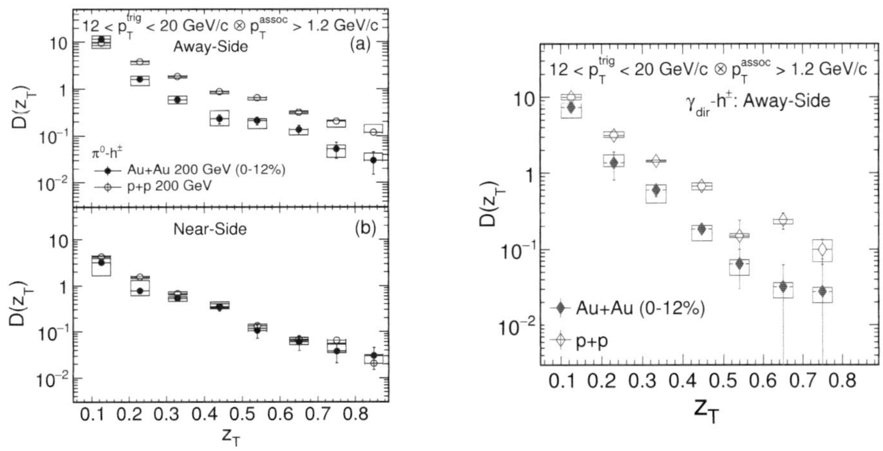

Figure 1. (Color online.) Left panel: The z_T dependence of π^0-$h^{\pm}$ away-side (a) and near-side (b) associated charged-hadron yields per trigger for Au+Au at 0-12% centrality (filled symbols) and $p+p$ (open symbols) collisions at $\sqrt{s_{NN}}$ = 200 GeV. Right panel: The z_T dependence of $\gamma_{\rm dir}$-$h^{\pm}$ away-side associated charged-hadron yields per trigger for Au+Au at 0-12% centrality (filled diamonds) and $p+p$ (open diamonds) collisions [6]. Vertical lines represent statistical errors, and the vertical extent of the boxes represents systematic uncertainties.

3. Results: $\gamma_{\rm dir}$- and π^0–hadron azimuthal correlation

The integrated away-side and near-side charged-hadron yields per π^0 trigger, $D(z_T)$, are plotted as a function of z_T, both for Au+Au (0-12% centrality) and $p+p$ collisions, in the left panel of Fig. 1. The away-side $D(z_T)$ for $\gamma_{\rm dir}$ triggers as a function of z_T for central Au+Au and minimum-bias $p+p$ collisions is shown in the right panel of Fig 1. Yields of the away-side associated charged hadrons are suppressed, in Au+Au relative to $p+p$, at all z_T except in the low z_T region both for $\gamma_{\rm dir}$ and π^0 trigger. On the other hand, no suppression is observed on the near-side in Au+Au, relative to $p+p$ collisions, due to the surface bias imposed by triggering on a high-p_T π^0. In order to quantify the medium modification for $\gamma_{\rm dir}$- and π^0-triggered recoil jet production as a function of z_T, the ratio defined as $I_{AA} = \frac{D(z_T)^{AuAu}}{D(z_T)^{pp}}$, of the per-trigger conditional yields in Au+Au to those in $p+p$ collisions is calculated.

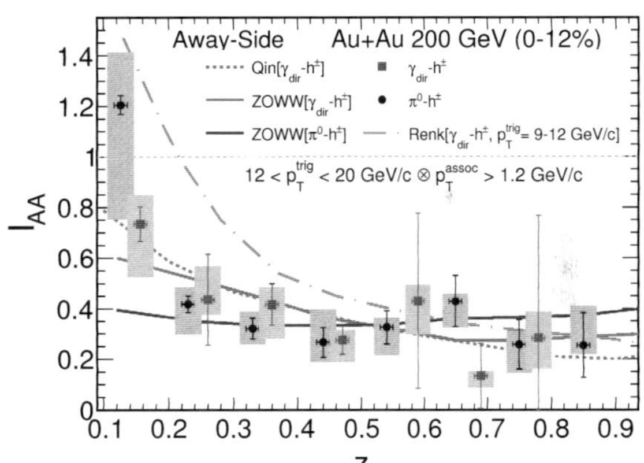

Figure 2. (Color online.) The $I_{AA}^{\gamma_{dir}}$ for $\gamma_{\rm dir}$ (red squares) and $I_{AA}^{\pi^0}$ for π^0 (blue circles) triggers are plotted as a function of z_T. The points for $I_{AA}^{\gamma_{dir}}$ are shifted by +0.03 in z_T for visibility. The vertical line and shaded boxes represent statistical and systematic errors, respectively [6]. The curves represent theoretical model predictions [3, 8, 9, 10].

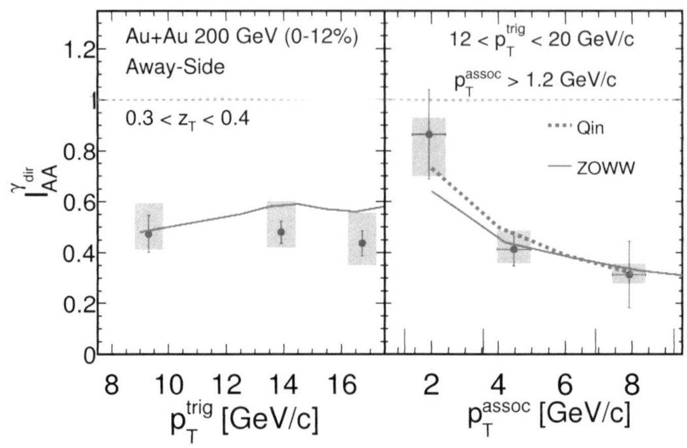

Figure 3. (Color online.) The values of $I_{AA}^{\gamma_{dir}}$ are plotted as a function of $p_T^{\rm trig}$ (left panel) and $p_T^{\rm assoc}$ (right panel) [6]. The vertical line and shaded boxes represent statistical and systematic errors, respectively. The curves represent theoretical model predictions [3, 8, 9].

Figure 2 shows the away-side medium modification factor for π^0 triggers ($I_{AA}^{\pi^0}$) and $\gamma_{\rm dir}$ triggers ($I_{AA}^{\gamma_{dir}}$), as a function of z_T. $I_{AA}^{\pi^0}$ and $I_{AA}^{\gamma_{dir}}$ show similar suppression within uncertainties. At low z_T ($0.1 < z_T < 0.2$), both $I_{AA}^{\pi^0}$ and $I_{AA}^{\gamma_{dir}}$ show an indication of less suppression than at higher z_T. This observation is not significant in the z_T-dependence of I_{AA} because the uncertainties in the lowest z_T bin are large. However, when I_{AA} is plotted vs. $p_T^{\rm assoc}$ (in Figure 3), the conclusion is supported with somewhat more significance. At high z_T, both $I_{AA}^{\pi^0}$ and $I_{AA}^{\gamma_{dir}}$ show a factor $\sim 3-5$ suppression. The ZOWW calculation also predicts $I_{AA}^{\gamma_{dir}}$ as a function of $p_T^{\rm trig}$ to be approximately flat in this range [3]. The YaJEM model predicts that at low $z_T = 0.2$, $I_{AA}^{\gamma_{dir}} = 1$ and rises above unity even at lower z_T, although at lower triggered p_T range 9-12 GeV$/c$.

The values of $I_{AA}^{\gamma_{dir}}$ are plotted as function of $p_T^{\rm assoc}$ in Fig. 3. It shows that the low-$p_T^{\rm assoc}$ hadrons on the away-side are not as suppressed as those at high $p_T^{\rm assoc}$. Both model predictions shown [3, 8], which do not include the redistribution of lost energy, are in agreement with the data. $I_{AA}^{\gamma_{dir}}$ shows no sensitivity as a function of $p_T^{\rm trig}$, for

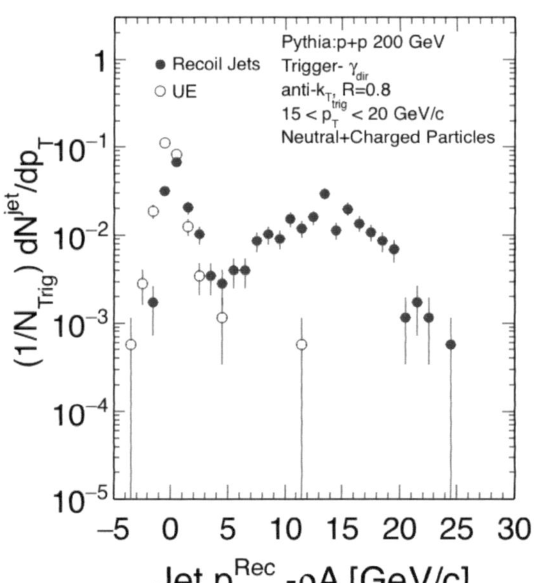

Figure 4. (Color online.) $\gamma_{\rm dir}$ triggered recoil (filled symbols) and uncorrelated (open symbols) reconstructed full jet transverse momentum spectra from PYTHIA simulation for p+p collisions at $\sqrt{s_{NN}} = 200$ GeV.

$0.3 < z_T < 0.4$, indicating that away-side parton energy loss is not sensitive to the initial energy of the parton in the range of 8-20 GeV/c.

4. Simulation study on $\gamma + \text{jet}$

We have performed a feasibility study for $\gamma + \text{jet}$ measurement in the kinematic acceptance for the STAR detector system using PYTHIA simulations. The γ-triggered events are generated within $15 < p_T^{\text{trig}} < 20$ GeV/c and all tracks are selected within $0.2 < p_{\text{T}}^{\text{track}} < 20$ GeV/c. Recoil full (including all charged and neutral particles) jets are reconstructed using the anti-k_T algorithm [11, 12] for a jet resolution parameter of R =0.8. Recoil jets are selected within $|\Delta\phi - \pi| < \pi/4$, where $|\Delta\phi|$ is the difference between triggered γ_{dir} and reconstructed jet azimuth. The uncorrelated reconstructed full jets (UE) are selected within $\Delta\phi \in [\pi/4, \pi/2]$ and $\Delta\phi \in [3\pi/2, 7\pi/4]$. The estimated background energy density (ρ) scaled by jet area (A) is subtracted from each reconstructed jet's raw transverse momentum, and the corrected value, $p_{T,jet}^{Rec} - \rho A$, is shown in Fig. 4. Similar analyses with respect to hadronic trigger objects are discussed in Refs [13, 14]. The triggered recoil jet peak, within $15 < p_{T,jet}^{Rec} - \rho A < 20$ GeV/c, can be seen around triggered p_T^{trig} of γ. In the STAR experiment, similar measurements for charged and full jets reconstruction using TPC and BEMC detectors is underway within wider acceptance in $|\eta| < 1$ and full azimuth.

5. Summary and outlook

Within experimental uncertainty, both $I_{AA}^{\pi^0}$ and $I_{AA}^{\gamma_{dir}}$ show similar levels of suppression with the expected differences due to the color-factor effect and the path-length dependence of in-medium energy loss not manifesting themselves. At high $z_T(p_T^{\text{assoc}})$, $I_{AA}^{\gamma_{dir}}$ shows high suppression than at low $z_T(p_T^{\text{assoc}})$. There is no trigger-energy dependence observed in the suppression of γ_{dir}-triggered yields, suggesting little dependence for energy loss on the initial parton energy, in the range of $p_T^{\text{trig}} =$ 8-20 GeV/c. A semi-inclusive study of jets correlated with high-p_T γ is underway in the STAR experiment to explore more on the jet energy loss in the medium created at RHIC.

References

[1] X.-N. Wang, Z. Huang, and I. Sarcevic, Phys. Rev. Lett. **77**, 231 (1996).
[2] T. Renk, Phys. Rev. C **74** (2006) 034906.
[3] H. Zhang, J.F. Owens, E. Wang, X.-N. Wang, Phys. Rev. Lett. **103** (2009) 032302.
[4] M. Anderson *et al.*, Nucl. Instrum. Methods A**499** (2003) 659.
[5] M. Beddo *et al.*, Nucl. Instrum. Methods A **499** (2003) 725.
[6] L. Adamczyk *et al.* (STAR Collaboration), Phys. Lett. B **760** (2016) 689.
[7] B. I. Abelev *et al.* (STAR Collaboration), Phys. Rev. C **82**, 034909 (2010).
[8] G.-Y. Qin, J. Ruppert, C. Gale, S. Jeon, G.D. Moore, Phys. Rev. C **80** (2009) 054909.
[9] X.-F. Chen, C. Greiner, E. Wang, X.-N. Wang, Z. Xu, Phys. Rev. C **81** (2010) 064908.
[10] T. Renk, Phys. Rev. C **80**, 014901 (2009).
[11] M.Cacciari, G. P. Salam, G. Soyez, JHEP **04**, 063 (2008).
[12] M.Cacciari, G. P. Salam, G. Soyez, JHEP **04**, 005 (2008).
[13] J. Adam, *et al.* (ALICE Collaboration), JHEP **09** (2015) 170.
[14] P. M. Jacobs and A. Schmah (for the STAR Collaboration), arXiv:1512.08784 [nucl-ex].

IOP Conf. Series: Journal of Physics: Conf. Series **832** (2017) 012004 doi:10.1088/1742-6596/832/1/012004

Jet structure modifications in heavy-ion collisions with JEWEL

Raghav Kunnawalkam Elayavalli

Department of Physics and Astronomy, Rutgers University, Piscataway, New Jersey 08854, USA

E-mail: raghav.k.e@cern.ch

Abstract. Key features of jet-medium interactions in heavy-ion collisions are modifications to the jet structure. Recent results from experiments at the LHC and RHIC have motivated several theoretical calculations and Monte Carlo models towards predicting these observables simultaneously. In this proceedings, the recoil picture in JEWEL is summarized and two independent procedures through which background subtraction can be performed in JEWEL are introduced. Information of the medium recoil in JEWEL significantly improves its description of several jet shape measurements.

1. Introduction

The qualitative effect of jet quenching through the measurements of the jet nuclear modification factor (R_{AA}) is confirmed with Run1 data at the LHC [1, 2, 3]. While the R_{AA} shows a clear and expected trend in the medium induced effects from central to peripheral events, the energy loss on a jet-by-jet level is not characterized. This motivated several measurements that probed the inner-structure of jets such as jet shapes [4, 5, 6], searches for the quenched energy away from the jet axis [7] and fragmentation functions [8, 9] to name a few. From these detailed measurements, jets that propagate the quark gluon plasma (QGP) were perceived as getting broader, losing energy inside the jet cone, having an increased multiplicity of low transverse momenta (p_T) particles around the periphery of the jet and several other consistent observations when compared to jets in pp collisions. The subjet groomed momentum fraction was recently measured as a function of the jet p_T and event centrality at CMS [10], concluding that jets were more asymmetrically split in heavy-ion collisions as opposed to those in pp collisions.

All these aforementioned results point to quantitative features of jet quenching, hence it is imperative that any phenomenological model describing the physics of QGP has to predict the general behavior and trend of these observables. Currently there are several theoretical calculations and Monte Carlo models available on the market (see [11] for a comprehensive review). These results offer a unique way of discriminating between these models.

JEWEL [12] is a Monte Carlo framework utilizing a perturbative quantum chromodynamical (pQCD) implementation of in-medium energy loss. The latest version of JEWEL is interfaced with PYTHIA6 [13] and the full framework operates as follows:

- PYTHIA produces the di-jet hard scattering and initial state radiation
- JEWEL takes these hard scattered partons and proceeds with the final state shower

IOP Conf. Series: Journal of Physics: Conf. Series **832** (2017) 012004 doi:10.1088/1742-6596/832/1/012004

- PYTHIA takes over for hadronization and hadron decays
- These events are then analyzed with the RIVET [14] analysis framework

The medium in JEWEL is characterized as a thermal distribution of scattering centers. The hard scattered partons undergo microscopic interactions that are estimated with perturbative matrix elements and the aforementioned parton shower modifications, giving rise to elastic and inelastic energy losses. At each interaction, a recoiled parton is created and in the current version of JEWEL, propagates without any further interaction. This is a limiting case, since in reality these recoiled partons can further interact with the medium. Detailed descriptions of the Monte Carlo implementation in JEWEL and corresponding studies are available in [12, 15].

The recoiled parton's energy is comprised of both the collisional part due from the hard scattered parton, and the thermal component of the scattering centers. This extraneous energy gets clustered in along with the other particles in the event and becomes part of the jets. Hence, when comparing data with JEWEL (including recoils), a slight mismatch appears for inter-jet observables due to the experimental implementation of the background subtraction. Since JEWEL does not simulate a full heavy ion event, the exact method utilized by different experiments cannot be reliably implemented. However, due to the microscopic nature of the interactions, the exact amount of background energy and momentum is easily estimated as the thermal component of the scattering centers before interaction. Any such subtraction techniques are only viable for infrared safe observables since the energy corresponding to the scattering centers are before hadronization effects.

2. Background subtraction in JEWEL

The information of the scattering centers before the interaction are first included in the event record with a separate tag, so as to not disturb the jet clustering. Simultaneously dummy particles with very small momenta and positions corresponding to the scattering centers are introduced to the final state particle collection. With this information, one can associate the scattering centers involved with the corresponding jet and hence proceed with background subtraction with either of the following methods:

- 4MomSub: Once the jet constituents are matched in position to their corresponding scattering centers, a simple four-momenta vectorial subtraction is performed to remove the background contribution to the respective jet.
- GridSub: A finite resolution grid is superimposed on the event confining the jet constituents and their scattering centers in grid cells. Inside each cell, the momenta of the scattering centers are vectorially subtracted and the jet is finally clustered with each cell as an input pseudo-particle. For the case when cells only contain scattering centers, their momentum is set to zero before the clustering procedure.

The 4MomSub is recommended when possible since it is closer to a true background subtraction. The GridSub method is employed in cases when observables require the information of jet constituents, and is very similar to experiments where the results are constrained by the detector's finite resolution. Detailed studies of the systematic uncertainties introduced by the background subtraction procedures are in preparation.

3. Comparisons with data

The ratio of the track yield in annuli around the jet axis in most central PbPb collisions at 2.76 TeV compared to pp collisions is shown in Fig. 1. The CMS data points [4] are shown in black markers while JEWEL+PYTHIA predictions are presented in blue dotted lines where the density is estimated with tracks and with green solid lines for density estimated with all final state particles. Due to the partonic nature of the background in JEWEL, the subtraction

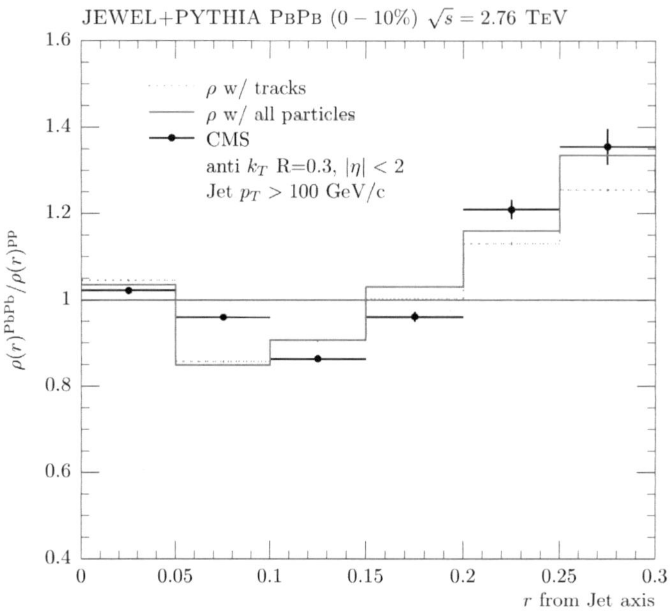

Figure 1. Differential jet shape ratio of PbPb to pp collisions comparing JEWEL+PYTHIA predictions with CMS data (black points). For the Monte Carlo predictions, the density estimated with only tracks vs. using all final state particles is shown in the blue dotted and green solid lines respectively. The data systematic uncertainties are shown in the yellow boxes around unity.

technique is efficient when considering both neutral and charged objects in the event. The data systematic uncertainty is presented in the yellow shaded region. The general trend of the data is reproduced by JEWEL+PYTHIA after subtraction using the 4MomSub method for anti-k$_T$ $R = 0.3$ jets clustered with the FASTJET [16] toolkit. As expected the agreement gets better with the data at large r when the density is estimated with all final state particles.

The JEWEL+PYTHIA predictions for the ratios of the subjet groomed momentum fractions in PbPb to pp collisions are compared with CMS data [10] as shown in Fig. 2. The subjet groomed momentum fractions are estimated via the softdrop framework [17], for low (left) and high (right) p_T anti-k$_T$ $R = 0.4$ jets respectively in Fig. 2. The ratio of JEWEL+PYTHIA to data is shown in the bottom panels where the yellow shaded region represents the total uncertainty in the data points. The systematic uncertainties in the JEWEL+PYTHIA predictions are shown by varying the grid resolution by a factor of two. JEWEL+PYTHIA also reproduces the general trend corresponding to more asymmetric jet splittings for jets in PbPb collisions at low p_T and more symmetric splittings as the jet p_T increases.

4. Conclusions

Including the recoils in JEWEL along with the background subtraction procedures allows the recovery of the jet energy acquired by the medium and its response to jets. This interplay makes JEWEL capable of reproducing the qualitative trend in data related to jet structure modifications. This new class of jet observables that probe the medium-jet interaction highlights the next era in jet tomography in heavy ion collisions.

This work was done in collaboration with Dr. Korinna Christine Zapp. RKE thanks the CERN theory department for its hospitality. This work was supported by Fundação para

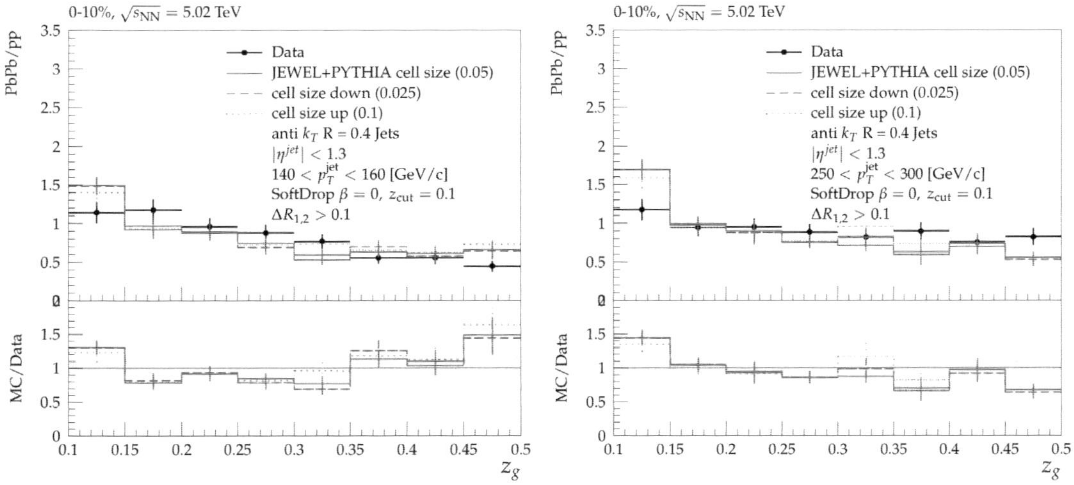

Figure 2. Comparison of JEWEL+PYTHIA predictions (colored lines) with CMS data (black markers) for the ratio of the subjet groomed momentum fraction distributions in central PbPb to pp events. The low and high p_T ranges are shown on the left and right respectively. The bottom panels present the ratio of the Monte Carlo predictions with data.

a Ciência e a Tecnologia (Portugal) under project CERN/FIS-NUC/0049/2015, postdoctoral fellowship SFRH/BPD/102844/2014 (KCZ) and by the European Union as part of the FP7 Marie Curie Initial Training Network MCnetITN (PITN-GA-2012-315877) (RKE). RKE also acknowledges support from the National Science Foundation under Grant No. 1352081, Claude Lovelace fellowship at Rutgers University and thanks Dr. Chun Shen for providing the initial hydrodynamics parameters for the event generation at 5.02 TeV.

References

[1] Aad G *et al.* (ATLAS) 2015 *Phys. Rev. Lett.* **114** 072302 (*Preprint* 1411.2357)
[2] Adam J *et al.* (ALICE) 2015 *Phys. Lett.* **B746** 1 (*Preprint* 1502.01689)
[3] Khachatryan V *et al.* (CMS) 2016 (*Preprint* 1609.05383)
[4] Khachatryan S *et al.* (CMS) 2014 *Phys. Lett.* **B730** 243 (*Preprint* 1310.0878)
[5] Cunqueiro L (ALICE) 2016 *Nucl. Phys.* **A956** 593 (*Preprint* 1512.07882)
[6] Khachatryan V *et al.* (CMS) 2016 (*Preprint* 1609.02466)
[7] Khachatryan V *et al.* (CMS) 2016 *JHEP* **01** 006 (*Preprint* 1509.09029)
[8] Khachatryan S *et al.* (CMS) 2014 *Phys. Rev.* **C90** 024908 (*Preprint* 1406.0932)
[9] Aad G *et al.* (ATLAS) 2014 *Phys. Lett.* **B739** 320 (*Preprint* 1406.2979)
[10] Collaboration C (CMS) 2016 *CMS-PAS-HIN-16-006*
[11] Mehtar-Tani Y 2016 *Nucl. Phys.* **A956** 168 (*Preprint* 1602.01047)
[12] Zapp K C 2014 *Eur. Phys. J.* **C74** 2762 (*Preprint* 1311.0048)
[13] Sjostrand T, Mrenna S and Skands P Z 2006 *JHEP* **05** 026 (*Preprint* hep-ph/0603175)
[14] Buckley A, Butterworth J, Lonnblad L, Grellscheid D, Hoeth H, Monk J, Schulz H and Siegert F 2013 *Comput. Phys. Commun.* **184** 2803 (*Preprint* 1003.0694)
[15] Zapp K C, Krauss F and Wiedemann U A 2013 *JHEP* **03** 080 (*Preprint* 1212.1599)
[16] Cacciari M, Salam G P and Soyez G 2012 *Eur. Phys. J.* **C72** 1896 (*Preprint* 1111.6097)
[17] Larkoski A J, Marzani S, Soyez G and Thaler J 2014 *JHEP* **05** 146 (*Preprint* 1402.2657)

IOP Conf. Series: Journal of Physics: Conf. Series **832** (2017) 012005 doi:10.1088/1742-6596/832/1/012005

Jet suppression measurement with the ATLAS detector

Radim Slovak on behalf of the ATLAS Collaboration

Charles University, Faculty of Mathematics and Physics, V Holesovickach 2, Prague 180 00, Czech Republic

E-mail: radim.slovak@cern.ch

Abstract. A hot medium with a high density of unscreened color charges is produced in relativistic heavy-ion collisions. Jets are produced at the early stages of these collisions and are known to become attenuated as they propagate through the hot matter. One manifestation of this energy loss is a lower yield of jets emerging from the medium than expected in the absence of medium effects. Another manifestation of the energy loss is the modification of the dijet balance and the modification of fragmentation functions. In these proceedings, the latest ATLAS results on single jet suppression, dijet suppression, and modification of the jet internal structure in Pb+Pb collisions are presented.

Relativistic heavy-ion collisions provide a unique opportunity to study the most intriguing aspects of the strong interaction in the laboratory. In high energy ion-ion collisions studied at RHIC and the LHC, the energy densities are sufficiently high to produce a new form of matter referred to as the quark-gluon plasma.

In the past few years, since the beginning of LHC Run 1, the ATLAS collaboration [1] has published several measurements showing strong modification of jet and hadron yields in lead-lead (Pb+Pb) collisions. Measurements of production of asymmetric dijets, single jets, measurement of jet fragmentation and production of neighbouring jets should provide new information on the mechanism of jet energy loss and modification of the jet substructure [2][3][4][5].

1. Nuclear modification factor of jets

The observable quantity that expresses the magnitude of suppression or enhancement of jets is the nuclear modification factor R_{AA}, which is defined as

$$R_{\mathrm{AA}} = \frac{\frac{1}{N_{\mathrm{evt}}} \frac{\mathrm{d}^2 N_{\mathrm{jet}}}{\mathrm{d}p_{\mathrm{T}}\mathrm{d}y}|_{\mathrm{central}}}{\langle T_{\mathrm{AA}} \rangle \frac{\mathrm{d}^2 \sigma^{pp}_{\mathrm{jet}}}{\mathrm{d}p_{\mathrm{T}}\mathrm{d}y}} \quad (1)$$

where T_{AA} is the nuclear overlap function.

For jet R_{AA} measurement, Pb+Pb data collected in 2011 (lumisosity $L_{\mathrm{int}}^{\mathrm{PbPb}}$=0.14 nb^{-1}) and pp data collected in 2013 ($L_{\mathrm{int}}^{\mathrm{pp}}$=4.0 pb^{-1}) at $\sqrt{s_{\mathrm{NN}}} = 2.76$ TeV were used. Minimum bias (MB) and jet triggered samples were combined to get jet spectra in the interval of $32 < p_{\mathrm{T}}^{\mathrm{jet}} < 500$ GeV. Jets were reconstructed using the anti-k_t algorithm [6] with parameter R=0.4. Background contribution to the jet energy from the underlying event (UE) was estimated and subtracted on the event-by-event basis by the iterative procedure. This procedure was used in both pp and

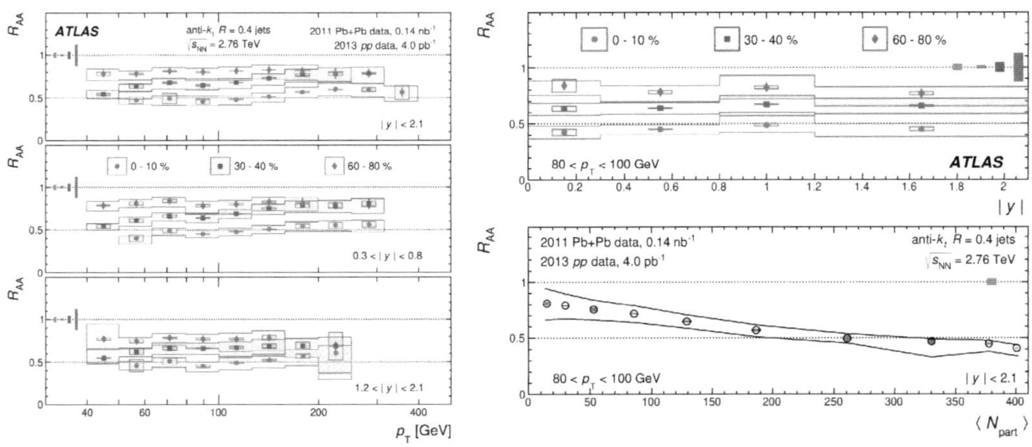

Figure 1. Left: Jet R_{AA} as a function of p_T in three centrality bins with each panel showing a different range in $|y|$. Right: The R_{AA} for jets with $80 < p_T < 100$ GeV as a function of y for different centrality bins (top) and as a function of N_{part} for $|y|$ <2.1 range (bottom) [7].

Pb+Pb collisions. Jet energies were calibrated by the procedure based on the MC. To account for the resolution effects an unfolding based on the Singular Value Decomposition (SVD) method was used.

Various dependencies of jet R_{AA} are shown in Fig.1. Left panel shows suppression of jet production in central Pb+Pb collisions by a factor of approximately two which is weakly dependent on the jet p_T. Size of suppression gradually decreases from central to peripheral collisions but is significant even in 60-80% centrality bin. This is depicted in the lower right panel where number of participants N_{part} dependence of R_{AA} in one p_T and one rapidity (y) bin is shown. The upper right panel shows rapidity dependence of R_{AA} for three different centralities in $80 < p_T < 100$ GeV bin. No rapidity dependence is observed in the measured y range.

2. Measurements of production of asymmetric dijets

ATLAS has also performed new measurements of dijet p_T correlations in Pb+Pb and pp collisions at $\sqrt{s_{NN}}$ =2.76 TeV [10]. The momentum balance of the dijet system was expressed in terms of the variable $x_J = p_{T_1}/p_{T_2}$, where $p_{T_1}$and $p_{T_2}$are the transverse momenta of the jets with the highest and second highest p_T in the event, respectively. Measurements of $\frac{1}{N}\frac{dN}{dX_J}$ were performed as a function of $p_{T_1}$and collision centrality. Previous experimental results [8][9][11] were not corrected to account for detector-specific experimental effects, in particular the resolution on the measured jet p_T, which lead to distortions of the distributions. To account for these effects two-dimensional Bayesian unfolding procedure was used. The new measurement of dijet asymmetry is free from such distortions and can be directly compared to theoretical calculations.

The $\frac{1}{N}\frac{dN}{dX_J}$ distributions obtained in central Pb+Pb and pp collisions are shown in Fig.2 for different intervals of p_{T_1}. The distributions in pp collisions have the most probable value near $x_J \sim 1$ corresponding to the dijet pairs having nearly the same p_T. In contrast, Pb+Pb event with $100 < p_T < 126$ GeV, the distributions develop a large peak at $x_J \sim 0.5$, corresponding to the subleading jet with half the p_T of the leading jet. The modifications observed in the Pb+Pb data are smaller with larger p_{T_1}. This evolution is much more abrupt than the p_T dependence of the single jet R_{AA}, which is approximately logarithmic [12]. The modifications are also smaller in more peripheral centrality bins.

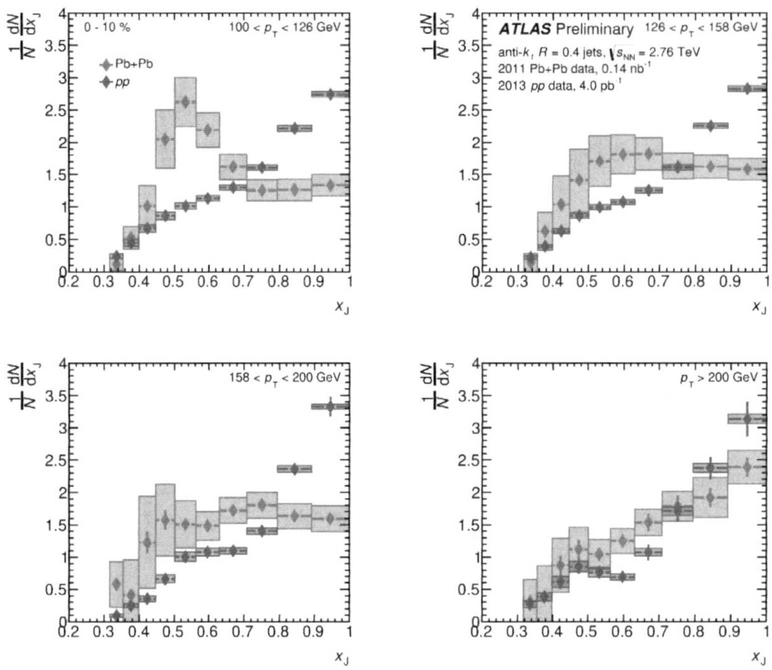

Figure 2. The $\frac{1}{N}\frac{dN}{dX_J}$ distributions for different selections on p_{T_1}, shown for the 0-10% centrality bin in Pb+Pb collisions (red) and *pp* (blue). Statistical uncertainties are indicated by the error bars while systematic uncertainties are shown with shaded boxes [10].

3. Fragmentation functions

Modifications of the jet internal structure can be directly accessed by measuring jet fragmentation functions [9]. ATLAS measured two sets of fragmentation functions, $D(p_T)$ and $D(z)$ which are defined as

$$D(p_T) = \frac{1}{N_{jet}}\frac{dN_{ch}}{dp_T} \qquad D(z) = \frac{1}{N_{jet}}\frac{dN_{ch}}{dz} \tag{2}$$

where p_T is transverse momentum of charged particles and longitudinal momentum fraction $z = (p_T \cos \Delta R)/p_T^{jet}$ and ΔR is the distance between jet axis and track in $\eta - \phi$ space. Jet triggered Pb+Pb and *pp* data taken in 2011 and 2013 were used in this analysis. Jets used in this analysis were clustered with anti-k_t algorithm, from UE background subtracted and fully calibrated. Jets with $100 < p_T < 398$ GeV were used in the analysis and tracks with $p_T > 1$ GeV that match the jet within $\Delta R < 0.4$ were used. Tracks were corrected for reconstruction efficiency. Because of the large contribution from UE η-dependent subtraction procedure was applied on the event-by-event basis to tracks with p_T less than 6 GeV. To correct for detector effects 2D Bayesian unfolding was used.

Ratios of jet fragmentation functions in Pb+Pb to *pp* jet fragmentation functions are shown in Fig.3. Jet fragmentation functions in Pb+Pb collisions exhibit enhancement in low and high p_T region and depletion at intermediate p_T compared to *pp*. Magnitude of the modification decreases for peripheral collisions. No significant difference between different η intervals at low p_T is observed but a hint of smaller enhancement in forward region ($1.2 < |\eta| < 2.1$) compared to barrel region ($|\eta| < 1.2$) is observed at high p_T.

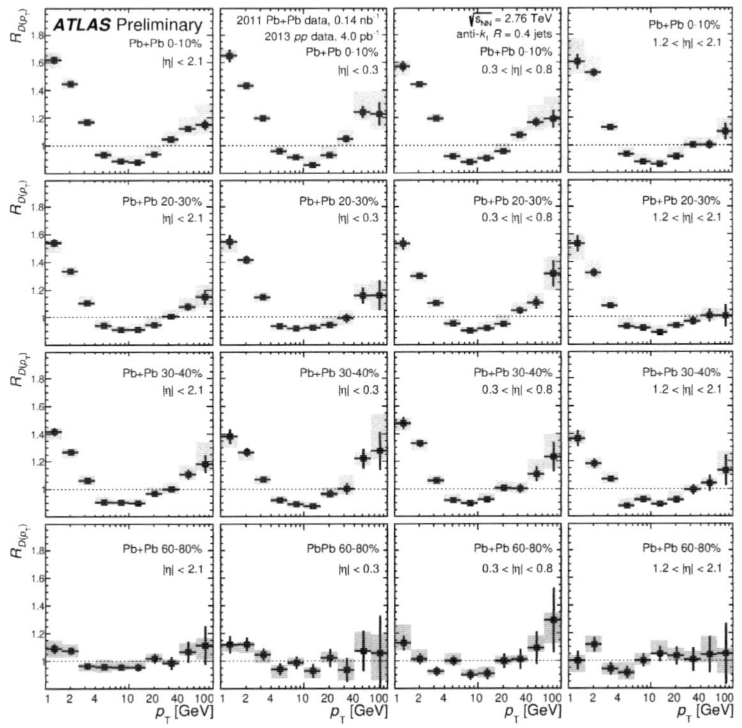

Figure 3. The ratios of $D(p_{\mathrm{T}})$ distributions measured in Pb+Pb collisions to $D(p_{\mathrm{T}})$ measured in *pp* collisions. Ratios are evaluated in four centrality bins (rows) and four selections on jet pseudorapidity (columns) [13].

4. Summary

Three presented measurements provide complex view on the jet quenching phenomenon. Measurements of single hadron and single jet R_{AA} show strong suppression of hadrons and jets in Pb+Pb collisions which is y independent. Dijets with $p_{\mathrm{T}_1} > 100$ GeV in the most central Pb+Pb collisions, the typical configurations are strongly asymmetric compared to that in *pp* collisions at the same collision energy. However, in peripheral collisions or at higher p_{T_1} (>200 GeV), the distributions are qualitatively similar to those in *pp* collisions. Measurement of jet fragmentation functions show modification of jet fragmentation in Pb+Pb collisions which is η independent except high p_{T} region where a hint of η dependence is observed.

References

[1] ATLAS Collaboration, JINST 3 (2008) S08003
[2] ATLAS Collaboration, Phys. Rev. Lett. 105 (2010) 252303, arXiv:1011.6182
[3] ATLAS Collaboration, Phys. Lett. B719 (2013) 220241, arXiv:1208.1967
[4] ATLAS Collaboration, Phys. Lett. B739 (2014) 320342, arXiv:1406.2979
[5] ATLAS Collaboration, Phys. Lett. B751 (2015) 376395, arXiv:1506.08656
[6] M. Cacciari, G. P. Salam, G. Soyez, JHEP 04 (2008) 063, arXiv:0802.1189
[7] ATLAS Collaboration, Phys. Rev. Lett. 114 (7) (2015) 072302, arXiv:1411.2357
[8] ATLAS Collaboration, Phys. Rev. Lett. 105 (2010) 252303, arXiv:1011.6182 [hep-ex]
[9] CMS Collaboration, Phys. Rev. C84 (2011) 024906, arXiv:1102.1957 [nucl-ex]
[10] ATLAS Collaboration, ATLAS-CONF-2015-052 (2015). http://cds.cern.ch/record/205567
[11] CMS Collaboration, Phys. Lett. B712 (2012) 176197, arXiv:1202.5022 [nucl-ex]
[12] ATLAS Collaboration, Phys. Rev. Lett. 114 (7) (2015) 072302, arXiv:1411.2357 [hep-ex]
[13] ATLAS Collaboration, ATLAS-CONF-2015-055

IOP Conf. Series: Journal of Physics: Conf. Series **832** (2017) 012006 doi:10.1088/1742-6596/832/1/012006

Cold nuclear matter effects on jet suppression in heavy-ion collisions

Eliane Epple for the ALICE Collaboration
Yale University, New Haven, Connecticut, USA
E-mail: eliane.epple@yale.edu

Abstract. These proceedings discuss the sensitivity of the nuclear modification factor R_{AA} of fully reconstructed jets to cold nuclear matter effects. To test the parton energy loss interpretation of the observed R_{AA} of $\sim$0.3, obtained from reconstructed jets in Pb–Pb collisions at $\sqrt{s_{\mathrm{NN}}}$=2.76 TeV, we measured the inclusive jet production in minimum bias p–Pb collisions at $\sqrt{s_{\mathrm{NN}}}$=5.02 TeV for resolution parameters R=0.2 and 0.4. The reconstructed jets incorporate the neutral and charged energy component and cover a momentum range $p_{\mathrm{T}}^{\mathrm{jet}}$=20-90 GeV. The comparison of the jet yield to PYTHIA simulations shows no significant depletion in the measured jet cross section attributed to cold nuclear matter effects present in p–Pb collisions.

1. Introduction

The depletion of the reconstructed jet yield in heavy-ion collisions, quantified by R_{AA} and R_{CP}, was demonstrated by several measurements at the LHC. The results of ALICE, ATLAS, and CMS showed that the suppression increases with centrality and decreases slightly with $p_{\mathrm{T}}^{\mathrm{jet}}$ [1, 2, 3, 4, 5]. To pin down the extent of suppression that can be attributed to the hot and dense phase in heavy-ion collisions, other possibilities for jet suppression must be studied as well. Cold nuclear matter effects have been suggested to possibly contribute to the measured suppression of jets in heavy-ion collisions. Cold nuclear matter effects incorporate several mechanisms by which the jet cross section in a nucleus-nucleus collision can be modified with respect to a pp reference, aside from energy loss due to a deconfined phase of the collision [6]. One of these mechanisms is that by taking pp collisions as a reference one does not preclude the fact that measurements in heavy-ion collisions may already deviate from the former due to a modified parton distribution function (PDF) of a nucleon as part of a nucleus with respect to a free proton [7]. The extent to which the nuclear modification factor R_{AA} is sensitive to a modified PDF can be tested by comparing the data with calculations including different PDFs. This is shown later in this proceeding. One way to test the influence of cold nuclear matter effects on R_{AA} is to study this observable in p–Pb collisions. These collisions should exhibit cold nuclear matter effects while inhibiting the creation of a hot and dense phase present in heavy-ion collisions.

R_{pPb} and R_{dAu} of reconstructed jets have been measured at LHC and RHIC [8, 9, 10, 11, 12]. Results by ATLAS indicate that R_{pPb} of jets deviates from unity for larger $p_{\mathrm{T}}^{\mathrm{jet}}$ (>100 GeV) and for forward-going rapidities (p-going side). Results from PHENIX further show that R_{dAu} exhibits a centrality dependence with R_{dAu} >1 for peripheral collisions and <1 for central collisions. These two effects combined produce an R_{dAu} ≈1 for a minimum bias event selection.

IOP Conf. Series: Journal of Physics: Conf. Series **832** (2017) 012006 doi:10.1088/1742-6596/832/1/012006

These effects challenge present understanding and may come from various mechanisms [8, 11]. The results reported by ALICE, on the other hand, demonstrated an R_{pPb} of unity for charged jets, independent of the selected centrality percentile. It was however noted that the type of centrality selection can introduce dynamical biases on this observable [10].

Comparing to full jets including the neutral component for the same reconstructed $p_{\mathrm{T}}^{\mathrm{jet}}$, charged jets on average stem from higher p_{T} partons. Thus, every effect in R_{pPb} that is dependent on the parton energy scale might result in differences in the p_{T} dependent R_{pPb} measurement for the two jet types. Furthermore, the majority of published results on reconstructed jet R_{AA} include the neutral energy component into their jet reconstruction analysis. These two considerations motivate a cross check of the results obtained by ALICE for charged jets [9] (including only charged tracks in the jet reconstruction) by an independent measurement of jets that include additionally the neutral energy component. This is discussed below.

2. Experimental setup and jet reconstruction in p–Pb collisions

The ALICE detector is a versatile spectrometer dedicated to study heavy-ion collisions [13]. Of primary importance for the present analysis is its inner tracking system (ITS) and the time projection chamber (TPC) for charged track reconstruction, as well as the electromagnetic calorimeter (EMCal) for determining the neutral energy component of reconstructed jets.

2.1. Event selection and jet reconstruction

This analysis used ~90 million p–Pb minimum bias events at $\sqrt{s_{\mathrm{NN}}}$=5.02 TeV, which were selected based on their primary vertex quality and position. Since this analysis focuses on jets including the neutral energy component, the η and φ acceptance of the EMCal ($|\eta| <0.7$ and φ=80°-180°) is a limiting factor for the jet reconstruction.

For the jet reconstruction an anti-k_{T} algorithm [14] was run over charged particle tracks with $p_{\mathrm{T}}>0.15$ GeV/c and EMCal clusters with $p_{\mathrm{T}}>0.3$ GeV/c. Jets were analyzed for a resolution parameter of R=0.2 and 0.4. Jets were rejected from the analysis, if they were not fully contained in the EMCal acceptance and if their reconstructed area did not fulfill A$> 0.6\pi R^2$.

2.2. Underlying event subtraction

The underlying event, produced in the p–Pb collision, shifts the reconstructed jet p_{T} to slightly higher values. This additional energy was estimated and subtracted from the jets on an event-by-event basis. In order to estimate the underlying event magnitude, a k_{T} jet finder was applied to charged tracks only. To avoid a contribution of signal jets to the background determination, the two k_{T}-background jets with the highest p_{T} were excluded. The remaining background jets were used to calculate the median jet momentum area density

$$\rho_{chrg.}^{BKG} = \text{median}\left\{\rho_j^{jet}\right\} = \text{median}\left\{\frac{p_T^{jet,j}}{A^{jet,j}}\right\}. \qquad (1)$$

To exclude pure ghost jets, $\rho_{chrg.}^{BKG}$ is calculated only for background jets that contain at least one charged track. The momentum density of the underlying event in p–Pb collisions is about two orders of magnitude lower than in Pb–Pb collisions, which causes the background to be scarcely populated in some areas. This "emptiness" of the event was accounted for by correcting $\rho_{chrg.}^{BKG}$ by an event occupancy factor C [15]. For each event C was defined as the the sum of background jet areas $A^{jet,j}$ normalized by the area of the TPC (which defines the acceptance of charged background jets). This factor C corrects the fact that $\rho_{chrg.}^{BKG}$ only determines the momentum density within reconstructed k_{T} jets. The anti-k_{T} signal jet however can lie anywhere in the acceptance, also in areas scarcely populated by tracks. The above procedure is performed only with charged tracks due to the larger acceptance of the TPC in comparison with the EMCal, resulting in a better control over local background fluctuations. The charged+neutral background density is scaled up from the charged background density with a scaling factor

$$SF = \frac{\sum E^{cl.-EMC} + p_T^{track-EMC}}{\sum p_T^{track-TPC}} \cdot \frac{A^{TPC}}{A^{EMC}}, \quad (2)$$

which builds the ratio of summed charged momenta and neutral energy in the EMCal acceptance to the sum of charged momenta in the TPC acceptance, corrected by the ratio of the two acceptances. Figure 1 combines the final corrected background momentum area density $\rho_{full}^{BKG,corr.}$ into a probability density for the event. The mean momentum background density is $\sim$3 GeV/c for central and $\sim$1 GeV/c for peripheral events.

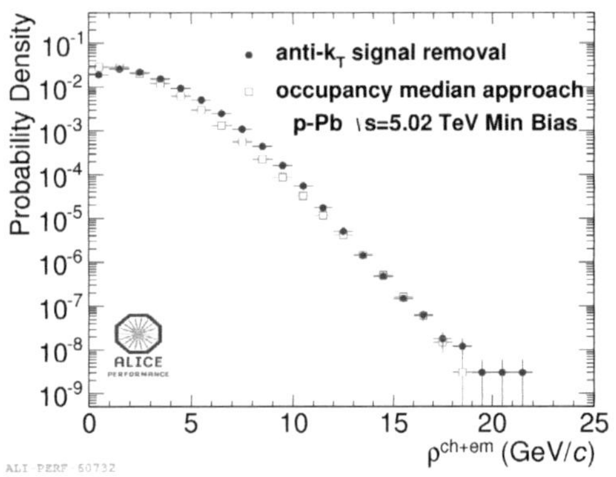

Figure 1. The probability density for extracting a certain momentum area density $\rho_{full}^{BKG,corr.}$ (labeled as ρ^{ch+em}), which describes the background of charged and neutral signals to the measured jet p_T in the event. The method described in the text is labeled "occupancy median approach". Another approach, not further discussed here, is the exclusion of overlap between a anti-k_T signal jet and a k_T background jet (blue dots).

2.3. Corrected jet spectra

The measured raw jet spectra for R=0.2 and 0.4 were corrected for their underlying event contribution by subtracting the obtained momentum area density of the event via:

$$p_T^{full\,jet} = p_T^{full\,jet+BKG} - A_{jet} \cdot \rho_{full}^{BKG,corr.}. \quad (3)$$

The background subtracted jet spectra were then corrected for detector and resolution effects by an unfolding procedure. For this, jets generated with PYTHIA6 were reconstructed at the particle and detector level and geometrically matched to each other. This way, a response matrix was generated that links the measured to the true jet momenta and thus encodes effects such as momentum resolution, and track and cluster reconstruction efficiency on the reconstructed jet momentum. The additional effect of local background fluctuations was also determined and included into the response matrix, as described in Ref. [16]. The unfolding procedure is based on the Singular Value Decomposition approach and uses the above-described response matrix to recover the true jet p_T spectrum. The unfolded jet spectra for resolution parameters R=0.2 and 0.4 are presented in Fig. 2. Since the EMCal acceptance does not cover the full azimuth, the measured yield was scaled up to $\varphi = 2\pi$ and is shown per unit η.

3. Results and conclusions

The fully-corrected jet p_T spectra were normalized to the number of events and the mean number of binary collisions in minimum bias p–Pb events (N_{coll}=6.87±0.56) [9] in order to compare the results to simulations and to other collision systems like pp and Pb–Pb. The absence of a pp reference spectrum of full jets at $\sqrt{s_{NN}}$=5.02 TeV prevents the direct comparison to measured

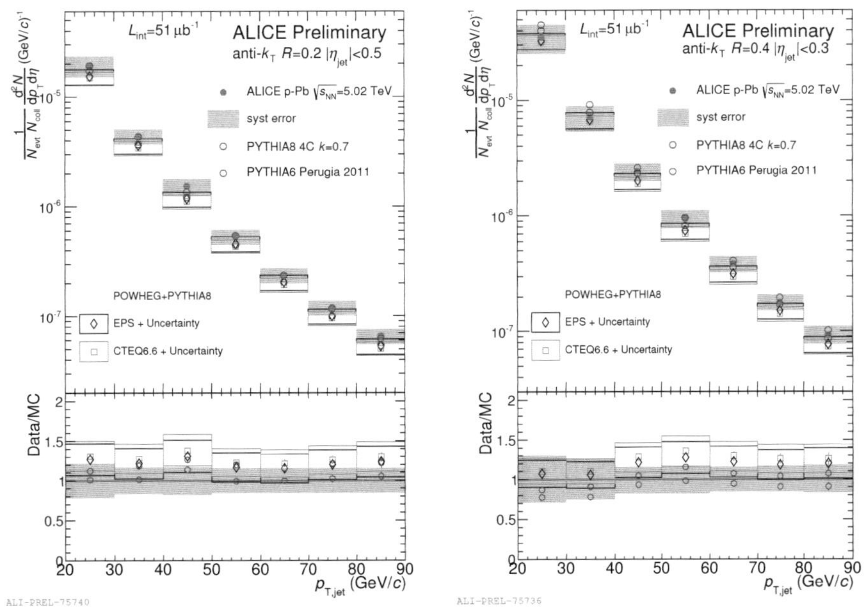

Figure 2. Fully corrected jet p_T spectra for R=0.2 (left) and R=0.4 (right). Lower panel show the ratio of data to different simulations.

data at the same energy. Therefore, PYTHIA simulations (Version 6, Perugia 2011 tune) at $\sqrt{s_{NN}}$=5.02 TeV were used as a reference. The data were divided by the PYTHIA simulation, which was scaled by the number of simulated events, to obtain a better visualization of the agreement between p–Pb data and the pp simulations. The ratio is shown in the lower panels of Fig. 2. The ratio is consistent with unity for both resolution parameters. Additionally, the measurements are compared to POWHEG calculations with either proton PDFs (CTEQ6.6) or nuclear PDFs (EPS09) included. Hadronization was in both cases modeled with PYTHIA8. The comparison of proton and nuclear PDFs shows that the influence of different PDFs on this observable is insignificant. Overall, the ratio illustrates that the low R_{AA} of ~0.3 measured in Pb–Pb collisions [2] cannot be explained by cold nuclear matter effects. The suppression can thus be attributed to a hot and dense phase, only believed present in heavy-ion collisions.

References

[1] Abelev B *et al.* (ALICE) 2014 *JHEP* **03** 013 (*Preprint* 1311.0633)
[2] Adam J *et al.* (ALICE) 2015 *Phys. Lett.* **B746** 1–14 (*Preprint* 1502.01689)
[3] Aad G *et al.* (ATLAS) 2013 *Phys. Lett.* **B719** 220–241 (*Preprint* 1208.1967)
[4] Aad G *et al.* (ATLAS) 2015 *Phys. Rev. Lett.* **114** 072302 (*Preprint* 1411.2357)
[5] Khachatryan V *et al.* (CMS) 2016 (*Preprint* 1609.05383)
[6] Kang Z B, Vitev I and Xing H 2015 *Phys. Rev.* **C92** 054911 (*Preprint* 1507.05987)
[7] Eskola K J, Paukkunen H and Salgado C A 2009 *JHEP* **04** 065 (*Preprint* 0902.4154)
[8] Aad G *et al.* (ATLAS) 2015 *Phys. Lett.* **B748** 392–413 (*Preprint* 1412.4092)
[9] Adam J *et al.* (ALICE) 2015 *Phys. Lett.* **B749** 68–81 (*Preprint* 1503.00681)
[10] Adam J *et al.* (ALICE) 2016 *Eur. Phys. J.* **C76** 271 (*Preprint* 1603.03402)
[11] Adare A *et al.* (PHENIX) 2016 *Phys. Rev. Lett.* **116** 122301 (*Preprint* 1509.04657)
[12] Khachatryan V *et al.* (CMS) 2016 *Eur. Phys. J.* **C76** 372 (*Preprint* 1601.02001)
[13] Abelev B *et al.* (ALICE) 2014 *Int. J. Mod. Phys.* **A29** 1430044 (*Preprint* 1402.4476)
[14] Cacciari M, Salam G P and Soyez G 2008 *JHEP* **04** 063 (*Preprint* 0802.1189)
[15] Chatrchyan S *et al.* (CMS) 2012 *JHEP* **08** 130 (*Preprint* 1207.2392)
[16] Abelev B *et al.* (ALICE) 2012 *JHEP* **03** 053 (*Preprint* 1201.2423)

IOP Conf. Series: Journal of Physics: Conf. Series **832** (2017) 012007 doi:10.1088/1742-6596/832/1/012007

Dijet asymmetries in Pb+Pb and pp collisions with the ATLAS detector

Laura Havener (on behalf of the ATLAS Collaboration)
Physics Department, Columbia University, New York, NY 10027 USA
E-mail: lbh2133@columbia.edu

Abstract. Measurements of dijet p_T correlations in both Pb+Pb and pp collisions at a nucleon–nucleon centre-of-mass energy of $\sqrt{s_{\mathrm{NN}}} = 2.76$ TeV are presented. The measurements were performed with the ATLAS detector at the Large Hadron Collider (LHC) using Pb+Pb and pp data samples with integrated luminosities of 0.14 nb^{-1} and 4.0 pb^{-1}, respectively. Jets were reconstructed using the anti-k_t algorithm with $R = 0.4$. A background subtraction procedure was applied to correct the jets for the large underlying event present in Pb+Pb collisions. The leading and subleading jet transverse momenta are denoted p_{T_1} and p_{T_2}. An unfolding procedure is applied to the two-dimensional p_{T_1}-p_{T_2} distributions to account for experimental effects in the measurement of both jets. Distributions of $\frac{1}{N}\frac{\mathrm{d}N}{\mathrm{d}x_{\mathrm{J}}}$, where $x_{\mathrm{J}} = \frac{p_{\mathrm{T}_2}}{p_{\mathrm{T}_1}}$, are presented as a function of p_{T_1} and collision centrality. The distributions are found to be similar in peripheral Pb+Pb collisions and pp collisions, but highly modified in central Pb+Pb collisions. The results are qualitatively consistent with expectations from partonic energy loss.

1. Introduction

Jets have been an important tool for probing the hot, dense medium produced in heavy-ion collisions known as the quark-gluon plasma (QGP). Partons are expected to lose energy in interactions with the QGP, resulting in "jet quenching". This phenomenon is expected in Pb+Pb collisions at the LHC where a direct observation can be made using reconstructed jets. Dijets are an effective probe of jet quenching because jets originating from the same hard scattering can lose different amounts of energy as they traverse the medium whereas in pp collisions the jets are expected to be balanced.

Previous measurements of the dijet asymmetry, $A_J = (p_{T1} - p_{T2})/(p_{T1} + p_{T2})$, where p_{T1} and p_{T2} are the transverse momentum of the leading and subleading jets in the event, respectively, show that the dijets become increasingly asymmetric in more central collisions [1-3]. However, these measurements were not corrected for experimental resolution so they can not be compared directly to theoretical calculations. These proceedings present a new measurement of the dijet asymmetry in Pb+Pb and pp collisions [4] at $\sqrt{s_{NN}}$ =2.76 TeV using the observable $x_{\mathrm{J}} = p_{T1}/p_{T2}$ that has been fully corrected for detector effects.

2. Experimental setup, data selection, and jet reconstruction

This measurement was performed using the ATLAS detector, which is described in Ref. [5]. This result made use of the inner detector, calorimeter, and trigger systems. The inner detector measures charged particles to $|\eta| < 2.5$. The calorimeter system consists of a steel-scintillator calorimeter ($|\eta| < 1.7$) and an electromagnetic ($|\eta| < 3.2$) and Liquid Argon (LAr) hadronic ($1.5 < |\eta| < 3.2$) calorimeter. The Pb+Pb data were selected by a combination of minimum-bias and jet triggers corresponding to a luminosity of 0.14 nb^{-1}. The centrality was determined by using the sum of the transverse energy in the forward calorimeter that covers $3.2 < |\eta| < 4.9$. The pp data were obtained from a combination of jet triggers that are used in different p_T ranges where the trigger is fully efficient. The pp data has a total luminosity of 4.0 pb^{-1}. A Monte Carlo (MC) sample was produced using the PYTHIA version 6.423 generator using AUET2D parameters. This was followed by a GEANT simulation of ATLAS [7, 8] and then embedded in real minimum bias heavy-ion data. The sample consisted of hard-scattering (not just dijet) events using multiple ranges of momentum transfer.

Jets are reconstructed using the anti-k_t algorithm [9] with $R = 0.4$ and utilize an event-by-event underlying event (UE) subtraction procedure described in Ref.[10] that includes the effects of harmonic modulation due to flow. Previous results accounted for the second harmonic modulation. In this measurement, that procedure has been extended to account for the third and fourth order harmonic modulation. Following the background subtraction, the jets are corrected using an energy scale calibration factor determined from MC simulations. Separately, an *in situ* calibration is applied to account for known differences between the the detector response in the data and the MC sample [12].

3. Data analysis, unfolding and systematics

In this analysis, dijet pairs were selected by finding the two highest jets in each event with $p_T > 25$ GeV and $|\eta| < 2.1$ where $p_{T1} > p_{T2}$. The pairs were required to have $\Delta\phi > 7\pi/8$. A two-dimensional $p_{T1} - p_{T2}$ distribution was filled in Pb+Pb and pp collisions using different trigger samples based on the leading jet p_T. The distributions are symmetrized to account for bin migration across the diagonal where the leading and subleading jets can switch roles due to the effects of experimental resolution. The 2D symmetrized distributions are shown in Fig. 1 for two different Pb+Pb centralities and for pp collisions.

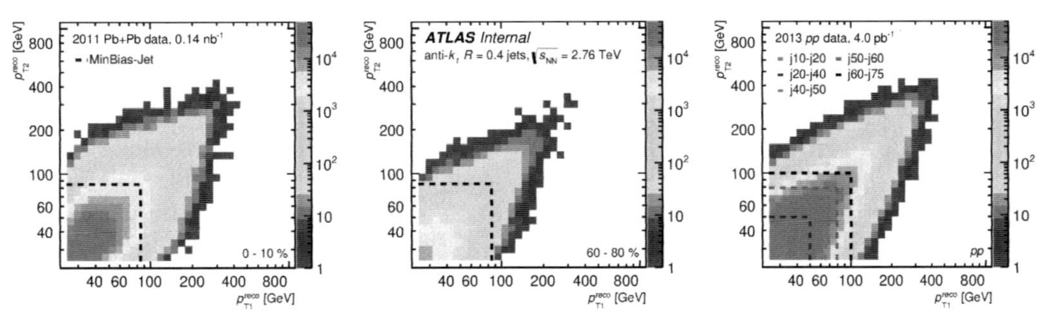

Figure 1. Symmetrized 2D $p_{T1} - p_{T2}$ distributions for Pb+Pb data 0-10 % (left), Pb+Pb data 60-80 % (middle), and pp (right). The dashed lines indicate the trigger boundaries [4].

The pairs have a combinatoric background contribution from uncorrelated pairs that arise from jets not originating from the same hard scattering. It is expected to be uniform in $\Delta\phi$ except for a small harmonic modulation due to imperfect removal of the background in the reconstruction. To account for residual flow harmonics, the background is assumed to be of the form $C(\Delta\phi) = Y(1 + 2c_3 \cos 3\Delta\phi + 2c_4 \cos \Delta\phi)$, which is motivated by flow harmonics. The coefficients c_3 and c_4 are determined in the region where $\Delta\eta > 1.0$, and the amplitude of the underlying event Y is estimated in the region $1.0 < \Delta\phi < 1.4$ using those coefficients. These estimates are used to determine the combinatoric contribution in the signal region which is subtracted bin-by-bin from the $p_{T1} - p_{T2}$ distribution.

The 2D $p_{T1} - p_{T2}$ distributions were corrected for bin migration due to experimental effects using a 2D unfolding procedure [11]. A four-dimensional response was generated from the MC sample in measured and true p_{T1} and p_{T2}. The Bayesian unfolding is an iterative procedure that requires a choice of number of iterations and prior. The prior is generated from the truth $p_{T1} - p_{T2}$ distribution in PYTHIA. The number of iterations is determined by balancing the amplification of statistical error at large number of iterations with bias due to the choice of the prior.

After unfolding, the 2D $p_{T1} - p_{T2}$ distribution is restored to a leading/subleading jet distribution by reflecting the distribution of the diagonal and projecting into the observable $\frac{1}{N}\frac{dN}{dx_J}$ as a function of x_J. Figure 2 shows these distributions before and after unfolding. The effect of unfolding is to move jets in *pp* and peripheral collisions to more balanced configurations at $x_J = 1$. In central Pb+Pb collisions the effect is to move jets to more asymmetric configurations at $x_J = 0.5$.

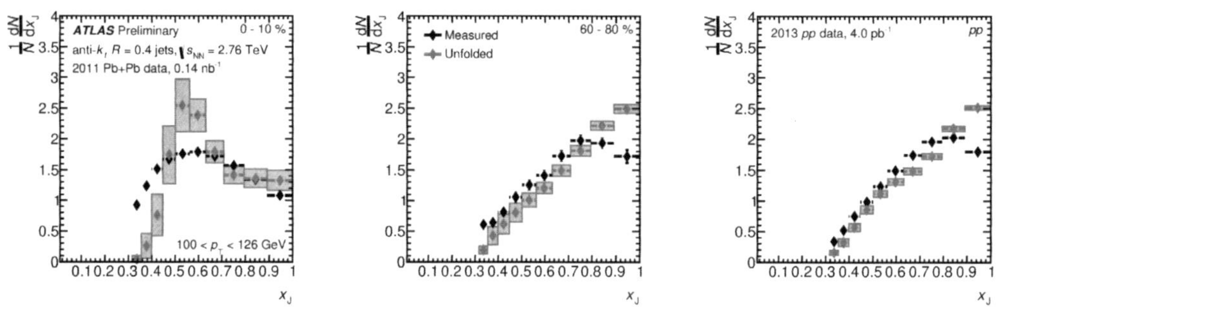

Figure 2. Distributions of $\frac{1}{N}\frac{dN}{dx_J}$ before (black) and after (red)unfolding for Pb+Pb data 0-10 % (left), Pb+Pb data 60-80 % (middle), and *pp* (right) [4].

Systematic uncertainties on the measured $\frac{1}{N}\frac{dN}{dx_J}$ distributions arise from uncertainties in the response matrix used in the unfolding. These are estimated by varying the jet energy scale (JES) and resolution (JER) in the MC sample. The JES uncertainty is applied as a function of p_T, η, and centrality and is the dominant uncertainty in the measurement. The JER has a dominant contribution due to the underlying event that is described well in the MC because of the data overlay. The analysis also has systematics due to the background subtraction and the unfolding procedure. The systematic due to the background subtraction was determined by varying the $\Delta\phi$ range used to estimate the amplitude of the background. The systematic uncertainties due to unfolding were determined by varying the number of iterations, reweighting the prior, and accounting for statistical effects in the response matrix.

4. Results

The final corrected results for Pb+Pb compared to *pp* collisions are shown in Figure 3. The figures on the left show the centrality dependence in the p_{T1} bin from 100-126 GeV. This demonstrates that as more central collisions are probed the dijets become more asymmetric. The most probable configuration for *pp* is $x_J = 1$ and for central Pb+Pb it is $x_J = 0.5$. As the Pb+Pb collisions become more peripheral the distribution becomes more similar to the *pp* distribution. The figures on the right show the leading jet p_T dependence of the most central collisions (0-10%) compared the *pp* collisions. A drastic leading jet p_T dependence is observed where the Pb+Pb collisions become like the *pp* at higher p_{T1}.

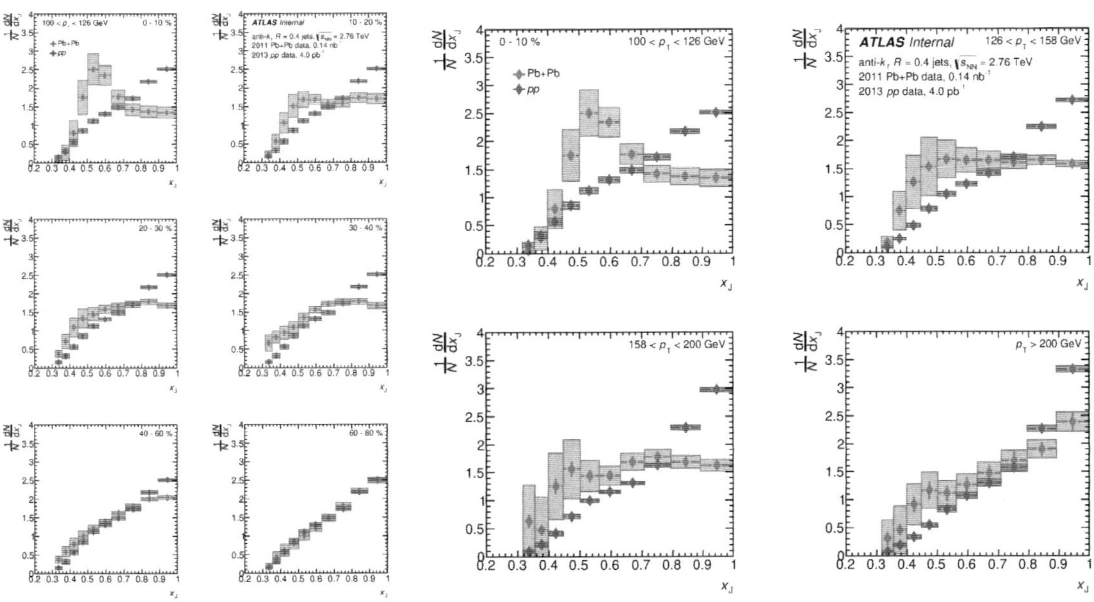

Figure 3. Distributions of $\frac{1}{N}\frac{dN}{dx_J}$ in Pb+Pb (red) and pp (blue) collisions for $100 < p_{T1} < 126$ GeV for different collision centralities (left six figures) and for 0-10% centrality for different leading jet p_T (right four figures). The error bars represent the statistical errors and the bands represent the systematic errors [4].

5. Conclusion

These proceedings presented a new measurement of the dijet asymmetry in Pb+Pb and pp collisions at $\sqrt{s_{NN}}$ =2.76 TeV. The result has been measured differentially in centrality and leading jet p_T and has been fully corrected by an unfolding procedure to account for experimental resolution. The results show more asymmetric dijet pairs in Pb+Pb data compared to pp data, which is consistent with jet quenching. The results demonstrate that the jets become more symmetric with increasing leading jet p_T. This measurement is qualitatively consistent with previous measurements of the dijet asymmetry, but since the measurement has been fully corrected for detector effects it can be directly compared to theoretical models of jet quenching.

Acknowledgments

This work was supported by U.S. Department of Energy grant DE-FG02-86ER40281.

References

[1] ATLAS Collaboration 2010 *Phys. Rev. Lett.* **105** 252303
[2] CMS Collaboration 2011 *Phys. Rev.* C**84** 024906
[3] CMS Collaboration 2012 *Phys. Lett.* B**712** 176
[4] ATLAS Collaboration 2015 ATLAS-CONF-2015-052, http://cdsweb.cern.ch/record/2055673
[5] ATLAS Collaboration 2008 *JINST* **3** S08003
[6] Sjostrand T, Mrenna S and Skands P 2006 *JHEP* **0605** 026
[7] Agostinelli S et al. 2003 *Nucl. Instrum. Meth.* A **506** 250
[8] ATLAS Collaboration 2010 *Eur. Phys. J.* C**70** 823
[9] Cacciari M, Salam G and Soyez G 2008 *JHEP* **0804** 063
[10] ATLAS Collaboration 2013 *Phys. Lett.* B**719** 220
[11] ATLAS Collaboration 2015 ATLAS-CONF-2015-016, http://cdsweb.cern.ch/record/2008677
[12] DAgostini G 1995 *Nucl. Instrum. Meth.* A**362** 487

IOP Conf. Series: Journal of Physics: Conf. Series **832** (2017) 012008 doi:10.1088/1742-6596/832/1/012008

Color coherence in multiple antenna medium radiation

Fabio Domínguez, Carlos A Salgado and Víctor Vila

Instituto Galego de Física de Altas Enerxías (IGFAE), Universidade de Santiago de Compostela, E-15706 Santiago de Compostela, Galicia, Spain

E-mail: `victor.vila@usc.es`

Abstract. We present the configuration in which a quark-antiquark pair with a fixed opening angle emits a hard gluon inside a medium, and an additional very soft emission afterwards (double antenna). We discuss the coherence effects in terms of the survival probability, which describes the interaction of the $q\bar{q}g$ system with the medium. We generalize previous studies of the antenna radiation to the case of more than two emitters and prove that this generalization provides further support to the picture of jet quenching with effective emitters in the parton cascade.

1. Jet quenching

Jet quenching has become an essential signal for the characterization of the medium formed in experiments of heavy ion collisions. Striking effects of energy loss of leading partons and reconstructed jets have been observed in heavy ion collisions at RHIC and LHC, see e.g. Ref. [1]. These results indicate that the spectra of high-p_T hadrons are strongly affected by the matter formed in the collisions as compared to those from proton-proton collisions. Color coherence is a key ingredient in understanding these effects from a theoretical point of view. It implies a reorganization of the cascade in terms of effective emitters for medium-induced radiation [2].

2. Color coherence in vacuum

The concept of color coherence is useful to understand the qualitative and quantitative features of the radiation in QCD, see e.g. Ref. [3]. In a parton shower, when a highly collinear parton is emitted during the propagation, the splitted collinear gluons are difficult to resolve unless they are sufficiently apart, and for large angles, their radiation interfere rather than add incoherently. A simple interpretation of this result is that, at certain angles, it is difficult to resolve the individual radiated partons, and their net behaviour is rather similar to the radiation from the total color charge of the parent parton. This problem becomes simple when the shower is implemented by using the angle of radiation as the ordering variable. The cascading of a jet, initiated by a hard parton, occurs in a coherent manner. In other words, subsequent parton branchings of the shower are not independent but depend on the characteristics of the previous branching.

The singlet antenna spectrum for $q\bar{q}$ production plus soft gluon emission is

$$dN = \frac{d\omega}{\omega}\frac{d\Omega}{2\pi}\frac{\alpha_s C_F}{2\pi}\Big[R_q + R_{\bar{q}} - 2\mathcal{J}\Big], \tag{1}$$

IOP Conf. Series: Journal of Physics: Conf. Series **832** (2017) 012008 doi:10.1088/1742-6596/832/1/012008

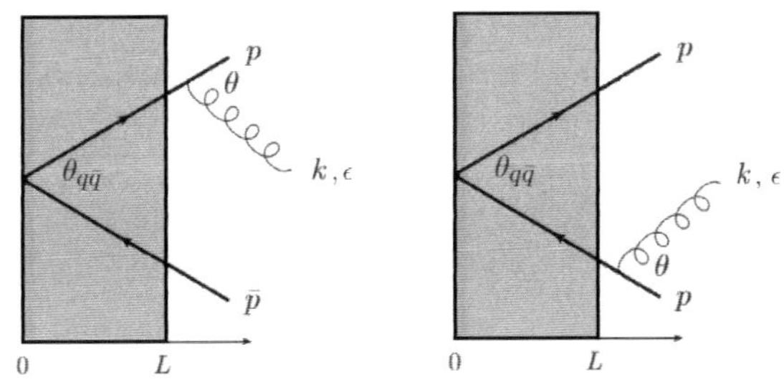

Figure 1. Out-out radiation taking a very soft gluon ($\omega \to 0$).

where R_q ($R_{\bar{q}}$) is the radiation spectrum off an independent constituent, and $\mathcal{J}$ describes the quark-antiquark interference (ω is the energy of emitted gluons, $d\Omega$ is the differential solid angle, α_s is the coupling constant in QCD and C_F[1] is a color factor).

This spectrum is divergent when the energy of the emitted gluon becomes soft, $\omega \to 0$, and when the gluon is emitted collinearly to the either the quark or the antiquark. The spectrum is suppressed at large angles due to the presence of destructive interferences ($\mathcal{R}_q \simeq \mathcal{J}$ at large angles). Thus, in vacuum, large-angle gluon emission is sensitive to the total charge of the system. This property relies on color conservation and leads to angular ordering of the vacuum cascade in QCD.

3. In-medium antenna radiation

Considering the case in which a very soft gluon is emitted from the antenna, the dominant contribution to the medium-induced radiation can be easily computed from the diagrams of Fig. 1. In this simple case, the spectrum of emitted gluons is a modification of eq. (1) [4]

$$dN = \frac{d\omega}{\omega}\frac{d\Omega}{2\pi}\frac{\alpha_s C_F}{2\pi}\Big[R_q + R_{\bar{q}} - (1-\Delta_{med})\, 2\mathcal{J}\Big]. \qquad (2)$$

The interaction of the $q\bar{q}$ pair with the medium is described by the survival probability

$$\mathcal{S} \equiv 1 - \Delta_{med} \equiv \frac{1}{N_c^2 - 1}\, Tr\, \Big\langle W(\vec{x}_\perp) W^\dagger(\vec{y}_\perp)\Big\rangle, \qquad (3)$$

where the Wilson lines describe the propagation of a gluon through a medium field $\mathcal{A}_-(x_+, \vec{x})$,

$$W(\vec{x}) = \mathcal{P}\ \exp\Big[ig\int dx_+ \mathcal{A}_-(x_+, \vec{x})\Big]. \qquad (4)$$

In the resulting spectrum we have introduced the Δ_{med} parameter, which is called the medium decoherence parameter. This factor determines a characteristic time scale for decoherence of the $q\bar{q}$ pair, and for the case of a static medium of size L,

$$\Delta_{med} \simeq 1 - \exp\Big[-\frac{1}{4}\hat{q}L(\vec{x}_\perp - \vec{\bar{x}}_\perp)^2\Big], \qquad (5)$$

where $\hat{q}$ is the transport coefficient, encoding all the information about the dynamical properties of the medium; this is the main parameter to be determined by fits to experimental data and to be compared with theoretical calculations.

We can consider two limits [2, 4]

[1] $C_F = \frac{N_c^2-1}{2N_c}$, with N_c the number of colors.

(i) when the color correlation length of the medium ($\simeq \frac{1}{\sqrt{\hat{q}L}}$) is larger than the size of the pairs, $\Delta_{med} \to 0$ ($dN \sim R_q + R_{\bar{q}} - 2\mathcal{J}$), the medium cannot resolve the individual emitters, thus acting like an individual object with the total charge of the pair (C_F for triplet, C_A for octet or 0 for singlet).

(ii) In the opposite case, $\Delta_{med} \to 1$ ($dN \sim R_q + R_{\bar{q}}$), the medium resolves the antenna (opaque medium: decoherence, two independent emitters) and it breaks the color coherence of the pair, behaving as two independent particles.

A way to estimate the coherence effects is through the definiton of a coherence time, $t_{coh} \sim \left[\frac{1}{12}\theta^2_{q\bar{q}}\hat{q}\right]^{1/3}$, so that the pair remains coherent (acting as an individual emitter) for path lenghts in the medium smaller than t_{coh}.

4. More than two emitters

The antenna provides a simple and intuitive picture, but does it hold for more than two emitters? In order to address this question, we repeated the previous calculations now including an additional hard gluon, with the final goal of extrapolating our results to n-gluon emissions. We restrict to the case of very soft gluon radiation ($\omega \to 0$), where only the out-out component needs to be computed [4].

So, let us consider now the case of three emitters (two hard splittings inside the medium). The direct terms R_q, $R_{\bar{q}}$ and R_g are proportional to a color factor, i.e., no medium effects appear as expected with the kinematics chosen.

$$|\mathcal{M}_1|^2 \propto C_F^2 \qquad |\mathcal{M}_2|^2 \propto C_F^2 \qquad |\mathcal{M}_3|^2 \propto N_c\, C_F^2 \tag{6}$$

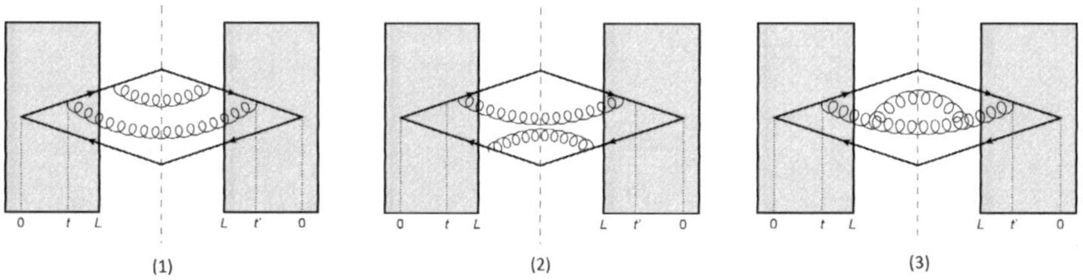

Figure 2. Direct terms for the case of three emitters.

More interesting results are obtained in the interference terms, where the survival probabilites $\mathcal{S}$ appear explicitly, providing information about the color coherence of the emitters.

$$\mathcal{M}_1 \otimes \mathcal{M}_3^* \propto \frac{1}{2}\, N_c^2\, \mathcal{S}(t,L)\, \mathcal{J} \tag{7}$$

$$\mathcal{M}_2 \otimes \mathcal{M}_3^* \propto N_c^2\, \mathcal{S}(0,t)\, \mathcal{S}(t,L)\, \mathcal{J} \tag{8}$$

The interpretation of equations (7) and (8) is very clear in the large-N_c limit (see Figs. 3 and 4). The survival probability, controlling the degree of color coherence in the soft gluon emission, corresponds to that of the dipole from which this gluon is emitted. In Fig. 3 this dipole is formed at time t during the evolution of the system, while it is produced at time 0 in the diagrams of Fig. 4. In this second case, the survival probability is the product of the two survival probabilities, from 0 to t and from t to L. If coherence is not preserved after the in-medium splitting, the antenna won't radiate coherently in the following emission.

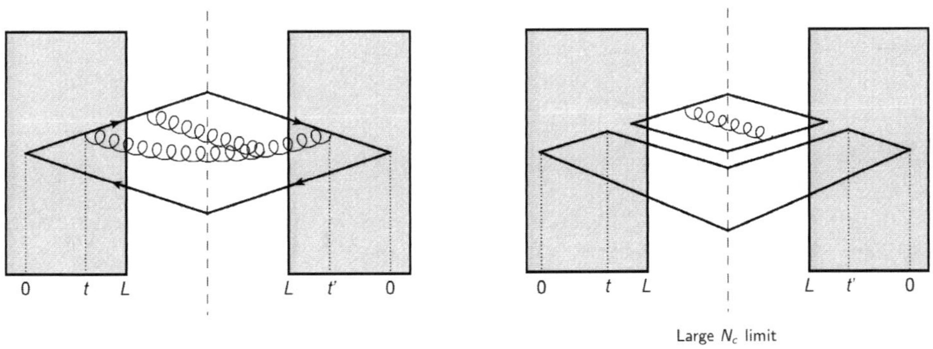

Figure 3. Interference term for the case of three emitters (soft gluon in amplitude).

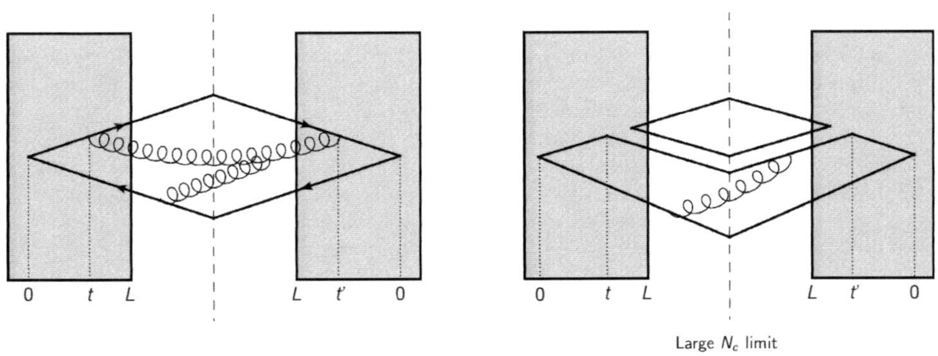

Figure 4. Interference term for the case of three emitters (soft gluon in conjugate amplitude).

5. Discussion and conclusions

Color coherence is essential to understand if multiple gluon emissions are correlated or not. Proceeding as in the vacuum case, in which coherence leads to angular ordering, antenna is a very convenient *laboratory*. An interesting goal is to incorporate the medium effects: the medium can destroy the coherence of the system. Our results definitely conclude that the results from the antenna setup (two emitters) can be easily generalized to the case of more than two emitters. These computations go a step forward to obtain a complete description of a QCD cascade.

Acknowledgments

This research was supported by the European Research Council grant HotLHC ERC-2011-StG-279579, Ministerio de Ciencia e Innovación of Spain under project FPA2014-58293-C2-1-P and Xunta de Galicia (Consellería de Educación) - the group is part of the Strategic Unit AGRUP2015/11.

References

[1] N. Armesto and E. Scomparin, Eur. Phys. J. Plus **131** (2016) no.3, 52 doi:10.1140/epjp/i2016-16052-4
[2] J. Casalderrey-Solana, Y. Mehtar-Tani, C. A. Salgado and K. Tywoniuk, Phys. Lett. B **725** (2013) 357
[3] Y. L. Dokshitzer, V. A. Khoze, A. H. Mueller and S. I. Troian, Gif-sur-Yvette, France: Ed. Frontieres (1991) 274 p.
[4] Y. Mehtar-Tani, C. A. Salgado and K. Tywoniuk, Phys. Lett. B **707** (2012) 156

IOP Conf. Series: Journal of Physics: Conf. Series **832** (2017) 012009 doi:10.1088/1742-6596/832/1/012009

Transverse momentum spectra and the nuclear modification factor of charged particles with ALICE

Edgar Perez Lezama for the ALICE Collaboration

GSI Helmholtz Centre for Heavy Ion Research, Planckstr. 1 Darmstadt, Germany

E-mail: eperezle@cern.ch

Abstract. The observed suppression of particle production at high transverse momentum in nucleus-nucleus collisions at high energies is an effect of the energy loss of partons as they propagate through the hot and dense deconfined QCD medium. The measurement of transverse momentum spectra of primary charged particles in Pb–Pb collisions and the comparison to the measurement in pp collisions are important to study the properties of the deconfined medium. In 2015, pp and Pb–Pb collisions have been recorded at the LHC at the energy of $\sqrt{s_{NN}} = 5.02$ TeV. The inclusive charged particle production is measured with the ALICE detector in the pseudo-rapidity range $|\eta| < 0.8$ and in the transverse momentum range $0.15 < p_{\mathrm{T}} < 40$ GeV/c. The spectra in Pb–Pb, determined for several centrality intervals, are compared to the reference spectrum in pp collisions by calculating the nuclear modification factor R_{AA}. The data are compared to measurements at the lower collision energy of 2.76 TeV and to model predictions.

1. Introduction

The charged particle production has been measured in different collisions systems and at several beam energies. Experiments at RHIC have found, from the measurements of the transverse momentum spectra in p+p and Au+Au collisions at $\sqrt{s_{NN}} = 130$ GeV [1, 2], that the high-p_{T} particle production in heavy-ion collisions is suppressed compared to the production measured in pp. Bjorken had earlier introduced the idea that such an observation is related to parton energy loss in the hot and dense QCD medium created in the collision of heavy ions [3], which leads to a modification of transverse momentum distributions of the final particles. The ALICE experiment has observed a stronger suppression of hadron production at high-p_{T} in central Pb–Pb collisions at $\sqrt{s_{NN}} = 2.76$ TeV [4], indicating the creation of a denser and hotter medium comparing to RHIC energies. This modification is typically quantified in terms of the nuclear modification factor:

$$R_{\mathrm{AA}} = \frac{1}{\langle T_{\mathrm{AA}} \rangle} \frac{\mathrm{d}N_{\mathrm{AA}}/\mathrm{d}p_{\mathrm{T}}}{\mathrm{d}\sigma_{pp}/\mathrm{d}p_{\mathrm{T}}} \tag{1}$$

where $\mathrm{d}N_{\mathrm{AA}}/\mathrm{d}p_T$ and σ_{pp} represent the yield in nucleus-nucleus collisions and the total inelastic cross section in pp collisions, respectively. $\langle \mathrm{T_{AA}} \rangle$ is the nuclear overlap function, expressed as the ratio of the number of binary nucleon-nucleon collisions $\langle \mathrm{N_{coll}} \rangle$ obtained from a Glauber model and the inelastic nucleon-nucleon cross section σ_{inel}^{NN}. If the Pb–Pb collisions are a simple superposition of elementary pp collisions the R_{AA} is expected to be equal to unity, while $R_{\mathrm{AA}} < 1$ indicates a suppression of charged-particle production compared to binary-collision scaling.

2. Analysis

The preliminary results reported here are based on an analysis of 25×10^6 pp collisions and 3.3×10^6 Pb–Pb collisions at $\sqrt{s_{\rm NN}} = 5.02$ TeV recorded with the ALICE detector during late 2015 that represents approximately 25% and 3% of the total data collected, respectively. Events are triggered using the V0A and V0C detectors, which have full azimuthal coverage and the pseudorapidity coverage of $2.8 < \eta < 5.1$ and $-3.7 < \eta < -1.7$, respectively. For the measurement of charged particle spectrum, secondary particles are rejected by requiring a minimum distance of the particle tracks to the reconstructed primary vertex. Primary particles are selected in Monte Carlo and they are defined as all prompt charged particles produced in the collision including all decay products, except the products from weak decays of light flavour hadrons. The charged particles are reconstructed using the Inner Tracking System (ITS) and the Time Projection Chamber (TPC). In Pb–Pb collisions the sum of amplitudes measured in V0A and V0C is used to estimate the collision centrality [5]. A detailed description of the ALICE detector, its sub-systems, and the performance have been reported in [6].

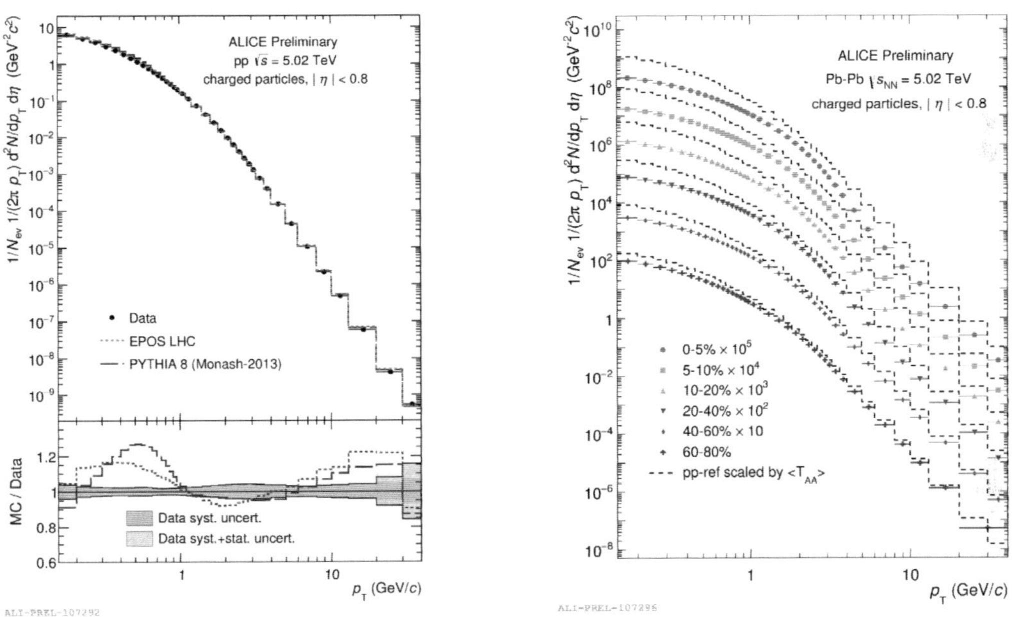

Figure 1. Transverse momentum distribution of charged particles. **Left:** For pp collisions at $\sqrt{s} = 5.02$ TeV, compared to PYTHIA8 and EPOS calculations [7, 13] and the ratio Monte Carlo to data in the lower panel. **Right:** For Pb–Pb collisions at $\sqrt{s_{\rm NN}} = 5.02$ TeV, for six centrality classes. The dashed curves correspond to the measured $p_{\rm T}$ distribution in pp collisions scaled by $\langle T_{\rm AA}\rangle$.

All corrections applied to data are calculated using Monte Carlo simulations from PYTHIA8 [7] for pp and HIJING [8] for Pb–Pb collisions with particle transport via a GEANT3 [9] simulation of the ALICE detector. To account for the differences in the particle composition between event generators and data, the charged-particle reconstruction efficiency ($\sim 70\%$) was calculated from the particle-dependent efficiencies weighted by the relative abundances of each particle measured in pp at 7 TeV[10] and Pb–Pb collisions at 2.76 TeV[11, 12]. The correction for contamination by secondary particles produced from weak decays or secondary interactions is important only at low transverse momentum ($\sim 10\%$), $p_{\rm T} < 1$ GeV/c, while the correction for $p_{\rm T}$-resolution is important at high $p_{\rm T}$ ($\sim 2\%$), $p_{\rm T} > 10$ GeV/c.

IOP Conf. Series: Journal of Physics: Conf. Series **832** (2017) 012009 doi:10.1088/1742-6596/832/1/012009

3. Results

The transverse momentum distribution has been measured for pp and Pb–Pb collisions at the same energy per nucleon for Pb–Pb at $\sqrt{s_{NN}} = 5.02$ TeV. Figure 1 (left) shows the distribution for pp collisions measured in the range of $0.15\ \text{GeV}/c \leq p_T < 40\ \text{GeV}/c$ and $|\eta| < 0.8$. The data is compared to Monte Carlo calculations performed with PYTHIA8 (Monash-2013 tune) [7] and EPOS LHC [13], both describing the data reasonably well at intermediate p_T. However, it is interesting to note that models overestimate the production at low and high transverse momentum by a similar amount $\sim$ 20%. The Pb–Pb transverse momentum distribution is shown in Figure 1 (right) for six centralities together with the pp cross section scaled by the nuclear overlap function.

Figure 3 shows the R_{AA} as a function of p_T for six centralities. The solid symbols correspond to the measurement at $\sqrt{s_{NN}} = 5.02$ TeV while the open symbols represent the published measurement at $\sqrt{s_{NN}} = 2.76$ TeV [4]. For the six centralities, the measurements for both energies show similar behavior within systematic uncertainties. In central collisions (0–5%) a strong suppression is observed, with a minimum at $p_T \sim 7$ GeV/c and a significant rise for $p_T >$ 7 GeV/c. For peripheral collisions (60–80%) R_{AA} exhibits a moderate suppression ($\sim$0.6-0.7). The similarity of the R_{AA} measured at 2.76 TeV and 5.02 TeV could be a hint of a stronger energy loss in the hotter and denser medium taking into account that the spectra are harder at 5.02 TeV than at 2.76 TeV.

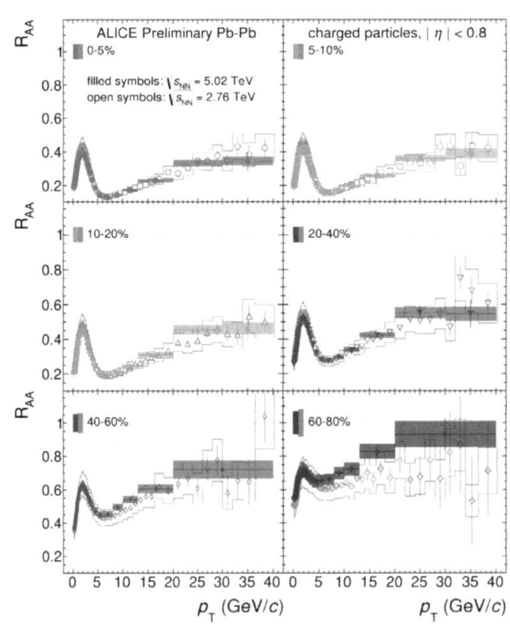

Figure 2. The nuclear modification factor (R_{AA}) as a function of p_T for six centralities. Filled symbols represent the measurement at $\sqrt{s_{NN}} = 5.02$ TeV and empty symbols correspond to R_{AA} at $\sqrt{s_{NN}} = 2.76$ TeV [4]. The normalisation uncertainty is shown for both energies as a box around unity for both energies.

Figure 3 (left) compares the R_{AA} at $\sqrt{s_{NN}} = 5.02$ TeV to model predictions from Djordjevic et al. [14] and Majumder et al. [15]. The two models succeed in describing the R_{AA} measurement at high-p_T, for the 0–5% centrality range. On the right panel of Figure 3, the measurement of the $R_{\rm AA}$ for the 0–10% centrality is compared to predictions by Vitev et at. [16] which achieve a good description of data.

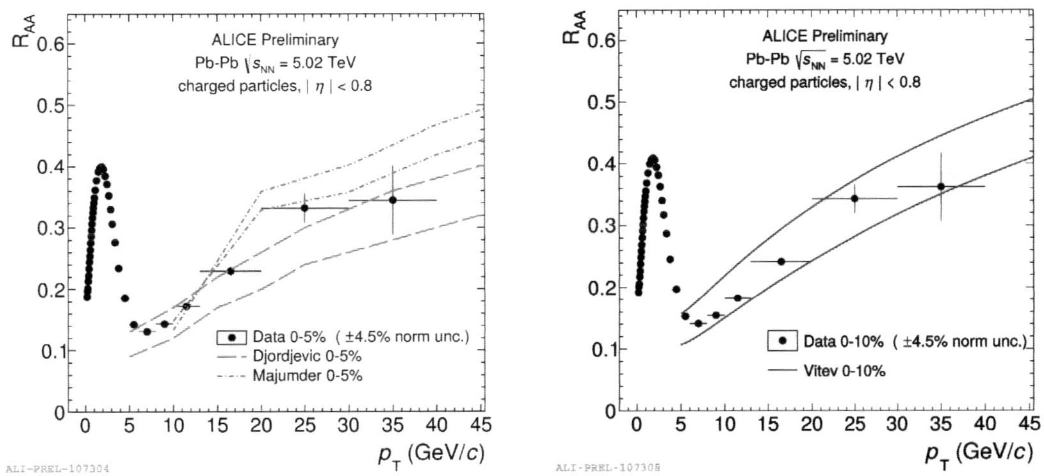

Figure 3. The nuclear modification factor measured at $\sqrt{s_{\rm NN}} = 5.02$ TeV for 0–5% (left) and 0–10% (right) compared with model calculations [14, 15, 16].

4. Summary

The nuclear modification factor in Pb–Pb collisions at $\sqrt{s_{\rm NN}} = 5.02$ TeV was found to be similar to the measurement at 2.76 TeV. A maximal suppression was observed around 6-7 GeV/c for the most central collisions (0–5%) which becomes smaller as centrality increases. The $R_{\rm AA}$ was compared with three models and they succeed in describing the general features of the measurement for a transverse momentum $p_{\rm T}$ >5 GeV/c. The full dataset is currently being analyzed. In addition, an improved analysis method will be applied to the $\sqrt{s_{\rm NN}} = 2.76$ TeV dataset, which is expected to reduce the systematic uncertainties and increase precision in the comparison with the higher collision energy.

References

[1] PHENIX Collaboration 2002 *Phys. Rev. Lett.* **88** 022301
[2] STAR Collaboration 2002 *Phys. Rev. Lett.* **89** 022301
[3] Bjorken J D 1982 *FERMILAB-PUB-82-059-THY*
[4] ALICE Collaboration 2013 *Phys. Lett.* **88** 52-62.
[5] ALICE Collaboration 2013 *Phys. Rev.* C **88** 044909.
[6] ALICE Collaboration 2014 *Int. J. Mod. Phys.* A **29** 1430044.
[7] Skands P, Carranza S and Rojo J 2014 *Eur. Phys. J* C **74** no.8 3024.
[8] Deng W T, Wang X N and Xu R 2011 *Phys. Rev.* C **83** 014915.
[9] Brun R, Bruyant F, Carminati F, Giani S, Maire M, McPherson A, Patrick G and Urban L 1994 *CERN-W5013*
[10] ALICE Collaboration 2014 *Phys. Lett.* B **736** 196-207.
[11] ALICE Collaboration 2016 *Phys. Rev.* C **93** no. 3 034913.
[12] ALICE Collaboration 2013 *Phys. Rev. Lett.* **111** 222301.
[13] Pierog T, Karpenko I, Katzy J M, Yatsenko E and Werner K 2015 *Phys. Rev.* C **92** no.3 034906.
[14] Djordjevic M, Blagojevic B and Zivkovic L 2016 *arXiv:* 1601.07852
[15] Majumder A and Shen C 2012 *Phys. Rev. Lett* **109** 202301.
[16] Vitev I, Chien Y T, Emerman A, Kang Z B and Ovanesyan G 2016 *Phys. Rev.* D **93** no.7 074030.

HQ16 IOP Publishing
IOP Conf. Series: Journal of Physics: Conf. Series **832** (2017) 012010 doi:10.1088/1742-6596/832/1/012010

Jet-hadron correlations relative to the event plane at the LHC with ALICE

Joel Mazer

Department of Physics, University of Tennessee, Knoxville, Tennessee 37996-1200, USA

E-mail: joel.mazer@cern.ch

Abstract. Jet quenching is observed at both RHIC and LHC energies. This suggests that partons lose energy as they traverse the medium. When the trigger jet is studied relative to the event plane, the path length dependence of medium modifications can be studied. We present measurements of the angular correlations relative to the event plane between reconstructed jets and charged hadrons in Pb–Pb collisions at $\sqrt{s_{\rm NN}}$ =2.76 TeV in ALICE. A newly implemented, robust background subtraction method to remove the complex, flow dominated, combinatorial background is used in this analysis.

1. Introduction

Jets are ideal probes of the Quark Gluon Plasma (QGP) because they originate from hard-scattered partons created early in the collision, prior to the formation of the medium. These partons are modified in the presence of a medium through collisional energy loss and induced gluon radiation. This modification is observed at both LHC and RHIC energies via the suppression of high-momentum particles [1–4]. A suppression is also seen for high p_T di-hadron correlations [5–8].

By reconstructing jets, ideally, a more complete picture of how lost energy is redistributed and how the effects of jet quenching emerge can be obtained. This article will discuss the status of azimuthal correlations of reconstructed jets binned relative to the event plane with charged hadrons in ALICE.

2. Experiment

Clusters measured in the Electromagnetic Calorimeter (EMCal) and charged hadrons measured with the central tracking system allow ALICE to study fully reconstructed jets [4]. Charged hadrons are reconstructed from their tracks using information from the Time Projection Chamber (TPC) and the Inner Tracking System (ITS).

Tracks are reconstructed over the full azimuthal range and come from mid-rapidity in a range of $|\eta_{lab}|$ <0.7 when reconstructed in a jet, and $|\eta_{lab}|$ <0.9 for the associated charged hadrons. The EMCal has an acceptance window of $|\eta_{lab}|$ <0.7 and $|\Delta\phi|$=107°. For a complete description of the ALICE detector, see [9–12].

3. Jet-hadron correlations

3.1. Data sample

The data used for the correlation analysis was collected by the ALICE Experiment in 2011 during the 2.76 TeV Pb–Pb collision data taking. This work uses events which fired the gamma trigger used by the ALICE EMCal [10]. In addition, a cluster constituent of the trigger jet was required to be matched to the fired trigger patch of the event.

3.2. Jet Reconstruction

Jets are reconstructed with a resolution parameter of R=0.2 using the anti-k_T jet-finding algorithm from the FastJet package [13]. Jets reconstructed in this analysis require a leading cluster to have E_T >6.0 GeV, which exceeds the trigger threshold. The jets are reconstructed with constituent tracks of p_T >3.0 GeV/c and clusters of E_T >3.0 GeV. Since the primary goal of jet finding is to reconstruct the jet axis accurately, this high constituent cut limits the influence of background on jet finding.

3.3. Measurement

We define the jet-hadron correlation function in heavy-ion collisions by Eq. 1.

$$\frac{1}{N_{trig}}\frac{d^2N_{assoc,jet}}{d\Delta\phi d\Delta\eta} = \frac{1}{\epsilon a N_{trig}}\frac{d^2N^{same}_{pairs}}{d\Delta\phi d\Delta\eta} - b_0(1+\sum v_n^{trig} v_n^{assoc}\cos(n\Delta\phi)) \tag{1}$$

Here the first term represents the same event pairs which are divided by an acceptance correction, a, provided by mixed events. Mixed events are used to correct for the limited detector efficiency and acceptance for track pairs. The single track reconstruction efficiency of associated particles is denoted by ϵ. The second term of Eq. 1 is the combinatorial heavy-ion background where b_0 is the background level and the v_n terms are the Fourier coefficients of the trigger jet and associated particles.

The trigger jets in this analysis are binned in angle relative to the event plane to explore the path length dependence of medium modifications. The orientations are defined such that in-plane is $0 < |\Delta\varphi| < \frac{\pi}{6}$, mid-plane is $\frac{\pi}{6} < |\Delta\varphi| < \frac{\pi}{3}$, and out-of-plane is $\frac{\pi}{3} < |\Delta\varphi| < \frac{\pi}{2}$, where $\Delta\varphi$ denotes the angular difference between the trigger jet and the reconstructed event plane.

3.4. Background subtraction

When the trigger jet is restricted relative to the event plane, both the background level and v_n are modified and will contain a dependence on the reaction plane resolution $\Re$. The derivation of the event plane dependent background equations are given in [14]. The reaction plane resolution corrects for the difference between the reconstructed event plane and the underlying symmetry plane, Ψ_n.

Various background subtraction methods were developed in [15] and applied to the azimuthal correlation functions. The primary method is the reaction plane fit (RPF). The RPF method works under the assumption that the signal is negligible in the large $\Delta\eta$ and small $\Delta\phi$ region. The 2D correlations are projected over 0.8< $|\Delta\eta|$ <1.2 to define the background dominated region, while the signal+background region is defined to be $|\Delta\eta|$ <0.6. To ensure as much information as possible is going into the fit by constraining the shape and level of the background, the in-plane, mid-plane, and out-of-plane orientations are simultaneously fit up to fourth order in v_n and required to have the same fit parameters.

The background dominated region is fit over the region $|\Delta\phi| < \pi/2$, shown in Fig. 1. We see from the blue band that the RPF fit models the data quite well, even at low p_T where the background is large. The higher order Fourier coefficents clearly matter, as the background for trigger jets mid-plane has four peaks, consistent with a v_4 dependence. This method does not require independent measurements of v_n and is able to extract the signal with smaller errors while requiring fewer assumptions and less bias than prior subtraction methods. This is especially useful since there are currently no v_3^{jet} or v_4^{jet} measurements.

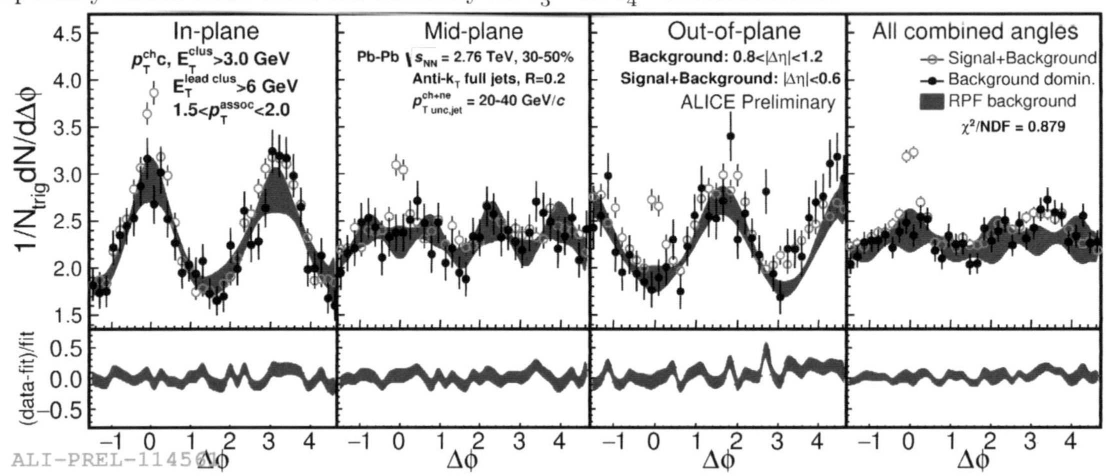

Figure 1. Signal+background region, background dominated region and RPF fit to background are shown for the 1.5-2.0 GeV/c associated p_T bin for full jets in Pb–Pb collisions at $\sqrt{s_{\rm NN}}$ = 2.76 TeV for the 30-50% most central events. The blue band shows the uncertainty of the background fit, which is non-trivially correlated point-to-point [15, 16].

3.5. Results

We extract the signal by subtracting the large correlated background from the correlation function. Figure 2 shows the signal for associated particles of 2.0-3.0 GeV/c. The uncertainties are dominated by statistics. With higher statistics, the uncertainties could be vastly reduced to allow for a more precise measurement.

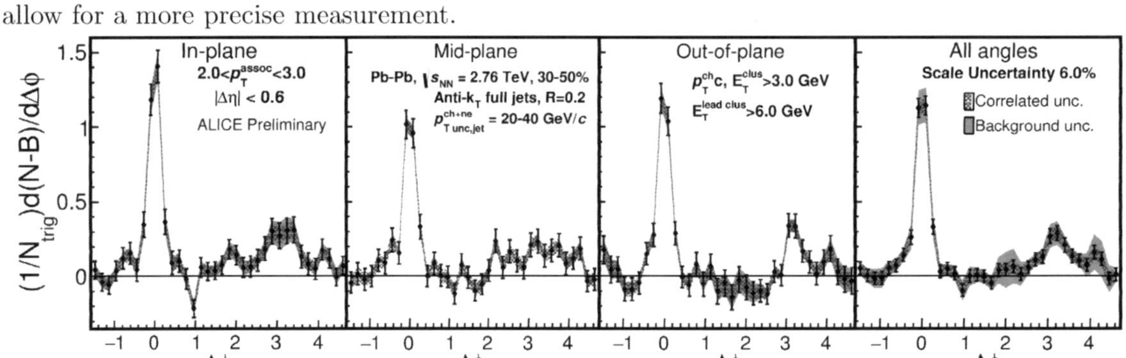

Figure 2. Corrected $\Delta\phi$ correlation function for 2.0< p_T <3.0 GeV/c associated hadrons for full jets in Pb–Pb collisions at $\sqrt{s_{\rm NN}}$ = 2.76 TeV for the 30-50% most central events. The blue band corresponds to the correlated uncertainty and the grey band corresponds to the background uncertainty which is non-trivially correlated point-to-point [15, 16]. There is an additional 6% global scale uncertainty.

We define the yields by Eq. 2.

$$Y = \frac{1}{N_{trig}} \int_c^d \int_a^b \frac{d(N_{meas} - N_{bkgd})}{d\Delta\phi} d\Delta\phi d\Delta\eta \qquad (2)$$

Where the integration limits a and b correspond to $\Delta\phi$ values of -1.047 and +1.047 on the near-side and +2.094 and +4.189 on the away-side respectively. In addition, the integration limits c and d correspond to $\Delta\eta$ values of -0.6 and 0.6 for both the near-side and away-side. The near-side (left) and away-side (right) jet yields for 1.0< p_T^{assoc} <10.0 GeV/c are shown in Fig. 3. The near-side yield is consistent with little or no modification. There is no clear dependence of the away-side peaks on orientation relative to the event plane. There are competing effects across different p_T ranges. Jet quenching could cause a decrease in yield going from in-plane to out-of-plane, while gluon radiation could cause an increase.

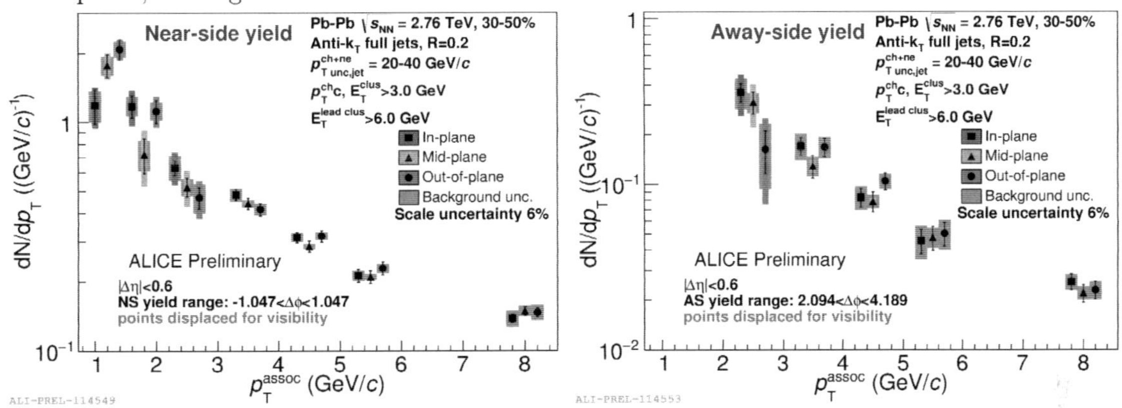

Figure 3. The near-side yield (left) and away-side yield (right) for full jets in Pb–Pb collisions at $\sqrt{s_{\rm NN}}$ = 2.76 TeV for the 30-50% most central events. The colored bands correspond to the correlated uncertainties and the grey band corresponds to the background uncertainty which is non-trivially correlated point-to-point [15,16]. There is an additional 6% global scale uncertainty.

4. Summary and Outlook

The jet-hadron correlation results were seen to have their uncertainties dominated by statistics. No significant event plane dependence was seen to within the current uncertainties on the extracted jet yield. The RPF background subtraction method applied to data was seen to have various advantages in that it does not require independent v_n measurements, has less bias and fewer assumptions than prior background subtraction methods, and can extract the signal with great precision. The current results are consistent with the re-analysis of STAR data seen in [16].

References

[1] Adare A *et al.* (PHENIX Collaboration) 2008 *Phys.Rev.* **C77** 011901 (*Preprint* 0705.3238)
[2] Adams J *et al.* (STAR Collaboration) 2003 *Phys. Rev. Lett.* **91** 172302 (*Preprint* nucl-ex/0305015)
[3] Adare A *et al.* (PHENIX Collaboration) 2011 *Phys.Rev.* **C84** 024904 (*Preprint* 1010.1521)
[4] Adam J *et al.* (ALICE Collaboration) 2015 *Phys. Lett.* **B746** 1–14 (*Preprint* 1502.01689)
[5] Adams J *et al.* (STAR Collaboration) 2005 *Phys. Rev. Lett.* **95**(15) 152301
[6] Abelev B I *et al.* (STAR Collaboration) 2009 *Phys. Rev. C* **80**(6) 064912
[7] Agakishiev G *et al.* (STAR Collaboration) 2012 *Phys. Rev. C* **85**(1) 014903
[8] Aggarwal M M *et al.* (STAR Collaboration) 2010 *Phys. Rev. C* **82**(2) 024912
[9] Abelev B B *et al.* (ALICE Collaboration) 2014 *Int. J. Mod. Phys.* **A29** 1430044 (*Preprint* 1402.4476)
[10] Abeysekara U *et al.* (ALICE EMCal Collaboration) 2010 (*Preprint* 1008.0413)
[11] Abelev B *et al.* (ALICE Collaboration) 2014 *J. Phys.* **G41** 087002
[12] Alme J *et al.* 2010 *Nuclear Instruments and Methods in Physics Research Section A: Accelerators, Spectrometers, Detectors and Associated Equipment* **622** 316 – 367 ISSN 0168-9002
[13] Cacciari M, Salam G P and Soyez G 2012 *Eur.Phys.J.* **C72** 1896 (*Preprint* 1111.6097)
[14] Bielcikova J, Esumi S, Filimonov K, Voloshin S and Wurm J 2004 *Phys.Rev.* **C69** 021901 (*Preprint* nucl-ex/0311007)
[15] Sharma N, Mazer J, Stuart M and Nattrass C 2016 *Phys. Rev.* **C93** 044915 (*Preprint* 1509.04732)
[16] Nattrass C, Sharma N, Mazer J, Stuart M and Bejnood A 2016 *Phys. Rev.* **C94** 011901 (*Preprint* 1606.00677)

IOP Conf. Series: Journal of Physics: Conf. Series **832** (2017) 012011 doi:10.1088/1742-6596/832/1/012011

Neutral pion and eta meson measurements with the ALICE detector

Adam Matyja[1,2,*]

[1] University of Tennessee, Knoxville, Tennessee, US
[2] Institute of Nuclear Physics PAN, Cracow, Poland
[*] On behalf of the ALICE collaboration

E-mail: adam.tomasz.matyja@cern.ch

Abstract. The ALICE experiment at the LHC is optimized to study the quark-gluon plasma (QGP), created in heavy-ion collisions. The medium-induced energy loss of particles can be investigated via the measurement of neutral meson spectra in heavy-ion collisions as well as via neutral meson-hadron correlations. Neutral mesons are identified from decay photon pairs via the invariant mass technique. Photons are measured in ALICE directly in the two electromagnetic calorimeters (PHOS and EMCal), as well as via the method of photon conversion (PCM) into electron-positon pairs, where the latter are measured in the inner tracking system (ITS) and the time projection chamber (TPC). Results obtained from EMCal, PHOS and PCM are consistent and allow measurements of spectra with high precision over a wide kinematical range. Suppression of the high-p_T meson production is observed through the mesasurement of nuclear modification factor (R_{AA}), which decreases with increasing the centrality of the collision. The suppression of the per-trigger yield on the away side in high-p_T π^0 hadron correlations as measured by the modification factor (I_{AA}) also shows evidence for parton energy loss in the medium.

1. Introduction

Neutral π^0 and η mesons are produced via parton fragmentation in pp collisions. The meson production differential cross section is a convolution of parton distribution functions (PDF) in the proton, fragmentation functions (FF) of the parton into a meson, and the inclusive cross section of partons production. Studies of neutral meson production constrain both the PDF and FF. The dense environment created in heavy-ion collisions allows observation of modification of the PDF or FF via linear and non-linear recombination effects caused by the dense matter [1]. Insight into modification can be quantified with particle ratio comparisons between ultra-relativistic heavy-ion collisions and pp collisions. Light mesons are mostly produced in gluon fragmentation at LHC energies. Due to the different color factor, gluons will suffer a larger energy loss in the medium than quarks. Both this and a different relative contribution of gluons and quarks to neutral meson production may lead to differences in the suppression pattern of π^0 and η mesons [2, 3]. High-p_T meson production spectra give information about the energy loss of the scattered parton in the dense medium.

Two-particle angular correlations between high-p_T particles and associated particles are also used to study jet-quenching phenomena. They provide information about the amount and distribution of the partonic energy loss in medium. Remnants of the radiated energy and the medium response to the initial parton can be studied in a soft and hard regime.

2. Neutral meson measurement

Photons were measured with two methods in the ALICE detector [4] at the LHC [5]. The first is via the measurement of the energy deposited by photons in the sensitive material on the EMCal [6] or PHOS [7] electromagnetic calorimeters. The second is based on the conversion of photons into e^+e^- pairs in the detector material (called the photon conversion method, PCM). Charged particles created by γ conversion are measured in the six layers of the silicon inner tracking system (ITS) and the gaseous time projection chamber (TPC). Neutral mesons are combined from photon pairs via the invariant mass technique.

ALICE has measured π^0 production spectra for four center-of-mass energies $\sqrt{s} = 0.9$ [8], 2.76 [9], 7 [8] and 8 TeV [10] in pp collisions. The p_{T} distributions presented in Fig. 1 are described by PYTHIA8 event generator and compared to the next-to-leading order perturbative QCD (NLO pQCD) calculations for two renormalization and factorization scales. The MSTW PDF and DSS14 FF [11] used in predictions describe the measured spectra better than previously used calculations [8]. Models describe data quite well for lower collision energies. A disagreement for higher energies starts at medium p_{T} and growing with p_{T} reaching almost factor two.

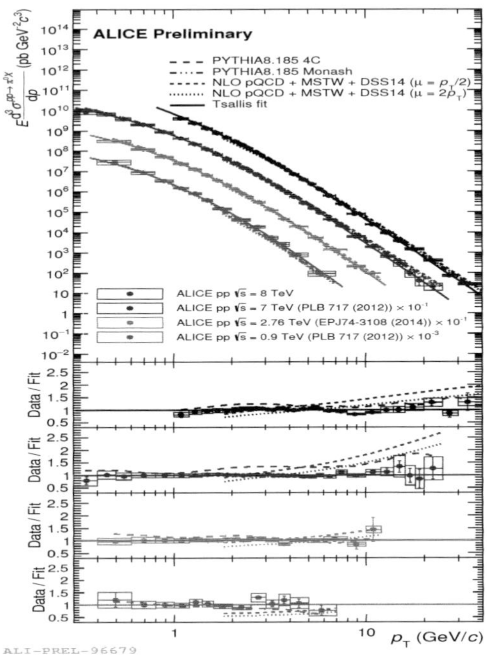

Figure 1: (On the left) π^0 production spectra in pp collisions for energies $\sqrt{s} = 0.9$, 2.76, 7 and 8 TeV compared to PYTHIA, NLO pQCD calculations and described by Tsalis fit.

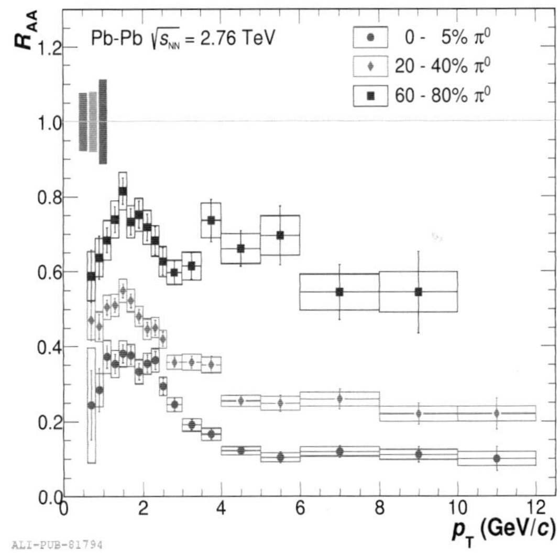

Figure 2: (Above) Neutral pion nuclear modification factor R_{AA} for three centrality classes in Pb-Pb collisions at $\sqrt{s_{\mathrm{NN}}} = 2.76$ TeV [9].

ALICE also measured p_{T} spectra for six centrality classes in Pb-Pb collisions at $\sqrt{s_{\mathrm{NN}}} = 2.76$ TeV [9]. Comparison of the data to theory predictions and Monte Carlo (MC) models can be found in [9]. Nuclear effects are quantified via the nuclear modification factor R_{AA} defined as: $R_{\mathrm{AA}}(p_T) = \frac{1}{N_{\mathrm{coll}}} \frac{dN_{\mathrm{AA}}/dp_{\mathrm{T}}}{dN_{\mathrm{pp}}/dp_{\mathrm{T}}}$, where dN/dp_{T} is the yield in nuclear and pp collisions scaled by the number of binary nucleon-nucleon collisions, N_{coll}, obtained from Glauber Monte Carlo simulations [12, 13]. The interplay of the initial and the final state effects contribute to the nuclear modification factor. There is no impact of medium when $R_{\mathrm{AA}} = 1$. Figure 2 shows the p_{T} dependence of R_{AA} for three centrality classes [9]. The largest suppression is observed for the most central Pb-Pb collisions with a value of $R_{\mathrm{AA}} \sim 0.1$ above $p_{\mathrm{T}} \approx 4$ GeV/c.

The increase of the integrated luminosity for [9] by a factor of 10 extended the p_{T} range of π^0 spectrum up to 20 GeV/c and allowed η meson p_{T} spectra measurements for the first time [10]

IOP Conf. Series: Journal of Physics: Conf. Series **832** (2017) 012011 doi:10.1088/1742-6596/832/1/012011

in Pb-Pb collisions at $\sqrt{s_{\mathrm{NN}}} = 2.76$ TeV. The η meson spectrum for pp and Pb-Pb collisions is shown in Fig. 3. The NLO pQCD prediction [11] is compatible with pp result. The same prediction scaled by N_{coll} compared to Pb-Pb results indicates a suppression of the yield.

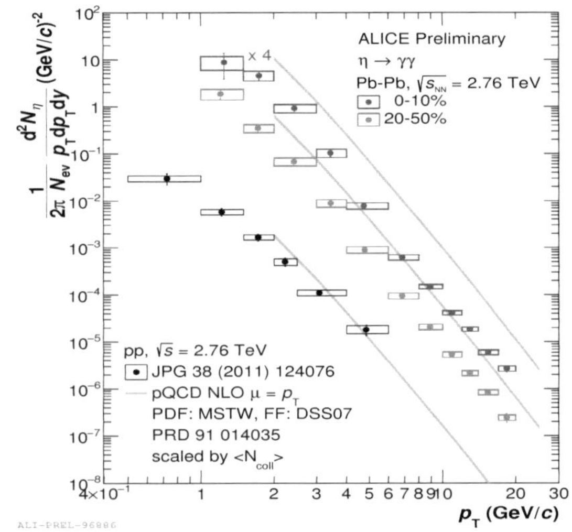

Figure 3: η meson production spectra in Pb-Pb collisions at $\sqrt{s_{\mathrm{NN}}} = 2.76$ TeV for two centrality classes compared to pp data and NLO pQCD predictions scaled by N_{coll}.

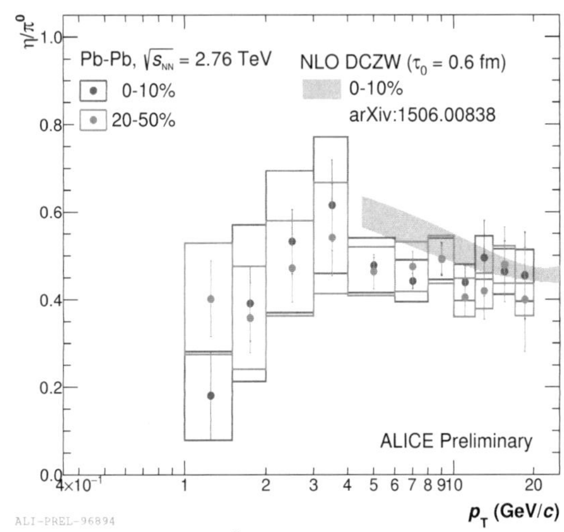

Figure 4: The η to π^0 meson production ratio in Pb-Pb collisions at $\sqrt{s_{\mathrm{NN}}} = 2.76$ TeV for two centrality classes compared to NLO theoretical calculation with quenching.

The measured η/π^0 production ratios in Pb-Pb collisions for two centrality classes compared to a theory prediction based on NLO and jet quenching [2] are shown in Fig. 4. The η/π^0 ratio is flat above $p_{\mathrm{T}} = 4$ GeV/c. An agreement with pQCD predictions within uncertainties is visible. A comparison to the pp result at $\sqrt{s} = 7$ TeV [8] and $K^{\pm}/\pi^{\pm}$ result [14] in Pb-Pb collisions for the same centrality classes indicates no significant differences within uncertainties. A comparison to the K/π ratio is of interest, even if strangeness content is different, because the kaon has approximately the same mass as the η meson.

3. π^0 - hadron correlations

Two particle correlations between high-p_{T} ($8 < p_{\mathrm{T}} < 16$ GeV/c) π^0 mesons and charged hadrons have been studied in the ALICE experiment in pp and central Pb-Pb collisions at $\sqrt{s_{\mathrm{NN}}} = 2.76$ TeV [15]. There are two jet origin structures in the azimuthal correlation $\Delta\varphi = \varphi^{\mathrm{trig}} - \varphi^{\mathrm{assoc}}$ of the trigger particle and associated particles. The near and away side peaks appear at $\Delta\varphi = 0$ and $\Delta\varphi = \pi$, respectively. The effect of medium is quantified with the medium induced per-trigger yield modification factor $I_{\mathrm{AA}}(p_T^{\pi^0}, p_T^{\mathrm{h}^{\pm}}) = \frac{Y^{\mathrm{AA}}(p_{\mathrm{T}}^{\pi^0}, p_{\mathrm{T}}^{\mathrm{h}^{\pm}})}{Y^{\mathrm{pp}}(p_{\mathrm{T}}^{\pi^0}, p_{\mathrm{T}}^{\mathrm{h}^{\pm}})}$, where $Y^{\mathrm{AA(pp)}}(p_{\mathrm{T}}^{\pi^0}, p_T^{\mathrm{h}^{\pm}})$ is the per-trigger yield for given trigger particle $p_{\mathrm{T}}^{\pi^0}$ and the associated particle $p_{\mathrm{T}}^{\mathrm{h}^{\pm}}$ in A-A (pp) collisions. The per-trigger yield modification factor for the near and away side are shown in the left and right hand side of Fig. 5, respectively. The enhancement on the near side ($1.2 < I_{\mathrm{AA}} < 1.8$) can be interpreted as the modification of the fragmentation function, quark jet to gluon jet ratio modification, or bias in the parton p_{T} due to the energy loss. A suppression is observed on the away side ($I_{\mathrm{AA}} \approx 0.6$) above $p_{\mathrm{T}}^{assoc} = 3$ GeV/c, which is a manifestation of energy loss in medium. The enhancement visible at low p_{T}^{assoc} may arise from k_{T} broadening, medium excitation or fragments from radiated gluons. Results are compared to different models. JEWEL [16] and NLO pQCD [17] models describe only suppression on the away side. AMPT [18] describes low-p_{T} part (quantitatively) and overpredicts high-p_{T} region. The π^0-hadron I_{AA} are compatible with the charged di-hadron correlation measurements in ALICE [19].

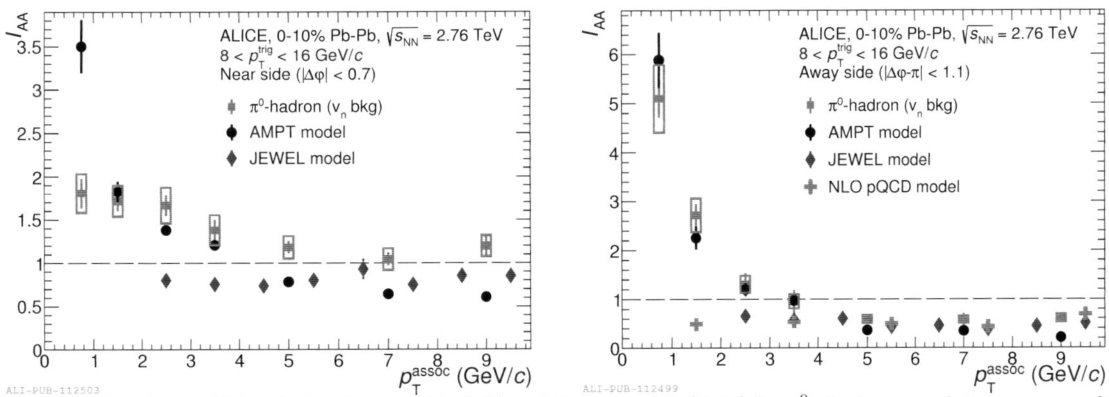

Figure 5: Per trigger yield ratio for the near side (left) and the away side (right) for π^0 - hadron correlations compared to several Monte Carlo predictions [15].

4. Summary

We presented measurements of the invariant π^0 yields for pp collisions at $\sqrt{s} = 0.9$, 2.76, 7 and 8 TeV as well as for Pb-Pb collisions at $\sqrt{s_{\rm NN}} = 2.76$ TeV over a wide $p_{\rm T}$ range. The NLO pQCD predictions describe pp results at lower energies quite well while there is an increasing discrepancy at high $p_{\rm T}$ for the highest energies. The nuclear modification factor $R_{\rm AA}$ for Pb-Pb collisions at $\sqrt{s_{\rm NN}} = 2.76$ TeV indicates a strong and $p_{\rm T}$ dependent π^0 suppression. The first η meson measurement at LHC in heavy-ion collisions has been shown. The η/π^0 ratio is rising towards $p_{\rm T} = 4$ GeV/c and then remaining essentially constant at 0.45 for higher $p_{\rm T}$ values. The π^0-hadron correlations have been measured in pp and the most central Pb-Pb collisions. The observed away side high-$p_{\rm T}$ suppression and low-$p_{\rm T}$ enhancement is partially described by different MC generators. Results constrain perturbative QCD inspired models, parametrization of parton distribution and fragmentation functions as well as phenomena governing the quark-gluon medium.

This work was supported by the U. S. Department of Energy under Grant No. DE-FG02-96ER40982 and National Science Centre, Poland under Grant No. 2013/08/M/ST2/00598.

References

[1] Lappi T and Mäntysaari H 2013 *Phys. Rev.* D **88** 114020
[2] Dai W, Chen X F, Zhang B W and Wang E 2015 *Phys. Lett.* B **750** 390
[3] Guo X and Wang X N 2000 *Phys. Rev. Lett.* **85** 3591
[4] Aamodt K *et al.* (ALICE Collaboration) 2008 *J. Instrum* **3** S08002
[5] Evans L *et al.* 2008 *J. Instrum* **3** S08001
[6] Abeysekara U *et al.* (ALICE EMCal Collaboration) 2010 ALICE EMCal PPR (*Preprint* 1008.0413)
[7] Aleksandrov D V *et al.* (ALICE PHOS Collaboration) 2005 *Nucl.Instrum. Meth.* A **550** 169–184
[8] Abelev B *et al.* (ALICE Collaboration) 2012 *Phys. Lett.* B **717** 162
[9] Abelev B *et al.* (ALICE Collaboration) 2014 *Eur. Phys. J.* C **74** 3108
[10] Morreale A (ALICE Collaboration) 2015 *Nucl. Phys.* A **00** 1 (*Preprint* 1512.05250v1)
[11] de Florian D, Sassot R, Epele M, Hernandez-Pinto R J and Stratmann M 2015 *Phys. Rev.* D **91** 014035
[12] Miller M L, Reygers K, Sanders S J and Steinberg P 2007 *Ann. Rev. Nucl. Part. Sci.* **57** 205
[13] Abelev B *et al.* (ALICE Collaboration) 2013 *Phys. Lett.* B **720** 52
[14] Abelev B *et al.* (ALICE Collaboration) 2014 *Phys. Lett.* B **736** 196
[15] Adam J *et al.* (ALICE Collaboration) 2016 *Phys. Lett.* B **763** 238
[16] Zapp K C, Krauss F and Wiedemann U A 2013 *JHEP* **03** 080
[17] Liu Z Q, Zhang H, Zhang B W and Wang E 2016 *Eur. Phys. J.* C **76** 20
[18] Ma G L and Wang X N 2011 *Phys. Rev. Lett.* **106** 162301
[19] Aamodt K *et al.* (ALICE Collaboration) 2012 *Phys. Rev. Lett.* **108** 092301

IOP Conf. Series: Journal of Physics: Conf. Series **832** (2017) 012012 doi:10.1088/1742-6596/832/1/012012

sPHENIX: The next generation heavy ion detector at RHIC

Sarah Campbell for the sPHENIX Collaboration
Columbia University, New York, NY USA
E-mail: sc3877@columbia.edu

Abstract. sPHENIX is a new collaboration and future detector project at Brookhaven National Laboratory's Relativistic Heavy Ion Collider (RHIC). It seeks to answer fundamental questions on the nature of the quark gluon plasma (QGP), including its coupling strength and temperature dependence, by using a suite of precision jet and upsilon measurements that probe different length scales of the QGP. This is possible with a full acceptance, $|\eta| < 1$ and 0-2π in ϕ, electromagentic and hadronic calorimeters and precision tracking enabled by a 1.5 T superconducting magnet. With the increased luminosity afforded by accelerator upgrades, sPHENIX is going to perform high statistics measurements extending the kinematic reach at RHIC to overlap the LHC's. This overlap is going to facilitate a better understanding of the role of temperature, density and parton virtuality in QGP dynamics and, specifically, jet quenching. This paper focuses on key future measurements and the current state of the sPHENIX project.

1. Introduction

The goal of the sPHENIX program [1] is to probe the Quark-Gluon Plasma (QGP) created in heavy ion collisions at multiple length scales. It does this through three avenues: studying the structure of jets as they evolve in the QGP, varying the jet parton (from the smallest gluon-jets, to light quark-, and then the larger, heavier charm and bottom quark-jets) and studying the sequential melting of the three upsilon states ($\Upsilon(1S)$, $\Upsilon(2S)$, $\Upsilon(3S)$). Initially jets created in heavy ion collisions are highly virtual, traveling through the QGP, but as they fragment they reach a scale close to that of the QGP and QGP interactions are more probable. As a result, the sensitivity of jet measurements to QGP effects is higher in lower energy jets and jets created at RHIC energies. Similarly, the partonic composition of jets differs at RHIC and LHC energies, with a higher fraction of quark-jets available at lower jet energies at RHIC compared to the LHC. Additional jet-parton studies require identifying heavy flavor jets. This allows further study of collisional versus radiative energy loss and the dead cone effect. Finally, measurements of the upsilon states at RHIC provide insight into bottom quarkonia behavior at lower QGP temperatures (T_{RHIC} is 77% lower than T_{LHC}) and with reduced Υ production from coalescence. For each of these signals, it is necessary to obtain complementary measurements at both RHIC and the LHC to disentangle their respective contributions and sensitivities. sPHENIX is the future RHIC detector capable of complementing the upgraded LHC heavy ion program.

2. The sPHENIX detector

The sPHENIX detector design, Figure 1, is driven by the goal of measuring the rare signals discussed above. To make the most of RHIC luminosities, the sPHENIX detector covers 2π in

IOP Conf. Series: Journal of Physics: Conf. Series **832** (2017) 012012 doi:10.1088/1742-6596/832/1/012012

azimuth (ϕ), ± 1 units in rapidity (η) and reads out data at a rate of 15 kHz. To efficiently resolve jets and their energies, sPHENIX has both hadronic and electromagnetic calorimeters. Efficient tracking of particles from 0.2 to 40 GeV/c in transverse momentum (p_T) is needed for jet fragmentation measurements. Heavy flavor jet identification requires precision vertexing with a distance of closest approach in the x-y plane (DCA_{xy}) resolution of better than 70 μm. To measure the three Υ states in the dielectron decay channel, hadron rejection of better than 99% and a mass resolution of 1% at the Υ mass are required. These precise and efficient tracking requirements are met using a 1.5 T magnet and three tracking subsystems: a time projection chamber (TPC), a Si strip inner tracker (INTT) and a precision MAPS detector based off of ALICE's ITS upgrade. The remainder of this section discusses the calorimeter and tracking detectors, Figures 2 and 3 respectively, in more detail.

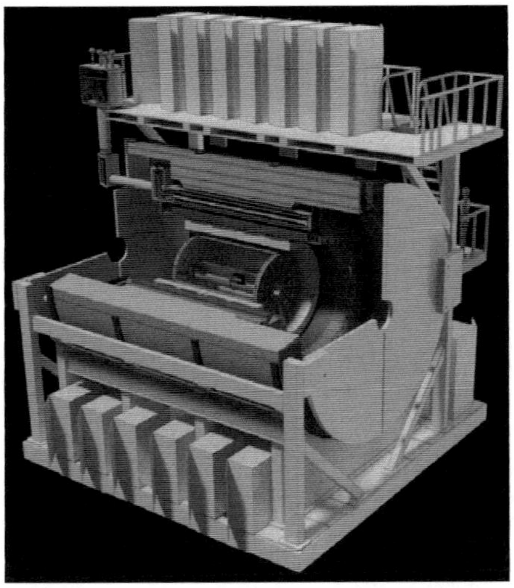

Figure 1. Engineering drawing of sPHENIX with its support structure, three calorimeters, the magnet, and three tracking detectors.

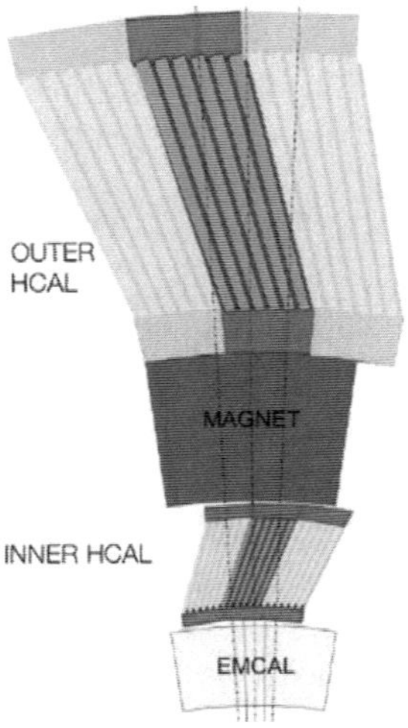

Figure 2. Angular section of the sPHENIX HCal and EMCal shown with the magnet.

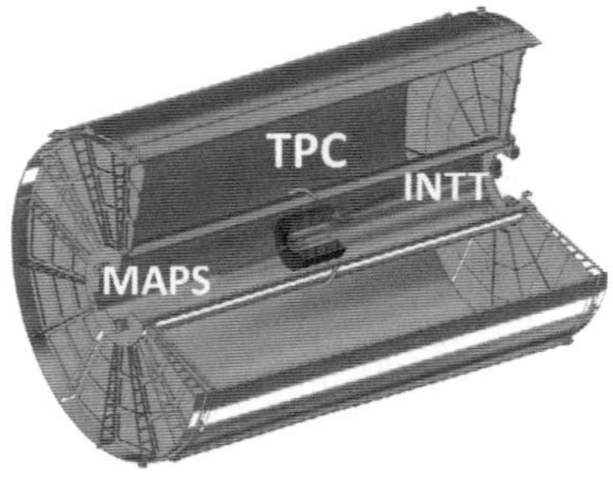

Figure 3. Engineering drawing of the sPHENIX tracking detectors, specifically the TPC, INTT, and MAPS detector.

2.1. Calorimeters

The hadronic calorimeters (HCal) consist of tilted steel plates interleaved with polystyrene panels embedded with 1 mm wavelength shifting fiber. It has a segmentation of 0.1×0.1 in ϕ and η. Inner and Outer HCal subsystems measure the hadronic energy deposited before and after the magnetic coil. The HCal design has a single particle energy resolution requirement of better than $100\%/\sqrt{E}$. The electromagnetic calorimeter (EMCal) is made up of W modules embedded with Si fibers evenly spaced in ϕ and η. The EMCal ϕ and η segmentation is 0.025×0.025 and it has an energy resolution requirement of better than $15\%/\sqrt{E}$. Research and development work on the EMCal is ongoing to determine whether modules should be projective in both ϕ and η or just in ϕ. Si-photomultipliers measure the light produced in both the HCal and EMCal systems and a common digitizer readout is planned.

In the winter of 2016, a prototype of the calorimeter systems was studied at the FermiLab Test Beam Facility. The measured energy distributions in the HCal are well reproduced by full GEANT simulations. This provides added confidence in our simulated detector response. Furthermore, preliminary analyses of the the combined HCal and EMCal single particle energy

IOP Conf. Series: Journal of Physics: Conf. Series **832** (2017) 012012 doi:10.1088/1742-6596/832/1/012012

resolution and electron energy resolution in the EMCal meet the design goals, with a projected single particle energy resolution in the range $(70.6\% - 95.7\%)/\sqrt{E}$ and an electron energy resolution of roughly $14.2\%/\sqrt{E}$ or $12.7\%/\sqrt{E}$ depending on the shower calibration method.

2.2. Tracking detectors

The outer-most tracking detector, the TPC, is located between 20 and 78 cm in radius and has an approximate 250 μm effective hit resolution. It provides the bulk of the pattern recognition and momentum resolution for the tracking of particles between 0.2 and 40 GeV/c in p_T. A continuous, non-gated, TPC readout is planned to be compatible with sPHENIX's high data acquisition rate. The inner-most tracking detector, the MAPS detector, consists of three layers of Si sensors following the ALICE ITS upgrade design [2]. It contributes both precision event vertex determination, $|z_{vtx}| < 10$ cm, and identification of off-vertex decays, $DCA_{xy} < 70$ μm. Located between the TPC and the MAPS layers, the INTT detector provides needed continuity in the tracking, redundancy in pattern recognition and DCA determination, and pile-up rejection. It consists of 4 layers of Si strips and will be read out out by reusing PHENIX FVTX electronics.

3. Simulated detector performance

The performance of the combined tracking systems is simulated by embedding pions in central HIJING events. Current results in this ongoing effort show efficient tracking out to 40 GeV/c in p_T and a distance of closest approach resolution of 40 μm at the lowest p_T values. The calorimetric jet reconstruction performance is characterized by the jet energy resolution, efficiency and purity. These quantities in sPHENIX are simulated using central HIJING events with an ATLAS-influenced jet reconstruction algorithm for jet radii of 0.2, 0.3, and 0.4 [3]. The resulting jet efficiency is better than 90% for jets with p_T greater than 20 GeV/c and the jet purity is better than 80% for jets with a p_T of greater than 25 GeV/c.

To estimate the available jet yields in sPHENIX, projected RHIC luminosities and perturbative QCD (pQCD) rates of hard processes are needed. Thanks to RHIC luminosity upgrades, over one hundred billion minimum bias events are expected in 22 weeks of $\sqrt{s_{NN}} =$ 200 GeV Au+Au collisions, of which approximately twenty billion are from 0-20% central events. These values, combined with pQCD rate calculations, provide estimates of the expected jet yields available to sPHENIX, presented in Table 1, confirming that rare jet probes can be measured with high statistics at sPHENIX. Figure 4 shows the projected statistical errors and kinematic reach of the nuclear modification factor, R_{AA}, for jets, b-jets, and direct photons, assuming 22 weeks of Au+Au collisions and 10 weeks of p+p collsions at RHIC. These projections extend the current RHIC capabilities shown here with PHENIX measurements as a reference. Specifically, projected inclusive jet measurements extend out to 80 GeV/c in p_T, projected b-jet measurements extend out to over 40 GeV/c and projected direct photon measurements extend out to over 50 GeV/c. With this increased kinematic range there is significant overlap in the projected p_T ranges for jet and heavy flavor measurements accessible at the future RHIC experiment, sPHENIX, and at future upgraded LHC experiments.

To complete the b-jet program, sPHENIX needs to be able to identify or tag b-jet events. Three methods of b-jet tagging are being explored: a) identifying multiple tracks with a large DCA, b) finding a secondary vertex, and c) tagging B-mesons by semi-leptonic decay or the baryon mass. While method c) is still under development for both p+p and A+A collisions, Pythia8 simulations find that methods a) and b) can identify b-jets with an estimated 30% purity and 70% efficiency. Given this level of b-jet tagging, the sPHENIX Collaboration expects to measure the R_{AA} of b-jets out to 40 GeV/c in p_T and potentially constrain b-jet transport coefficients in models.

For the Υ program, Figure 5 shows the projected upsilon yields in 10 weeks of p+p collisions measured by sPHENIX. Over 8800 $\Upsilon(1S)$, 2200 $\Upsilon(2S)$, and 1160 $\Upsilon(3S)$ are expected.

Table 1. Projected yields for 0-20% $\sqrt{s_{NN}} = 200$ GeV Au+Au collisions in 22 weeks at RHIC.

Signal	p_T range	pQCD	Yields
Light q + g jets	$p_T > 20$ GeV/c	NLO	10^7
Light q + g jets	$p_T > 30$ GeV/c	NLO	10^6
Direct photons	$p_T > 20$ GeV/c	NLO	10^4
c-jets	$p_T > 20$ GeV/c	FONLL	10^4
b-jets	$p_T > 20$ GeV/c	FONLL	10^4

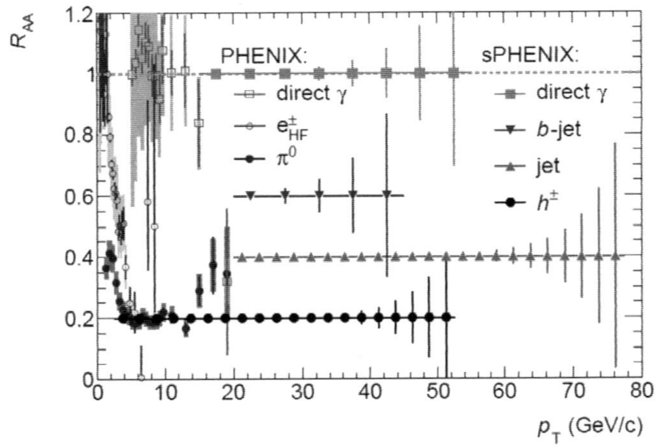

Figure 4. Statistical projections of the R_{AA} for various hard processes measured with sPHENIX given 22 week of Minimum Bias Au+Au collisions with 10 weeks of p+p collisions.

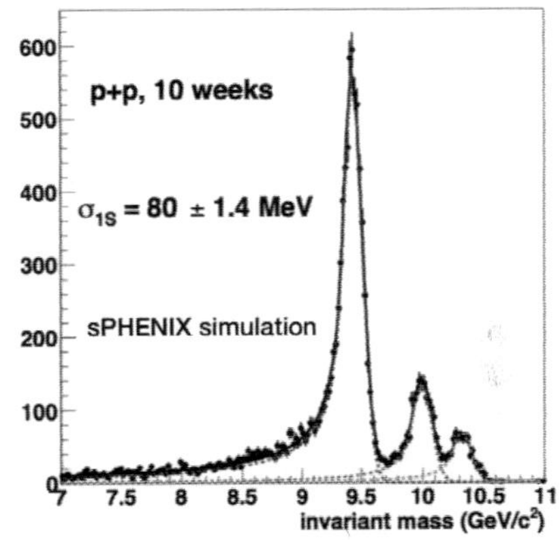

Figure 5. Simulated results of sPHENIX's ability to measure the three upsilon states from the dielectron decay channels in 10 weeks of p+p collisions.

Furthermore, sPHENIX can clearly separate the three upsilon states with a $\Upsilon(1S)$ width of 80 ± 1.4 MeV/c^2. The future sPHENIX detector provides, for the first time, the ability to separate all three upsilon states at RHIC. This mass resolution is maintained in Au+Au collisions, allowing centrality-dependent R_{AA} measurements for each of the upsilon states.

4. Conclusions

The future sPHENIX detector will extend RHIC results beyond PHENIX and STAR's current capablilites, and will provide necessary complementary measurements to the future LHC experiments. This complementarity is needed to form a complete picture of the properties and behavior of the QGP created in heavy ion collisions. sPHENIX plans to achieve this by focusing on jet, upsilon and b-jet observables. Having recently obtained CD-0 designation, with first beam expected in 2022 [4], the future for heavy ion physics with sPHENIX is bright.

References

[1] Adare A *et al.* 2015 An upgrade proposal from the PHENIX Collaboration (*Preprint* `arXiv:1501.06197`)
[2] Abelev B *et al.* 2014 *J. Phys. G* **41** 087002
[3] Hanks J A *et al.* 2012 *Phys. Rev. C* **86** 024908
[4] Hallman T 2016 RHIC/AGS User's Meeting talk

IOP Conf. Series: Journal of Physics: Conf. Series **832** (2017) 012013 doi:10.1088/1742-6596/832/1/012013

The Initial Stages of Colliding Nuclei and Hadrons

Prithwish Tribedy
Physics Department, Brookhaven National Laboratory, Upton, NY 11973, USA
E-mail: ptribedy@bnl.gov

Abstract. The final day of the Hot Quarks 2016 conference was focused on the discussions of the initial stages of colliding nuclei and hadrons. In this conference proceedings we give a brief overview of a few selective topics discussed at the conference that include latest developments in the theoretical description of the initial state towards understanding a number of recent experimental results from RHIC and LHC.

1. Introduction

In the standard model of heavy ion collisions, by "initial stages", one commonly refers to two Lorentz contracted sheets of saturated partonic matter that collide to form an ensemble of highly occupied gluon fields. This initial stage undergoes a pre-equilibrium evolution and eventually evolves to a phase describable by viscous hydrodynamics. Many aspects of the standard model of heavy collisions have been verified by a number of experimental observations. One powerful observable in this context is the two-dimensional (2D) di-hadron correlation function (per-trigger-particle associated yield distribution) expressed in terms of relative pseudo-rapidity ($\Delta\eta$) and azimuthal angle ($\Delta\phi$) of the emitted particles as shown in the left panel of Fig.1. Such correlations include two major components, the di-jet and the ridge. The short range di-jet correlations give rise to a narrow near-side peak at ($\Delta\eta = 0, \Delta\phi = 0$) but can extend over the entire $\Delta\eta$ range on the away side ($\Delta\phi \sim \pi$) whereas the long-range ridge-like correlations can persist up to large $\Delta\eta$ on both near and away sides.

The production of back-to-back di-jets are constrained by momentum conservation and the strength of such correlations gets diminished when going from peripheral to central events [1]. Such observation is a consequence of the well-known scenario of jet-quenching, a key signature to support the formation of a strongly correlated Quark Gluon Plasma (sQGP) in A+A collisions. On the other hand, the strength of the long range ridge-like correlations are often characterized by the Fourier coefficients $V_{\Delta n}$ obtained after the harmonic decomposition of the correlation functions integrated over $\Delta\eta$. These two particle $V_{\Delta n}$ coefficients can be related to single-particle azimuthal anisotropy harmonics v_n, a possible source of which is the hydrodynamic expansion of the medium in response to initial spatial anisotropy that is nearly boost-invariant, once again consistent with the scenario of sQGP formation in A+A collisions.

For years collisions of small systems such as p+p and p+A have been providing baselines for measurements in A+A collisions. Such consensus has been strongly challenged by a number of striking recent measurements that resemble features commonly observed in A+A collisions. As shown in Fig.1, the long range ridge like structure seen in peripheral Pb+Pb collisions also appears in the high multiplicity events in p+p and p+Pb collisions for hadrons with $1<p_T<3$ GeV.

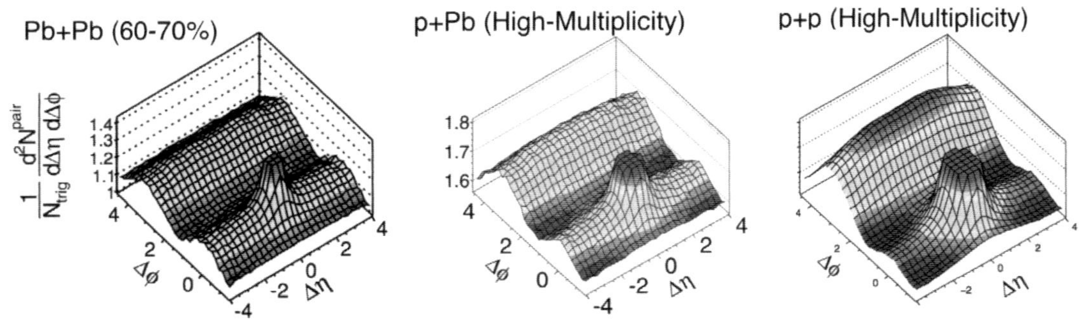

Figure 1. Di-hadron correlations in relative pseudo-rapidity and azimuthal angle in peripheral Pb+Pb and high multiplicity p+Pb and p+p collisions [1–3].

2. Initial state correlations

By looking at Fig. 1 a natural question arises : does the same underlying mechanism drive these long-range correlations in all collision systems ? From causality arguments it is known that any form of correlation that spans over a wide range of rapidity must originate at the early stages of collisions [4]. This makes long-range di-hadron correlations an initial state driven phenomenon for any collision systems. However what short of initial state correlations dominate the experimental observations could be a matter of debate. In recent years, two possible sources of initial state correlations have been identified : 1) intrinsic momentum space correlations of partons that survive the process of hadronization, 2) position space correlations (geometric) of partons that are converted to momentum space correlations by the final state interactions. In principle both should contribute to the observed experimental correlations although their relative contribution will depend on the collision system.

In a scenario like A+A collisions where the system size and the number of produced initial partons is large, the mean free path of a typical semi-hard parton ($\sim$ GeV) is expected to be much smaller than the system size. Therefore, the initial-state momentum space correlations between such partons may be destroyed due to the phenomenon of quenching. This is a natural consequence for a system approaching thermalization and undergoing hydrodynamic evolution in the subsequent phase [5, 6]. As a result of which the azimuthal correlations of final hadrons particularly at small transverse momentum in A+A collisions will be dominated by hydrodynamic response to the initial state geometry caused by the position space correlations of initial partons.

In the collisions of small systems, however, the scenario is different as the number of initial (semi-) hard partons is small. In a typical low multiplicity event of small collision systems, the dominant correlation will be due to initial state momentum space correlations of back-to-back jets that escape the interaction region without loosing a fraction of their energy by interacting with the other soft partons produced in the collision. Going to relatively higher multiplicity events, additional multi-parton processes will start to contribute to such momentum space correlations as a consequence of gluon saturation. Going to even higher multiplicity events, interaction among the produced particles will lead to dilution of initial state correlations. Eventually in the limit of very high multiplicity, when all semi-hard partons are quenched, a complete loss of momentum space correlations will be accompanied by formation of a thermalized medium and one will recover the picture of A+A collisions. Therefore, for a fixed size system there may be a critical density of partons leading to a cross over from a regime dominated by initial-sate dynamics to one dominated by final state interaction such as hydrodynamic expansion (see Ref [7] for a detailed discussion). So far a first-principles description of a given system over

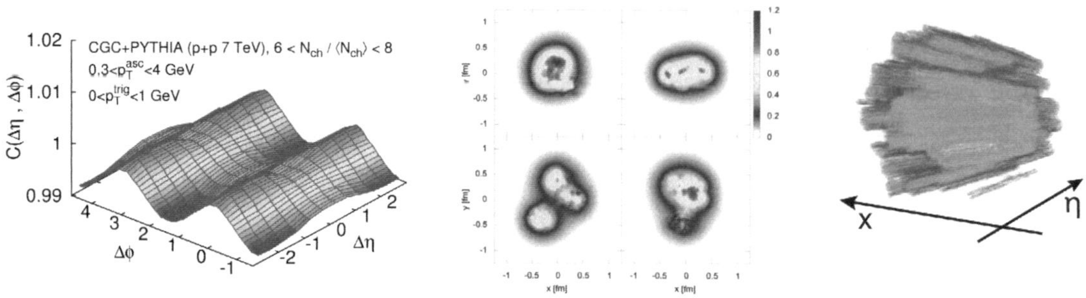

Figure 2. (left) Ridge-like structure of the di-hadron correlations obtained in CGC-PYTHIA (CGC-Lund) framework [10]. (Center) Distribution of gluon density around valance quarks inside a proton constrained by HERA data [11]. (Right) Distribution of initial energy density obtained from the 3D-Glasma initial conditions [12].

the entire regime of parton densities in a single framework is not feasible. Current theoretical studies of the initial state therefore mainly focus in a regime where either 1) final state effects can be neglected, e.g. to describe the systematics of di-hadron correlations in small collision systems or 2) the initial state momentum space correlations can be neglected, e.g. to describe the initial conditions for fluid-dynamic simulations in A+A collisions. We will discuss these approaches in the rest of the proceedings.

3. Angular correlations in the collisions of small systems

The interesting observations in small collision systems such as p+p are made in high multiplicity events. It is therefore necessary to understand how such events originate in the first place. This would require an understanding of multi-particle productions in QCD at high energies ($\sqrt{s} \to \infty, x \to 0$). At such energies the occupation number of gluons below the saturation momentum $k_\perp < Q_S$ in the hadron wave functions becomes large ($O(1/\alpha_S)$) as a consequence of gluon saturation. In the collisions of two such hadrons, the weak coupling description of particle production at high energies is best captured in the framework of Color-Glass-Condensate (CGC). In this framework the origin of high-multiplicity events is a consequence of initial state fluctuations that lead to rare configurations of the parton distribution in the colliding hadrons and the mechanism of correlated n-particle production in CGC [8, 9]. The same underlying mechanism is also responsible for azimuthally collimated two particle production.

The angular collimation in CGC can be attributed to the details of the wave functions of the colliding hadrons that lead to enhancement in the probability of finding two gluons with similar quantum numbers and transverse momentum close to the saturation scale Q_S. Such angular collimation extends over a long range in rapidity due to the boost-invariant nature of the classical gluon fields that dominate the hadronic wave function. Such fields lead to rapidity independent gluon production over an interval of $\Delta y \lesssim 1/\alpha_s$ that can generate ridge like structure. The systematics of such correlations in CGC have been analyzed under different approximation schemes e.g. the glasma-graph contribution to two gluon production, Bose-enhancement of two idential momentum partons in the projectile, spatial density variations of partons, local anisotropy of color-domains in the target, multiple interactions of the classical fields [4, 13–23], etc. For a recent review on this we refer the reader to Ref [24].

The different sources of fluctuations and the mechanisms of multi-particle production that lead to initial partonic correlations within the framework of CGC discussed above are well incorporated in the phenomenological approach of the IP-Glasma model [25]. In this model one can compute the momentum space distributions of gluons dN_g/dyd^2k_t after collisions

IOP Conf. Series: Journal of Physics: Conf. Series **832** (2017) 012013 doi:10.1088/1742-6596/832/1/012013

that contain the information of $n-$particle correlations. Averaged over many events, these lead to quantities of experimental interests like two-particle azimuthal correlation functions, multiplicity distributions, etc. However such correlations exist at the parton and in order to make comparisons to experimental data, a realistic hadronization scheme is required. The recent major progress in this direction has been the development of CGC-PYTHIA (CGC-Lund) model [10] which combines the event-by-event distribution dN_g/dyd^2k_t from IP-Glasma and the Lund string fragmentation of PYTHIA. CGC-PYTHIA naturally produces the probability distribution of multiplicity $P(N_{ch})$ for charged hadrons that is consistent with the experimental distributions. The tails of such distributions are populated by high multiplicity events ($N_{ch} > 6 \times \langle N_{ch} \rangle$) for which one can calculate the 2D di-hadron correlation function as shown in Fig.2 (left). One can clearly see the long-range[1] ridge like structure as seen in the experiment (Fig.1). In addition, the CGC-PYTHIA model, purely from initial state dynamics, can also reproduce the mass ordering of average transverse momentum $\langle p_T \rangle_p > \langle p_T \rangle_K > \langle p_T \rangle_\pi$ and elliptic flow coefficients $v_2(\pi) > v_2(K) > v_2(p)$ at $p_T \lesssim 3$ GeV that are often regarded as signatures of collectivity driven by hydrodynamics [10].

4. Initial state geometry, fluctuations in heavy ion collisions

Significant amount of research in recent years has been focused on the development of fluctuating initial conditions for A+A collisions. The commonly used Monte-Carlo models of A+A collisions can be classified into many categories e.g. 1) the models purely based on collision geometries such as different implementations of MC-Glauber, 2) models that combine collision geometry with string-fragmentations such as NeXus [26], EPOS [27], UrQMD [28] and AMPT [29] 3) models that include gluon saturation such as DIPSY [30], MC-KLN [31, 32], MC-rcBK [33] and IP-Glasma [25]. These models provide the distributions of initial energy, entropy density, flow etc. that can be matched to hydrodynamic simulations under certain approximations [34]. Due to a large number of unknown parameters involved in the hydrodynamic simulations, in principle it is difficult to directly constrain these models from the measurements in A+A collisions. However recently a few techniques and observables have been identified. For example the studies in Ref. [35] have demonstrated that, in the regime where non-linearities from hydrodynamics are negligible, the correlation between v_3 and v_2 can directly constrain the correlation of ε_3 and ε_2. Such studies have largely excluded different models expect DIPSY, IP-Glasma and MC-Glauber with binary collision scaling. Among other observables, the probability distributions of flow harmonics $P(v_n)$ measured by the ATLAS collaboration [36] have been found to be largely insensitive to the hydrodynamic evolution, particularly the effects of viscosity [37] and therefore can be directly compared to the corresponding initial state eccentricity distributions $P(\varepsilon_n)$. Both MC-KLN and MC-Glauber models have failed to describe $P(v_n)$ data and are therefore been largely constrained at LHC energies. In addition, recently the STAR collaboration [38] has measured the correlation of v_2 with multiplicity N_{ch} from RHIC in ultra-central Au+Au and deformed U+U collisions. Such measurements indicate that models of initial conditions with binary collision scaling of multiplicity such as the two-component Glauber model can not describe the trend seen in data. A good description to such data can be obtained from IP-Glasma model which indicates that color-coherence play important role in the particle production at RHIC [39].

In this context, there have been three major recent developments for Glauber-like models of initial conditions that include, TRENTO [40], Quark-Glauber [41, 42] and Shadowed Glauber [43]. These improvements to Glauber initial conditions modify the two-component model that relate collision geometry to multiplicity or energy density by including additional coherence and lead to a successful description of ultra-central U+U data at RHIC. Recently a p-QCD based model of initial conditions "EKRT" has been developed [37] that includes NLO

[1] The experimental distribution of the di-hadron correlations also includes the di-jet peak which is not present in the current implementation of CGC-PYTHIA.

IOP Conf. Series: Journal of Physics: Conf. Series **832** (2017) 012013 doi:10.1088/1742-6596/832/1/012013

cross section of mini-jet production using nuclear PDF, geometry of A+A collisions and an implementation of gluons saturation. EKRT combined with viscous hydrodynamic simulations successfully describes a large number of data including v_n, $P(v_n)$ at RHIC and LHC.

The IP-Glasma is the only model of initial conditions that naturally includes sub-nucleonic fluctuations of initial energy densities. However importance of additional sources of sub-nucleonic fluctuations in the IP-Glasma model was realized while analyzing p+A collisions [44]. In the original implementation of IP-Glasma the proton shape was assumed to be round. A recent modification to the IP-Glasma model includes eccentric shape of protons with gluon density distributed around three valence quarks as shown in Fig. 2(center). Such shape has been constrained by incoherent diffractive $e + p$ data of HERA [11] and will be important for studying light-heavy ion collisions at RHIC and LHC.

Over the past years most of the models of initial conditions have assumed boost-invarince and have not included the full three dimensional (3D) fluctuating structure of the initial energy density. Although such assumptions are somewhat acceptable for observables at mid-rapidity at the LHC and top RHIC energy, at lower energies it is essential to incorporate the longitudinal dependence of initial energy density to perform a full 3+1 dimensional hydrodynamic simulations. Such calculations have been performed by initializing the hydrodynamic simulations with inputs from UrQMD [28], AMPT [29] and EPOS [45] initial conditions. Recently, the inclusion of longitudinal structure in Glauber like initial conditions was performed in the 3D-Glauber model introduced in Ref [46] by combining the LeXus model [47] of scattering with the geometry of heavy ion collisions. 3D-Glauber provides both energy density and net-baryon density needed to perform full 3D+1 hydrodynamic simulations for the Beam Energy Scan (BES) program at RHIC. Many CGC models of initial conditions based on k_T-factorization can provide full 3D initial energy density. In case of the IP-Glasma model, in the original implementation, boost-invariance was assumed. Recently the rapidity dependence of the initial energy density has been included in the IP-Glasma model (Fig.2(right)) by incorporating the JIMWLK evolution of the classical fields in rapidity and this new implementation is referred to as the 3D-Glasma model [12]. Among other models of initial conditions that can provide rapidity dependent initial conditions are the strong coupling based Holographic initial conditions that include the numerical relativity solutions to AdS/CFT to match with hydrodynamic simulations [48,49]. In such models the transverse geometry is needed to be introduced separately, using a MC-Glauber implementation. Such initial conditions can also be generalized to lower energies.

It must be noted that in most cases the current state of hydrodynamic modeling of A+A collisions comes with a big caveat of directly matching of different models of initial conditions under the assumption of instant thermalization or isotropization. For example, in the weak coupling approach, the initial state consisting of highly occupied gluons $f \sim \mathcal{O}(\frac{1}{\alpha_S}) \gg 1$ is far from being isotropic. However a recent work [5] demonstrates that one can use an effective kinetic theory (EKT) description of QCD to such initial distribution after a subsequent pre-equilibrium evolution when the occupation number becomes $f \ll \mathcal{O}(\frac{1}{\alpha_S})$. The EKT description can then lead the system towards isotropization that can be smoothly matched to a hydrodynamical description when the occupancies are $f \sim \mathcal{O}(1)$ [5].

5. Conclusion

The study of the initial state is a continuously progressing field of research and a large number of exciting new developments have improved our understanding of the structures of colliding hadrons and nuclei. Recent measurements in small systems have provided many challenges towards understanding of the interplay between initial and final state effects in different collisions systems. A comprehensive simulation of heavy ion collisions by smoothly matching initial conditions to fluid-dynamic simulations is very close to reality with recent progress in the understanding of thermalization.

IOP Conf. Series: Journal of Physics: Conf. Series **832** (2017) 012013 doi:10.1088/1742-6596/832/1/012013

Aknowledgement

This work was supported under Department of Energy Contract No. DE-SC0012704. We thank Björn Schenke for his comments on the manuscript.

References

[1] Chatrchyan S *et al.* (CMS) 2012 *Eur. Phys. J.* **C72** 2012 (*Preprint* 1201.3158)
[2] Chatrchyan S *et al.* (CMS Collaboration) 2013 *Phys.Lett.* **B718** 795–814 (*Preprint* 1210.5482)
[3] Khachatryan V *et al.* (CMS) 2016 *Phys. Rev. Lett.* **116** 172302 (*Preprint* 1510.03068)
[4] Dumitru A, Gelis F, McLerran L and Venugopalan R 2008 *Nucl.Phys.* **A810** 91–108 (*Preprint* 0804.3858)
[5] Kurkela A and Zhu Y 2015 *Phys. Rev. Lett.* **115** 182301 (*Preprint* 1506.06647)
[6] Berges J, Boguslavski K, Schlichting S and Venugopalan R 2014 *Phys. Rev.* **D89** 074011 (*Preprint* 1303.5650)
[7] Schlichting S and Tribedy P 2016 *Advances in High Energy Physics* **8460349** 17 (*Preprint* 1611.00329)
[8] Schenke B, Tribedy P and Venugopalan R 2014 *Phys. Rev.* **C89** 024901 (*Preprint* 1311.3636)
[9] McLerran L and Tribedy P 2015 (*Preprint* 1508.03292)
[10] Schenke B, Schlichting S, Tribedy P and Venugopalan R 2016 *Phys. Rev. Lett.* **117** 162301 (*Preprint* 1607.02496)
[11] Mntysaari H and Schenke B 2016 *Phys. Rev. Lett.* **117** 052301 (*Preprint* 1603.04349)
[12] Schenke B and Schlichting S 2016 *Phys. Rev.* **C94** 044907 (*Preprint* 1605.07158)
[13] Dusling K and Venugopalan R 2012 *Phys.Rev.Lett.* **108** 262001 (*Preprint* 1201.2658)
[14] Altinoluk T, Armesto N, Beuf G, Kovner A and Lublinsky M 2015 *Phys. Lett.* **B751** 448–452 (*Preprint* 1503.07126)
[15] Levin E and Rezaeian A H 2011 *Phys. Rev.* **D84** 034031 (*Preprint* 1105.3275)
[16] Kovner A and Lublinsky M 2011 *Phys.Rev.* **D83** 034017 (*Preprint* 1012.3398)
[17] Kovchegov Y V and Wertepny D E 2013 *Nucl. Phys.* **A906** 50–83 (*Preprint* 1212.1195)
[18] McLerran L and Skokov V V 2014 (*Preprint* 1407.2651)
[19] Dumitru A and Giannini A V 2014 (*Preprint* 1406.5781)
[20] Dumitru A, McLerran L and Skokov V 2015 *Phys. Lett.* **B743** 134–137 (*Preprint* 1410.4844)
[21] Lappi T 2015 *Phys. Lett.* **B744** 315–319 (*Preprint* 1501.05505)
[22] Schenke B, Schlichting S and Venugopalan R 2015 *Phys.Lett.* **B747** 76–82 (*Preprint* 1502.01331)
[23] Dusling K, Tribedy P and Venugopalan R 2016 *Phys. Rev.* **D93** 014034 (*Preprint* 1509.04410)
[24] Lappi T, Schenke B, Schlichting S and Venugopalan R 2016 *JHEP* **01** 061 (*Preprint* 1509.03499)
[25] Schenke B, Tribedy P and Venugopalan R 2012 *Phys. Rev. Lett.* **108** 252301 (*Preprint* 1202.6646)
[26] Drescher H J, Ostapchenko S, Pierog T and Werner K 2002 *Phys. Rev.* **C65** 054902
[27] Werner K, Liu F M and Pierog T 2006 *Phys. Rev.* **C74** 044902 (*Preprint* hep-ph/0506232)
[28] Petersen H, Steinheimer J, Burau G, Bleicher M and Stocker H 2008 *Phys. Rev.* **C78** 044901
[29] Pang L G, Qin G Y, Roy V, Wang X N and Ma G L 2015 *Phys. Rev.* **C91** 044904 (*Preprint* 1410.8690)
[30] Flensburg C, Gustafson G and Lonnblad L 2011 *JHEP* **08** 103 (*Preprint* 1103.4321)
[31] Kharzeev D, Levin E and Nardi M 2005 *Phys. Rev.* **C71** 054903 (*Preprint* hep-ph/0111315)
[32] Drescher H J and Nara Y 2007 *Phys. Rev.* **C76** 041903 (*Preprint* 0707.0249)
[33] ALbacete J L and Dumitru A 2010 (*Preprint* 1011.5161)
[34] Noronha-Hostler J 2016 (*Preprint* 1610.05341) URL https://inspirehep.net/record/1492549/files/arXiv:1610.05341.pdf
[35] Retinskaya E, Luzum M and Ollitrault J Y 2014 *Phys. Rev.* **C89** 014902 (*Preprint* 1311.5339)
[36] Aad G *et al.* (ATLAS) 2013 *JHEP* **11** 183 (*Preprint* 1305.2942)
[37] Niemi H, Eskola K J and Paatelainen R 2016 *Phys. Rev.* **C93** 024907 (*Preprint* 1505.02677)
[38] Adamczyk L *et al.* (STAR) 2015 *Phys. Rev. Lett.* **115** 222301 (*Preprint* 1505.07812)
[39] Schenke B, Tribedy P and Venugopalan R 2014 *Phys. Rev.* **C89** 064908 (*Preprint* 1403.2232)
[40] Moreland J S, Bernhard J E and Bass S A 2015 *Phys. Rev.* **C92** 011901 (*Preprint* 1412.4708)
[41] Adare A *et al.* (PHENIX) 2016 *Phys. Rev.* **C93** 024901 (*Preprint* 1509.06727)
[42] Eremin S and Voloshin S 2003 *Phys. Rev.* **C67** 064905 (*Preprint* nucl-th/0302071)
[43] Chatterjee S, Singh S K, Ghosh S, Hasanujjaman M, Alam J and Sarkar S 2016 *Phys. Lett.* **B758** 269–273 (*Preprint* 1510.01311)
[44] Schenke B and Venugopalan R 2014 *Phys. Rev. Lett.* **113** 102301 (*Preprint* 1405.3605)
[45] Werner K, Karpenko I, Pierog T, Bleicher M and Mikhailov K 2010 *Phys. Rev.* **C82** 044904 (*Preprint* 1004.0805)
[46] Monnai A and Schenke B 2016 *Phys. Lett.* **B752** 317–321 (*Preprint* 1509.04103)
[47] Jeon S and Kapusta J I 1997 *Phys. Rev.* **C56** 468–480 (*Preprint* nucl-th/9703033)
[48] van der Schee W, Romatschke P and Pratt S 2013 *Phys. Rev. Lett.* **111** 222302 (*Preprint* 1307.2539)
[49] van der Schee W and Schenke B 2015 *Phys. Rev.* **C92** 064907 (*Preprint* 1507.08195)

IOP Conf. Series: Journal of Physics: Conf. Series **832** (2017) 012014 doi:10.1088/1742-6596/832/1/012014

Femtoscopy in $\sqrt{s_{\mathrm{NN}}} = 5.02$ TeV p-Pb collisions with *ATLAS*

Michael Clark on behalf of the ATLAS Collaboration
Columbia University, 116th St & Broadway, New York, NY 10027, USA
E-mail: michael.ryan.clark@cern.ch

Abstract. Bose-Einstein correlations between identified charged pions are measured for p+Pb collisions at $\sqrt{s_{\mathrm{NN}}} = 5.02$ TeV with the ATLAS detector using a total integrated luminosity of 28 nb^{-1}. Pions are identified using ionisation energy loss measured in the pixel detector. Two-particle correlation functions and the extracted source radii are presented as a function of average transverse pair momentum (k_{T}) and rapidity ($y^{\star}_{\pi\pi}$) as well as collision centrality. Pairs are selected with a rapidity $-2 < y^{\star}_{\pi\pi} < 1$ and with an average transverse momentum $0.1 < k_{\mathrm{T}} < 0.8$ GeV. The effect on the two-particle correlation function from jet fragmentation is studied, and a new method for constraining its contributions to the measured correlations is described. The measured source sizes are substantially larger in more central collisions and are observed to decrease with increasing pair k_{T}. A correlation with the local single-particle multiplicity $\mathrm{d}N_{\mathrm{ch}}/\mathrm{d}y^{\star}$ is demonstrated. The scaling of the extracted radii with the mean number of participants is also used to compare a selection of initial-geometry models. The cross term R_{ol}, which couples radial and longitudinal expansion, is measured as a function of rapidity, and a departure from zero is observed with 4.8 σ combined significance for $y^{\star}_{\pi\pi} > -1$ in the most central events.

Multi-particle correlations in proton-lead (p+Pb) [1, 2, 3, 4, 5] and proton-proton ($p+p$) [6] collisions exhibit long-range azimuthal correlations similar to those observed in lead-lead (Pb+Pb) collisions, in which they are attributed to collective expansion of the quark-gluon plasma. Hydrodynamic models can describe the observed correlations in p+Pb [7, 8, 9], but the extent to which they are appropriate in "small" systems remains an open question.

Bose-Einstein correlations between charged pion pairs can be used to measure the spacetime extent of particle production with a technique called femtoscopy (see Reference [10] and references therein). The width of the correlations in relative momentum $\mathbf{q}$ is studied as a function of pair transverse momentum k_{T}. The measured Hanbury Brown and Twiss (HBT) radii, which are interpreted as the size of the particle-emitting region at freezeout, are expected to decrease with rising transverse momentum if the source expands [11]. Thus, HBT measurements can be used to address the questions raised by angular correlation results. While femtoscopic methods have already been applied to p+Pb systems at the LHC [12, 13], these results are obtained using a new data-driven technique to constrain the significant background contribution from jet fragmentation, and they provide the first measurements of the dependence of these source radii on the rapidity $y^{\star}_{\pi\pi}$.

The data used in this analysis were collected by the ATLAS detector [15] during the 2013 p+Pb run at the Large Hadron Collider (LHC) with a nucleon-nucleon centre-of-momentum energy of $\sqrt{s_{\mathrm{NN}}} = 5.02$ TeV. The inner detector is used to reconstruct charged particles.

IOP Conf. Series: Journal of Physics: Conf. Series **832** (2017) 012014 doi:10.1088/1742-6596/832/1/012014

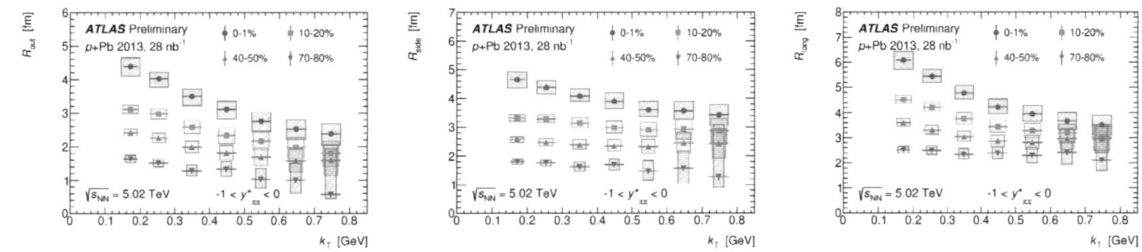

Figure 1. Three-dimensional HBT radii as a function of the average transverse momentum k_T of pion pairs [14]. The widths of the boxes vary between centrality intervals only for visual clarity.

It consists of a silicon pixel detector, a semiconductor tracker made of double-sided silicon microstrips, and a transition radiation tracker made of straw tubes. Pions are identified using a measurement of the charge deposited in the pixel detector. All three subdetectors are composed of a barrel and two symmetrically placed end-cap sections. The forward calorimeters, located in a pseudorapidity range of $3.1 < |\eta| < 4.9$, are used to estimate the centrality of each collision.

The two-particle correlation functions are taken to be of the form

$$C_{\mathbf{k}}(\mathbf{q}) = [(1-\lambda) + \lambda K(\mathbf{q}) C_{\mathrm{BE}}(\mathbf{q})]\, \Omega(\mathbf{q}), \tag{1}$$

where K is a correction factor for final-state Coulomb interactions and Ω represents non-femtoscopic background which arises due to jet fragmentation. The Bose-Einstein part of the three-dimensional correlation functions are fit to functions of the form

$$C_{\mathrm{BE}}(\mathbf{q}) = 1 + e^{-\|R\mathbf{q}\|}, \tag{2}$$

where R is a symmetric matrix in the Bertsch-Pratt coordinate system:

$$R = \begin{pmatrix} R_{\mathrm{out}} & 0 & R_{\mathrm{ol}} \\ 0 & R_{\mathrm{side}} & 0 \\ R_{\mathrm{ol}} & 0 & R_{\mathrm{long}} \end{pmatrix}. \tag{3}$$

In these coordinates, q_{out} is the projection along $\mathbf{k}_T$, q_{side} is the projection along $\hat{\mathbf{z}} \times \mathbf{k}_T$ (with the z-axis along the beamline), and q_{long} is the longitudinal component. The relative momentum of the pair is evaluated in the longitudinally comoving frame (LCMF), i.e., the frame boosted such that $k_z = 0$.

The jet fragmentation background, $\Omega(\mathbf{q})$, contributes an enhancement at low $|\mathbf{q}|$, though usually over a wider range in $|\mathbf{q}|$ and with a smaller amplitude than the Bose-Einstein correlation. This background is constrained in same-charge correlation functions by measuring the opposite-sign correlation functions, which do not contain a Bose-Einstein enhancement. Opposite-charge correlations contain contributions from resonance decays and the most prominent of these are removed by cuts on the opposite-charge pion pair mass. Even with resonances removed, due to charge conservation the fragmentation contribution is not identical between same- and opposite-charge correlations, but the shape is similar. The relationship between the two is studied in detail using PYTHIA 8, and a mapping from opposite- to same-charge correlation functions is derived in simulation for use with the data. For details on this and other aspects of the analysis, see Reference [14].

Figure 1 shows the HBT radii as a function of k_T in four centrality intervals. A decrease of the radii with rising k_T, interpreted as a signature of collective expansion, is observed in central events. This trend is diminished in peripheral events.

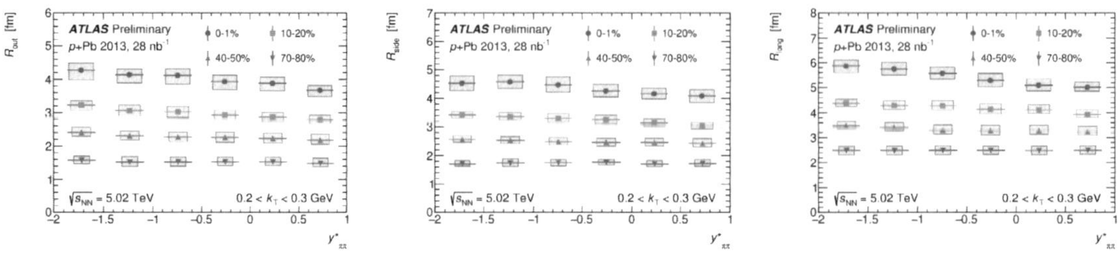

Figure 2. Three-dimensional HBT radii as a function of the rapidity $y^{\star}_{\pi\pi}$ of pion pairs [14]. The widths of the boxes vary between centrality intervals only for visual clarity.

The HBT radii are presented as a function of pair rapidity $y^{\star}_{\pi\pi}$ in Figure 2. Central collisions exhibit larger radii on the backward (Pb-going) side of the event, while peripheral events show no distinguishable variation of the radii with rapidity.

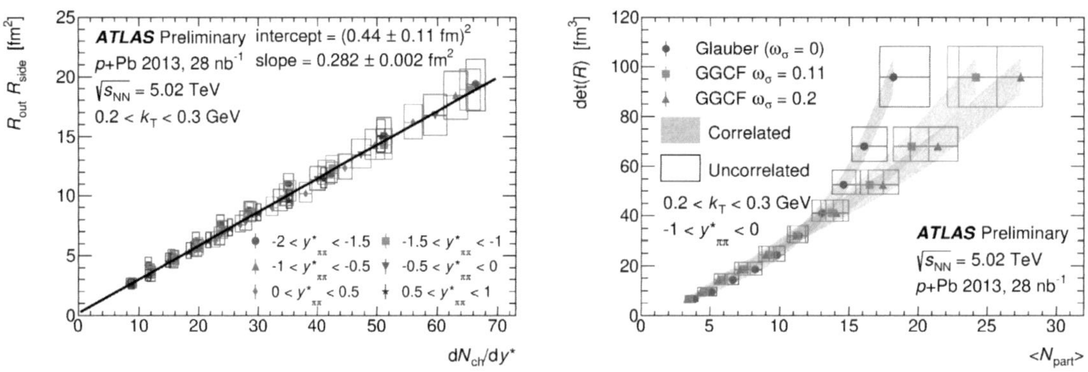

(a) (b)

Figure 3. The transverse area $R_{\text{out}}R_{\text{side}}$ is shown in (a) as a function of local multiplicity in several intervals of centrality and $y^{\star}_{\pi\pi}$ [14]. The volume element of the HBT matrix $\det(R)$ is shown in (b) as a function of N_{part} for three Glauber-Gribov colour fluctuation models.

Source radii are observed to correlate strongly with the single-particle local multiplicity $\mathrm{d}N_{\text{ch}}/\mathrm{d}y^{\star}$, as illustrated in Figure 3(a), which presents the transverse area $R_{\text{out}}R_{\text{side}}$ at $0.2 < k_{\text{T}} < 0.3$ GeV in several centrality and rapidity intervals as a function of $\mathrm{d}N_{\text{ch}}/\mathrm{d}y^{\star}$. The linear behaviour with multiplicity indicates a constant areal density at low k_{T}.

Figure 3(b) compares the volume scale $\det(R)$ with the average number of nucleon participants $\langle N_{\text{part}}\rangle$, both for the standard Glauber model and for two choices of the Glauber-Gribov colour fluctuation (GGCF) extension [16]. This extension of the Glauber initial-geometry model account for fluctuations in the nucleon-nucleon cross section. The volume scale increases at a rate greater than linearly for all choices shown for N_{part}. While the final-state volume indicated by $\det(R)$ should not necessarily scale linearly with the initial size N_{part}, extreme departure from a linear behaviour is difficult to describe. These results indicate that fluctuations in the size of the proton are crucial for the understanding of the initial geometry of p+Pb collisions.

The cross-term, R_{ol}, which couples to the lifetime of the source [17], is shown in Figure 4. In hydrodynamic models, a nonzero R_{ol} necessitates both longitudinal and transverse expansion.

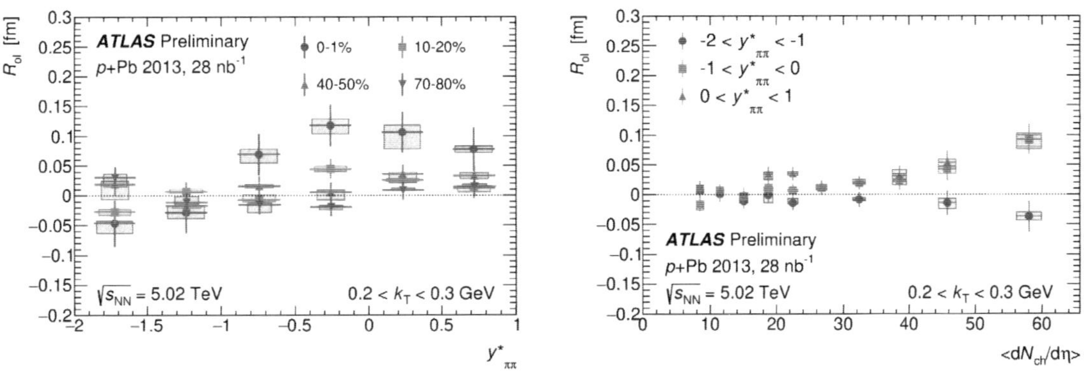

(a) (b)

Figure 4. The cross-term $R_{\rm ol}$, which couples radial and longitudinal expansion, is shown as a function of rapidity (a) and average global multiplicity $\langle {\rm d}N_{\rm ch}/{\rm d}\eta \rangle$ (b) [14].

A significant departure from zero is observed in this parameter in central events, but only for rapidities $y^{\star}_{\pi\pi} \gtrsim -1$. This indicates that the particle production at mid-rapidity and in the proton-going direction is sensitive to the local z-asymmetry of the system.

These results provide a detailed description of the freeze-out geometry of proton-lead collisions. The large pseudorapidity coverage of the ATLAS detector has been utilised to probe the rapidity dependence of the asymmetric p+Pb collisions. Clear signatures of collective behaviour are observed in central collisions, but they are significantly diminished in peripheral collisions.

This work was supported by U.S. Department of Energy grant DE-FG02-86ER40281.

References

[1] ALICE Collaboration 2013 *Phys. Lett. B* **719** 29
[2] ATLAS Collaboration 2013 *Phys. Rev. Lett.* **110** 182302
[3] CMS Collaboration 2013 *Phys. Lett. B* **724** 213
[4] ATLAS Collaboration 2014 *Phys. Rev. C* **90** 044906
[5] CMS Collaboration 2015 *Phys. Rev. Lett.* **115** 012301
[6] ATLAS Collaboration 2016 *Phys. Rev. Lett.* **116** 172301
[7] Bożek P 2012 *Phys. Rev. C* **85** 014911
[8] Shuryak E and Zahed I 2013 *Phys. Rev. C* **88** 044915
[9] Bożek P and Broniowski W 2013 *Phys. Rev. C* **88** 014903
[10] Lisa M A, Pratt S, Soltz R and Wiedemann U 2005 *Ann. Rev. Nucl. Part. Sci.* **55** 357
[11] Kolb P F and Heinz U W 2003 (*Preprint* `nucl-th/0305084`)
[12] ALICE Collaboration 2015 *Phys. Rev. C* **91** 034906
[13] ALICE Collaboration 2014 *Phys. Lett. B* **739** 139
[14] ATLAS Collaboration 2016 ATLAS-CONF-2016-027 URL `https://cds.cern.ch/record/2157691`
[15] ATLAS Collaboration 2008 *JINST* **3** S08003
[16] Alvioli M and Strikman M 2013 *Phys. Lett. B* **722** 347
[17] Chapman S, Scotto P and Heinz U W 1995 *Phys. Rev. Lett.* **74** 4400

IOP Conf. Series: Journal of Physics: Conf. Series **832** (2017) 012015 doi:10.1088/1742-6596/832/1/012015

Light flavour hadron production in pp collisions at $\sqrt{s} = 13$ TeV with ALICE

Raúl Tonatiuh Jiménez Bustamante for the ALICE Collaboration
GSI Helmholtzzentrum für Schwerionenforschung GmbH, Planckstraße 1, 64291 Darmstadt, Germany and Ruprecht-Karls Heidelberg University, Seminarstraße 2, 69117 Heidelberg, Germany.

E-mail: raul@cern.ch

Abstract.
The ALICE detector has excellent Particle IDentification (PID) capabilities in the central barrel ($|\eta| < 0.9$). This allows identified hadron production to be measured over a wide transverse momentum (p_{T}) range, using different sub-detectors and techniques: their specific energy loss ($\mathrm{d}E/\mathrm{d}x$), the velocity determination via time-of-flight measurement, the Cherenkov angle or their characteristic weak decay topology. Results on identified light flavour hadron production at mid-rapidity measured by ALICE in proton-proton collisions at $\sqrt{s} = 13$ TeV are presented and compared with previous measurements performed at lower energies. The results cover a wide range of particle species including long-lived hadrons, resonances and multi-strange baryons over the p_{T} range from 150 MeV/c up to 20 GeV/c, depending on the particle species.

1. Introduction

In June 2015, the Large Hadron Collider (LHC) has resumed its physics program with proton-proton (pp) collisions at $\sqrt{s} = 13$ TeV. Currently, this is the highest centre-of-mass energy reached in the laboratory. A full description of particle production in pp collisions cannot be obtained from perturbative Quantum Chromodynamics and hence modelling efforts typically employ various empirical components that have to be adjusted based on experimental data. The measurement of light flavour particles production in pp collisions provides important input for event generators to model the soft parton interactions and the hadronization processes. Furthermore, pp collisions are used as reference to isolate medium effects present in larger colliding systems. Previous ALICE measurements [1], [2], [3] on identified particles in pp collisions at lower energies show that the models usually give a fair description of the shapes of the p_{T} spectra but fail on the description of the pion, kaon and proton yields.

2. Particle Identification

$\pi^{\pm}$, $\mathrm{K}^{\pm}$, p and $\bar{\mathrm{p}}$ are identified in the rapidity window $|y| < 0.5$ using several Particle IDentification (PID) techniques in the different ALICE sub-detectors: the Inner Tracking System (ITS), the Time Projection Chamber (TPC), the Time Of Flight detector (TOF) and High Momentum Particle Identification Detector (HMPID). The combined information of these detectors in different p_{T} regions allows us to measure the particle production in the region starting from 150 MeV/c up to 20 GeV/c.

IOP Conf. Series: Journal of Physics: Conf. Series **832** (2017) 012015 doi:10.1088/1742-6596/832/1/012015

The K^{*0} and ϕ mesons are measured with invariant-mass analyses via the reconstruction of their hadronic decays. The combinatorial background is subtracted using either a like-sign or event mixing technique. The resulting invariant-mass distributions are then fitted. Previous measurements by ALICE and details on the analysis procedure can be found in [4].

The multi-strange Ω^+, Ω^-, Ξ^+, Ξ^- baryons are reconstructed via their weak hadronic decay products. Candidates are selected with restrictions on the topology of the decay and the signal is extracted from the resulting invariant mass distributions. The full analysis procedure is described in [5].

3. Results

The p_T-dependent $(p + \bar{p})/(\pi^+ + \pi^-)$ and $(K^+ + K^-)/(\pi^+ + \pi^-)$ ratios measured in central rapidity ($|y| < 0.5$) at $\sqrt{s} = 2.76$, 7.0 and 13 TeV are shown in Fig. 1. The comparison to PYTHIA 8 [6] (Monash-2013 [7]) for the different energies is also shown.

Within the systematic uncertainties the $(K^+ + K^-)/(\pi^+ + \pi^-)$ ratio is consistent for the three different energies. The $(p+\bar{p})/(\pi^+ + \pi^-)$ ratio for the low-intermediate p_T region shows a slight shift towards higher p_T when increasing $\sqrt{s}$. Within our systematic uncertainties the ratio is compatible for the three different energies in the p_T region above 10 GeV/c. This shift is also observed on the Monte Carlo predictions from the PYTHIA 8 model. The origin of the peak in PYTHIA 8 is attributed to the colour reconnection mechanism [13], and is expected to become more important at higher centre-of-mass energy due to the increase of number of multiparton interactions [14].

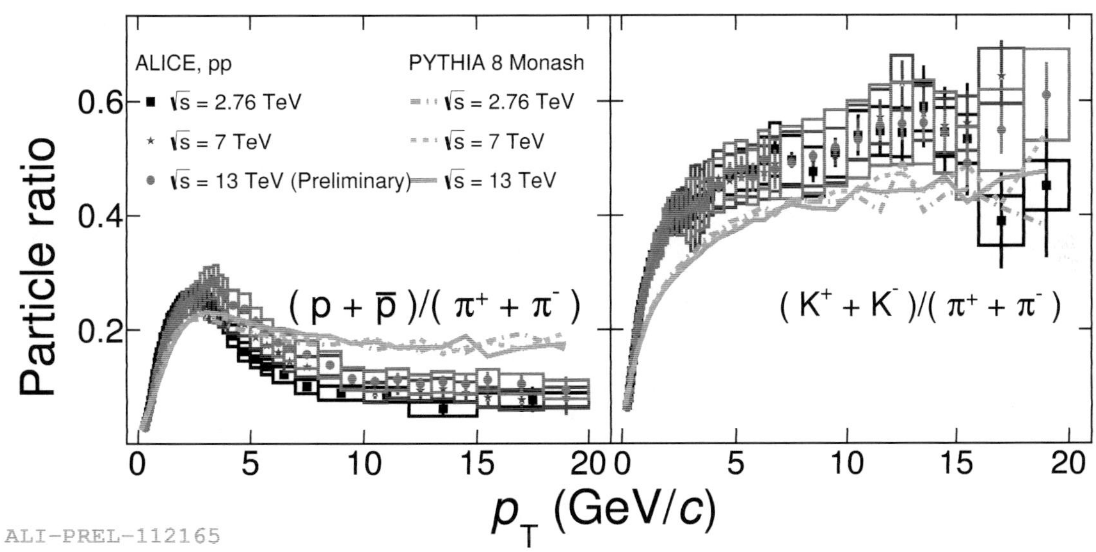

Figure 1. Ratios $(p+\bar{p})/(\pi^+ + \pi^-)$ and $(K^+ + K^-)/(\pi^+ + \pi^-)$ in pp collisions at $\sqrt{s} = 2.76$, 7.0 and 13 TeV. The lines show PYTHIA 8 Tune Monash-2013 [7] predictions for the 3 different energies.

The p_T-integrated $(K^+ + K^-)/(\pi^+ + \pi^-)$ and $(p+\bar{p})/(\pi^+ + \pi^-)$ ratios in pp collisions as a function of $\sqrt{s}$ are shown in Fig.2. Results at different energies for different experiments are also shown. Results from E735 at $\sqrt{s} = 0.3$, 0.54, 1.0 and 1.8 TeV [8], [9], from PHENIX at $\sqrt{s} = 62.4$ and 200 GeV [10] and CMS at $\sqrt{s} = 0.9$, 2.76 and 7 TeV [11] are compared for both particle ratios. The last added point at 13 TeV confirms the similar trend and saturation for $\sqrt{s} > 900$ GeV observed in ALICE previous measurements.

The hyperon-to-pion ratio as a function of $\sqrt{s}$ measured in ALICE is shown in Fig. 3. In comparison with the kaon- and proton-to-pion ratio the strange- and multi-strange-to-pion ratios show a hint of increase at 13 TeV with respect to the lower collision energies. This enhanced production may be related to the enhancement observed as a function of multiplicity in pp collisions at $\sqrt{s} = 7$ TeV [12], as the charged particle multiplicity density increases with the collision energy.

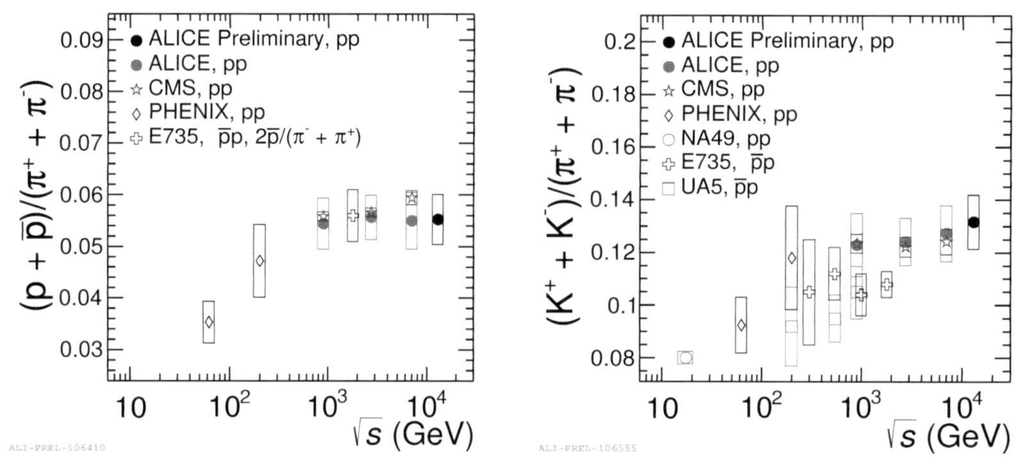

Figure 2. $(p + \bar{p})/(\pi^+ + \pi^-)$ and $(K^+ + K^-)/(\pi^+ + \pi^-)$ in pp and p$\bar{\text{p}}$ collisions as a function of $\sqrt{s}$. Errors bars represent the quadratic sum of statistical and systematics uncertainties.

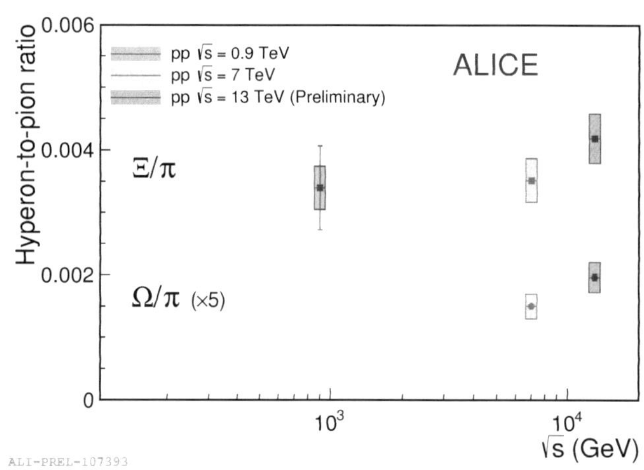

Figure 3. Hyperon to pion ratio in pp collisions as a function of the collision energy.

The K^{*0} and ϕ meson production has been also measured in pp collisions at $\sqrt{s} = 13$ TeV. The p_{T}-integrated K^{*0}/K and ϕ/K ratios are shown in Fig. 4. However, no energy evolution is observed in the range from RHIC to the highest LHC energy.

IOP Conf. Series: Journal of Physics: Conf. Series **832** (2017) 012015 doi:10.1088/1742-6596/832/1/012015

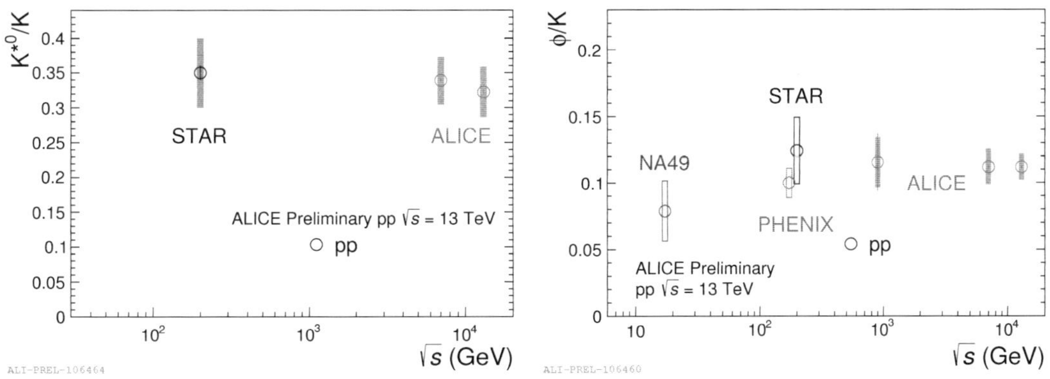

Figure 4. K^{*0}/K and ϕ/K ratio in pp collisions as a function of the collision energy. Bars represent statistical uncertainties. Boxes represent the total systematic uncertainties.

4. Conclusions

The p_{T} dependence of the $(p+\bar{p})/(\pi^{+}+\pi^{-})$ and $(K^{+}+K^{-})/(\pi^{+}+\pi^{-})$ is the same at LHC energies within our systematic errors. PYTHIA 8 describes the shape but exhibits large deviations from the data for transverse momenta larger than 2 GeV/c. Above $\sqrt{s}=900$ GeV the $(p+\bar{p})/(\pi^{+}+\pi^{-})$ and $(K^{+}+K^{-})/(\pi^{+}+\pi^{-})$ ratios are energy independent within systematic uncertainties. The relative production of strange resonances (K^{*0}/K and ϕ/K) remains constant within a wide range of centre of mass energy. Multiplicity dependent studies in pp collisions at 13 TeV will help disentangle the role of collision energy and event multiplicity in the identified particle production of pp collisions.

References

[1] ALICE Collaboration, *Eur. Phys. J. C* **71** (2011) 1655.
[2] ALICE Collaboration, *Phys. Lett. B* **736** (2014) 196-207.
[3] ALICE Collaboration *Eur. Phys. J.C* **75** (2015) 226.
[4] ALICE Collaboration, *Eur. Phys. J. C* **72** (2012) 2183.
[5] ALICE Collaboration, *Phys. Lett. B* **712** (2012) 309-318.
[6] T. Sjöstrand, S. Mrenna, and P. Z. Skands *Comput. Phys. Commun.* **178** (2008) 852-867.
[7] P. Z. Skands, S. Carrazza, and J. Rojo, *Eur. Phys. J. C* **74** (2014) 3024.
[8] E735 Collaboration *Phys. Rev. D* **48** (1993) 984-997.
[9] E735 Collaboration, *Phys. Rev. D* **46** (1992) 2773-2786.
[10] PHENIX Collaboration, *Phys. Rev. C* **83** (2011) 064903.
[11] CMS Collaboration, *Eur. Phys. J. C* **72** (2012) 2164.
[12] ALICE Collaboration (2016) *arXiv*:1606.07424 [nucl-ex].
[13] A. O. Velasquez, P. Christiansen, E. C. Flores, I. M. Cervantes, and G. Paic, *Phys. Rev. Lett.* **111** (2013) 042001.
[14] E. Cuautle, A. O. Velazquez, G. Paic, *Nucl. Phys. A* **956** (2016) 749-752.

IOP Conf. Series: Journal of Physics: Conf. Series **832** (2017) 012016 doi:10.1088/1742-6596/832/1/012016

Initial conditions for hydrodynamics from weakly coupled pre-equilibrium evolution

Aleksas Mazeliauskas

E-mail: aleksas.mazeliauskas@stonybrook.edu
Department of Physics and Astronomy, Stony Brook University, Stony Brook, New York 11794, USA

Abstract. We use leading order effective kinetic theory to simulate the pre-equilibrium evolution of transverse energy and flow perturbations in heavy-ion collisions. We provide a Green function which propagates the initial perturbations of the energy-momentum tensor to a time when hydrodynamics becomes applicable. With this map, the pre-thermal evolution from saturated nuclei to hydrodynamics can be modeled in the framework of weakly coupled QCD.

1. Introduction

Viscous relativistic hydrodynamic simulations of heavy ion collisions at the BNL Relativistic Heavy Ion Collider and the CERN Large Hadron Collider have shown tremendous success in describing simultaneously many of the soft hadronic observables, however initial conditions for hydrodynamics remain one of the largest uncertainties in hydrodynamic modeling of heavy ion collisions [1, 2, 3, 4]. In this work I use the effective kinetic theory of weakly coupled quasi-particles to study the equilibration and the onset of hydrodynamics in heavy ion collisions [5, 6]. In particular, I focus on the transverse perturbations, which are expected to initiate flow during the equilibration process [7].

2. Separation of scales

In the weak coupling limit, kinetic theory describes the evolution of the system from the microscopic formation time $\tau_0 \sim Q_s^{-1}$ to the onset of hydrodynamics at a much later time $\tau_{\rm init} \sim \tau_{\rm equilibrium}$ [8]. For realistic values of coupling constant $\alpha_s \sim 0.3$, the equilibration time is short and the causally connected region $c(\tau_{\rm init} - \tau_0) \sim 1\,{\rm fm}$ is much smaller than the transverse nuclear geometry $R_{\rm Pb} \sim 5\,{\rm fm}$, but comparable to a single nucleon scale $R_p \sim 1\,{\rm fm}$ (see Fig. 1)

$$R_{\rm Pb} \gg c(\tau_{\rm init} - \tau_0) \sim R_{\rm p}. \quad (1)$$

For this reason, the global nuclear geometry contributes a small gradient to a locally constant background, while event-by-event nucleon fluctuations are suppressed by $1/\sqrt{N_{\rm part}}$, where $N_{\rm part}$ is the number of participant nucleons. Therefore initial energy density can be expanded locally as

$$e(\boldsymbol{x}, \tau_0) = e(\tau_0) + \delta e(\boldsymbol{x}, \tau_0), \quad (2)$$

where $e(\tau_0) = \langle e(\boldsymbol{x}, \tau_0)\rangle_{|\boldsymbol{x}-\boldsymbol{x}_0|\leq c(\tau_{\rm init}-\tau_0)}$ is the average energy density in the causal region and $\delta e(\boldsymbol{x}, \tau_0)$ is a small perturbation. We will use effective kinetic theory described below

IOP Conf. Series: Journal of Physics: Conf. Series **832** (2017) 012016 doi:10.1088/1742-6596/832/1/012016

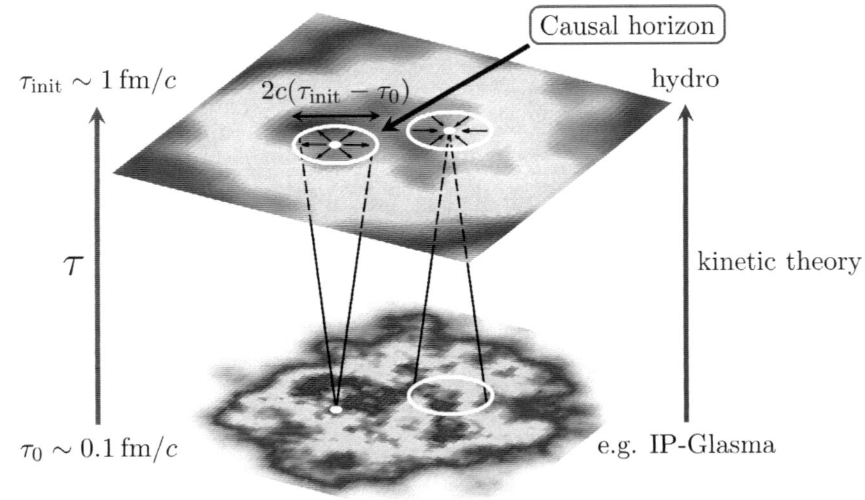

Figure 1. Kinetic theory describes the evolution from the microscopic formation time τ_0 to the equilibration time $\tau_{\rm init}$, when hydrodynamics becomes applicable [8]. By causality, for a given point in the transverse plane it is sufficient to analyze the pre-equilibrium evolution within the causal neighborhood of that point.

to simulate a translationally invariant background with linearized perturbations in the causal region $|\boldsymbol{x} - \boldsymbol{x}_0| < c(\tau_{\rm init} - \tau_0)$.

3. Effective kinetic theory

We use the effective kinetic theory of high temperature QCD at leading order in α_s to model the pre-equilibrium evolution in heavy ion collisions [9]. At early times gluons dominate over fermions in the plasma, so we solve the Boltzmann equation for boost invariant gluon distribution function $f(\tau, \boldsymbol{x}, \boldsymbol{p})$ with leading order elastic $2 \leftrightarrow 2$ and inelastic $1 \leftrightarrow 2$ collision processes [5, 6, 9]

$$\partial_\tau f + \frac{\boldsymbol{p}}{|p|} \cdot \nabla f - \underbrace{\frac{p_z}{\tau}\partial_{p_z} f}_{\text{Bjorken expansion}} = - \underbrace{\mathcal{C}_{2\leftrightarrow 2}[f]}_{} - \underbrace{\mathcal{C}_{1\leftrightarrow 2}[f]}_{} . \tag{3}$$

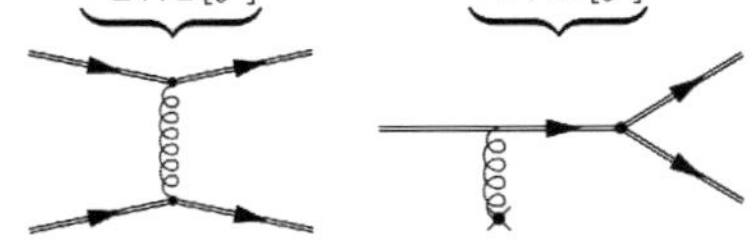

We split the distribution function into translationally invariant background $\bar{f}(\tau, \boldsymbol{p})$ and a linear perturbation with a wavenumber $\boldsymbol{k}_\perp$ in the transverse plane $\delta f_{\boldsymbol{k}_\perp}(\tau, \boldsymbol{p}) e^{i\boldsymbol{k}_\perp \cdot \boldsymbol{x}}$. Then we solve the Boltzmann equation as a system of coupled differential equations with constant $\boldsymbol{k}_\perp$.

We use a Color Glass Condensate (CGC) inspired initial background distribution function $\bar{f}(\boldsymbol{p})$, which possesses large initial pressure anisotropy, $P_T \ll P_L$, and take the functional form of transverse perturbations to be $|\delta f_{\boldsymbol{k}_\perp}| \sim p\partial_p \bar{f}(\boldsymbol{p})$ [5, 6]. In Fig. 2(a) we show the evolution of the background energy momentum tensor components relative to their asymptotic values in scaled time $\tau T/(\eta/s)$. We see that at sufficiently late times the longitudinal pressure P_L in kinetic theory approaches the constitutive equation of viscous conformal hydrodynamics for Bjorken expansion [10]

$$\frac{P_L}{e} = \underbrace{\frac{1}{3}}_{\text{ideal}} - \underbrace{\frac{16}{9}\frac{\eta/s}{\tau T}}_{\text{1st order}} \underbrace{- \frac{32}{27}\frac{\tau_\pi}{\eta/(sT)}\left(1 - \frac{\lambda_1}{\tau_\pi \eta}\right)\left(\frac{\eta/s}{\tau T}\right)^2}_{\text{2nd order}} + \ldots \tag{4}$$

IOP Conf. Series: Journal of Physics: Conf. Series **832** (2017) 012016 doi:10.1088/1742-6596/832/1/012016

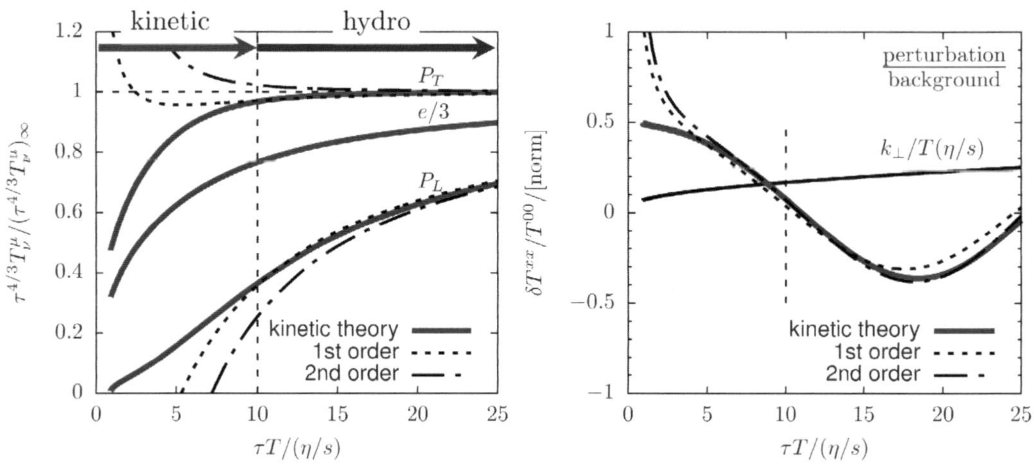

Figure 2. (a) Equilibration of background energy momentum tensor in kinetic theory relative to the asymptotic values. Dashed lines correspond to the asymptotic first and second order hydrodynamic constitutive equations. (b) Evolution of energy momentum tensor perturbation δT^{xx} relative to the background energy density T^{00} due to wavenumber $k_\perp$ energy perturbation in the transverse plane (in x-direction).

Therefore at late times the system evolution can be smoothly passed from kinetic theory to hydrodynamics. In Fig. 2(b) we show that the linear perturbations of the energy momentum tensor $\delta T^{\mu\nu}$ can also be described with hydrodynamic constitutive equations similar to Eq. 4 at sufficiently late times and sufficiently small wavenumbers $k_\perp$.

4. Linear response functions

The goal of the pre-equilibrium evolution is to construct the initialization conditions for hydrodynamics at $\tau_{\rm init}$ from a given initial state at τ_0. Close to equilibrium the full energy momentum tensor $T^{\mu\nu}$ can be constructed via hydrodynamic constitutive equations from the local energy $e + \delta e$ and momentum $\vec{g}$ densities, therefore we only need to know energy and momentum response functions to initial conditions.

In kinetic theory the linear energy response to the initial energy perturbations can be written as a convolution

$$\frac{\delta e(\tau_{\rm init}, \boldsymbol{x}_0)}{e(\tau_{\rm init})} = \int d^2\boldsymbol{x}\; E(|\boldsymbol{x}_0 - \boldsymbol{x}|; \tau_{\rm init}, \tau_0) \times \frac{\delta e(\tau_0, \boldsymbol{x})}{e(\tau_0)}, \tag{5}$$

where $E(r; \tau_{\rm init}, \tau_0)$ is the coordinate space Green function for energy perturbations. In Fig. 3(a), we show the radial profile of $E(r; \tau, \tau_0)$ at $\tau T/(\eta/s) = 10$ and compare it with the free streaming response. The coordinate space Green function has the meaning of a system response to initial δ-like perturbation (see Fig. 3(b)). In the absence of collisions, disturbances propagate at the speed of light and the free streaming response function shown in Fig. 3(a) is centered on the causality circle $r = c(\tau - \tau_0)$. As seen in Fig. 3(a) collision processes modify the system response in kinetic theory and eventually it will become identical to the hydrodynamic response (not shown).

Spatial Green functions are obtained by simulating the kinetic theory response to several values of wavenumber $\boldsymbol{k}_\perp$ perturbations and then taking Fourier-Hankel transform to the coordinate space. Similarly, we find momentum response function to the initial energy gradients in the transverse plane [6].

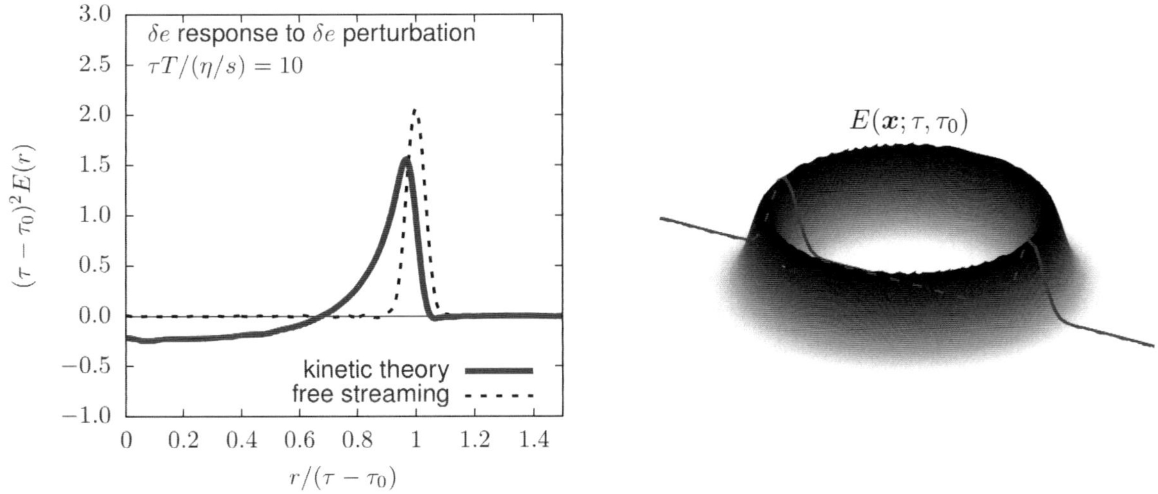

Figure 3. (a) Radial profile of energy response function $E(r; \tau, \tau_0)$ due to initial energy perturbation at time $\tau T/(\eta/s) = 10$. (b) Illustration of coordinate space response function $E(\boldsymbol{x}; \tau, \tau_0)$ to initial δ-like perturbation in two dimensions.

5. Summary

We used effective kinetic theory to study equilibration and approach to hydrodynamics of linearized transverse energy perturbations around an initially anisotropic but boost invariant background. At the hydrodynamic initialization time all components of energy momentum tensor can be initialized from the local energy and momentum densities, which can be determined from the kinetic theory response to initial perturbations. Using the kinetic theory pre-equilibrium evolution in heavy ion collision models could reduce the dependence on hydro initialization time and better account for the pre-equilibrium flow.

Acknowledgments

I gratefully acknowledge my collaborators on this work: Liam Keegan, Aleksi Kurkela and Derek Teaney. I would like to thank Jean-François Paquet and Sören Schlichting for useful comments. Special thanks to the APS Forum on Graduate Student Affairs for supporting my participation in the Hot Quarks 2016 conference. Finally, I would like to thank the organizers and participants of the Hot Quarks 2016 conference for the stimulating week of talks and discussions. This work was supported in part by the U.S. Department of Energy under Contracts No. DE-FG-88ER40388.

References

[1] Heinz U and Snellings R 2013 *Ann. Rev. Nucl. Part. Sci.* **63** 123–151 (*Preprint* 1301.2826)
[2] Luzum M and Petersen H 2014 *J. Phys.* **G41** 063102 (*Preprint* 1312.5503)
[3] Teaney D A 2009 (*Preprint* 0905.2433)
[4] Antinori F *et al.* 2016 (*Preprint* 1604.03310)
[5] Kurkela A and Zhu Y 2015 *Phys. Rev. Lett.* **115** 182301 (*Preprint* 1506.06647)
[6] Keegan L, Kurkela A, Mazeliauskas A and Teaney D 2016 *JHEP* **08** 171 (*Preprint* 1605.04287)
[7] Vredevoogd J and Pratt S 2009 *Phys. Rev.* **C79** 044915 (*Preprint* 0810.4325)
[8] Kurkela A 2016 *25th International Conference on Ultra-Relativistic Nucleus-Nucleus Collisions (Quark Matter 2015) Kobe, Japan, September 27-October 3, 2015* (*Preprint* 1601.03283)
[9] Arnold P B, Moore G D and Yaffe L G 2003 *JHEP* **01** 030 (*Preprint* hep-ph/0209353)
[10] Baier R, Romatschke P, Son D T, Starinets A O and Stephanov M A 2008 *JHEP* **04** 100 (*Preprint* 0712.2451)

IOP Conf. Series: Journal of Physics: Conf. Series **832** (2017) 012017 doi:10.1088/1742-6596/832/1/012017

Study of Drell-Yan pair production on nuclear targets

Michal Krelina

Czech Technical University in Prague, FNSPE, Brehova 7, 11519 Prague, Czech Republic

E-mail: michal.krelina@fjfi.cvut.cz

Victor P. Goncalves

High and Medium Energy Group, Instituto de Fisica e Matematica, Universidade Federal de Pelotas, Pelotas, RS, 96010-900, Brazil

Department of Astronomy and Theoretical Physics, Lund University, SE-223 62 Lund, Sweden

Jan Nemchik

Czech Technical University in Prague, FNSPE, Brehova 7, 11519 Prague, Czech Republic

Institute of Experimental Physics SAS, Watsonova 47, 04001 Kosice, Slovakia

Roman Pasechnik

Department of Astronomy and Theoretical Physics, Lund University, SE-223 62 Lund, Sweden

Abstract. Drell-Yan pair production off nuclei is an ideal tool to test the cold nuclear effects occurring before a hard collision since no interaction in the final state is expected, neither energy loss or absorption. We present for the first time a comprehensive study of the nucleus-to-nucleon production ratio (the nuclear modification factor) within the color dipole approach using the Green function formalism which naturally incorporates for the color transparency and quantum coherence effects. We study a different onset of nuclear shadowing in various kinematical regions. At large values of the Feynman variable x_F and dilepton invariant mass M we include also a suppression factor due to restrictions caused by the energy conservation induced by multiple initial state interactions (ISI effects). We present a variety of predictions for the nuclear suppression as a function of x_F and M that can be verified by experiments at RHIC and LHC. The mixing of coherence effects with ISI effects can be eliminated going to large values of the dilepton invariant mass. Then predictions for the nuclear suppression is a direct manifestation for the onset of net ISI effects that can be verified by the future measurements.

1. Introduction

The aim of this contribution is to present a study of cold nuclear effects in proton-nucleus interactions at intermediate and low energies (RHIC energies and lower). For this purpose, we use the Drell-Yan (DY) pair production process representing a clean and precise tool for dynamics of the cold nuclear effects not only in proton-nucleus interactions but also in heavy-ion collisions. Moreover, besides no final state energy loss or absorption, the variability of the invariant mass of the dilepton pair allows to reach kinematical regions where the coherence or non-coherence effects are dominant [1].

IOP Conf. Series: Journal of Physics: Conf. Series **832** (2017) 012017 doi:10.1088/1742-6596/832/1/012017

The dynamics of the quantum coherence effects is controlled by the coherence length (CL) that corresponds to the lifetime of the lowest Fock component $|q\gamma^*\rangle$ for the production of the Drell-Yan dilepton pair in the light-cone formalism. Here, the coherence length for the DY process reads

$$l_c = \frac{1}{x_2 m_p} \frac{(M_{l\bar{l}}^2 + p_T^2)(1-\alpha)}{(1-\alpha)M_{l\bar{l}}^2 + \alpha^2 m_q^2 + p_T^2}, \tag{1}$$

where p_T is the transverse momentum of the dilepton pair, m_q is the quark effective mass, and α is the fraction of the light-cone momentum of the projectile quark carried out by the virtual photon.

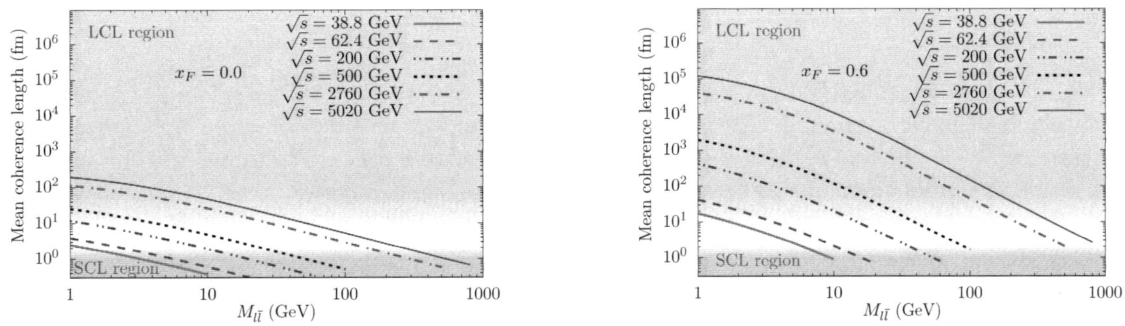

Figure 1. The mean CL as a function of the invariant dilepton mass for several values of the c.m. collision energy $\sqrt{s}$ and Feynman x_F variable. The LCL and SCL regions are highlighted by the yellow and green bands, respectively.

Studying the kinematical dependence of the CL we can distinguish two limits. First, the short coherence length (SCL) limit, where the CL is too short, $\langle l_c \rangle \leq 1 \div 2$ fm, and excludes any coherence effects. This regime corresponds to the so-called Bethe-Heitler regime [3]. Second, for the long coherence length (LCL) limit, $\langle l_c \rangle \gg R_A$ where R_A is the nuclear radius, coherence effects, i.e. nuclear shadowing, are maximal. LCL limit is analogy of the so-called Landau-Pomeranchuk-Migdal effect [4] in QED. For both limits, the formalism for calculation of cross sections on nuclear targets are well-known, see Refs. [5, 1] for the LCL and Ref. [6] for the SCL limit.

In Fig. 1, the mean coherence length is shown for several CMS energies and for two values of Feynman $x_F = 0.0, 0.6$ where regions of SCL and LCL are highlighted by the green and yellow bands, respectively. Especially for $x_F = 0.0$ (mid-rapidity), one can see that the most common measured values of the dilepton pair mass occurs in the white band between these limits. Here, the more general and rigorous Green function formalism that treats the CL exactly has to be used.

The Green function formalism was formulated in Ref. [7] and exact numerical solution was provided in Ref. [8] for nuclear DIS and in Ref. [2] for proton-nucleus collisions. This formalism describes the propagation of $q\gamma^*$ fluctuation through the nucleus where the transverse separation of the fluctuation is varied and interacts with the local bounded nucleons via the dipole cross section as is illustrated in Fig. 2. The Green function effectively resums over all possible trajectories of the $q\gamma^*$ fluctuation. For more details, see Ref. [2].

The considered Green function formalism includes the exact quark shadowing only due to $|q\gamma^*\rangle$ Fock state. At CMS energies, where we use this formalism, the gluon shadowing is very small to negligible. Despite, we include the gluon shadowing in the same way as in Ref. [1].

Besides the coherence effects, we also study the effective energy loss induced by the ISI effects. The latter are expected to cause a significant suppression of the nuclear DY cross section when

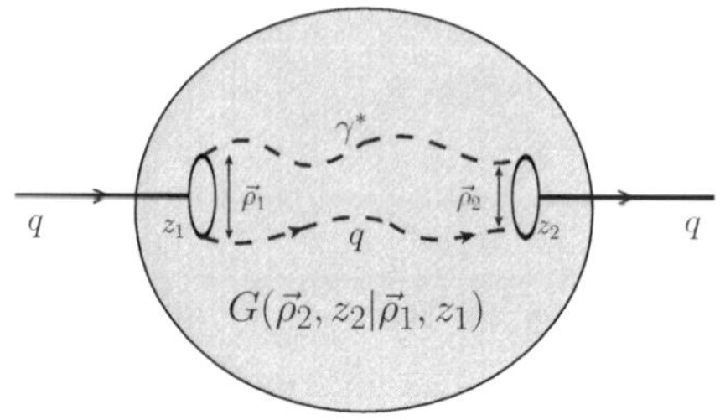

Figure 2. An illustration of propagation of the $q\gamma^*$ fluctuation through a nucleus is described by the Green function $G(\vec{\rho}_2, z_2|\vec{\rho}_1, z_1)$ as a result of summation over different paths of the $q\gamma^*$ state in the medium.

reaching the kinematical limits, $x_L = 2p_L/\sqrt{s} \to 1$ and/or $x_T = 2p_T/\sqrt{s} \to 1$, and were analyzed in Ref. [9]. The ISI effects explain the nuclear suppression at large p_T as well as at forward rapidities in fixed-target experiments where the coherence effects are not allowed due to a low collision energy. For more details, see Ref. [1].

2. Results

In Fig. 3 we present predictions for the future fixed-target experiment AFTER@LHC as functions of $M_{l\bar{l}}$. In the left part one can see that the variation of dilepton mass covers the whole space between LCL and SCL limits. The middle part compares the calculations with different dipole cross sections, and the right part compares the results with quark shadowing (red solid line), quark + gluon shadowing (blue dashed line) and shadowing in combination with ISI (green dashed line). Here, the coherence effects (nuclear shadowing) dominate at small values of $M_{l\bar{l}}$ while the non-coherence (ISI) effects dominate at large values of $M_{l\bar{l}}$.

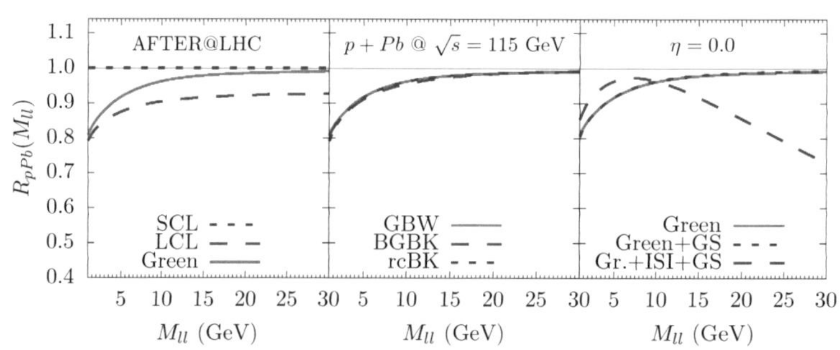

Figure 3. The dilepton mass dependence at midrapidity of the lead-to-proton ratio R_{pPb} at the AFTER@LHC c.m. collision energy $\sqrt{s} = 115$ GeV.

Fig. 4 includes predictions for the LHCb-gas experiment where four different nuclei were tested. In the left part, the quark shadowing is counted only (the gluon shadowing is neglected), and the right part includes both, the quark shadowing and ISI effects together. Here, it is demonstrated that the suppression R_pA as a function of rapidity is a superposition of the coherence and ISI effects whereas R_pA as a function of dilepton mass allows to separate the coherence and ISI effects.

3. Conclusions

In this paper, we have considered the Drell-Yan process as an ideal tool for study of nuclear effects occurring before a hard collision. We showed that the coherence effects (nuclear shadowing) are

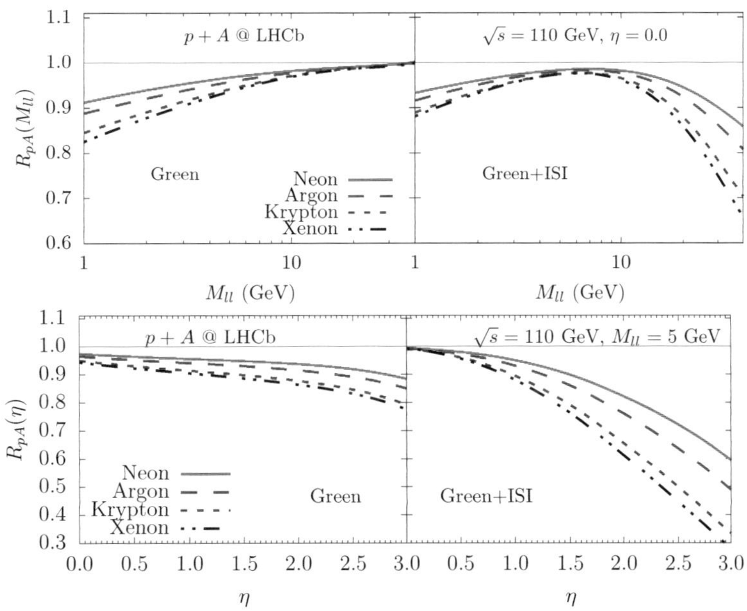

Figure 4. The dilepton invariant mass (upper panels) and pseudorapidity (lower panels) dependences of the nucleus-to-proton ratio R_{pA} at LHCb fixed-target c.m. collision energy $\sqrt{s} = 110$ GeV for several different nuclear targets.

controlled by the coherence length which depends on energy, rapidity and dilepton invariant mass. Also, we demonstrated the necessity of the Green function formalism for RHIC energies and lower where the coherence length corresponds to the transition region between LCL and SCL limits. Using this formalism, the predictions for the future experiment AFTER@LHC and current LHCb-gas experiment were presented. We advice to study the nuclear modification factor as function of $M_{l\bar{l}}$ that is a good probe for both the coherence and non-coherence sources of suppression allowing to reduce or eliminate the shadowing-ISI mixing.

4. Acknowledgements

V.P.G. has been supported by CNPq, CAPES and FAPERGS, Brazil. R.P. is supported by the Swedish Research Council, contract number 621-2013-428. J.N. and M.K. are partially supported by the grant 13-20841S of the Czech Science Foundation (GACR) and by the Grant MSMT LG15001. J.N. is supported by the Slovak Research and Development Agency APVV-0050-11 and by the Slovak Funding Agency, Grant 2/0020/14.

References

[1] E. Basso, V. P. Goncalves, M. Krelina, J. Nemchik and R. Pasechnik, Phys. Rev. D **93**, no. 9, 094027 (2016)
[2] V. P. Goncalves, M. Krelina, J. Nemchik and R. Pasechnik, arXiv:1608.02892 [hep-ph].
[3] H. Bethe and W. Heitler, Proc. Roy. Soc. Lond. A **146**, 83 (1934).
[4] L. D. Landau and I. Pomeranchuk, Dokl. Akad. Nauk Ser. Fiz. **92**, 535 (1953).
A. B. Migdal, Phys. Rev. **103**, 1811 (1956).
[5] B.Z. Kopeliovich, in *Proceedings of the international workshop XXIII on Gross Properties of Nuclei and Nuclear Excitations, Hirschegg, Austria, 1995*, edited by H. Feldmeyer and W. Nörenberg (Gesellschaft Schwerionenforschung, Darmstadt, 1995), p. 385.
[6] M. B. Johnson *et al.*, Phys. Rev. C **65**, 025203 (2002)
[7] B.Z. Kopeliovich, A. Schafer, and A.V. Tarasov, Phys. Rev. C**59**, 1609 (1999).
[8] J. Nemchik, Phys. Rev. C **68**, 035206 (2003)
[9] B.Z. Kopeliovich, J. Nemchik, I.K. Potashnikova, I. Schmidt, Int. J. Mod. Phys. E**23**, 1430006 (2014).

IOP Conf. Series: Journal of Physics: Conf. Series **832** (2017) 012018 doi:10.1088/1742-6596/832/1/012018

Event-By-Event Initial Conditions for Heavy Ion Collisions

S Rose and R J Fries

Cyclotron Institute, Texas A&M University, College Station, TX, 77843-1411, US

E-mail: zenth @ physics.tamu.edu

Abstract. The early time dynamics of heavy ion collisions can be described by classical fields in an approximation of Quantum ChromoDynamics (QCD) called Color Glass Condensate (CGC). Monte-Carlo sampling of the color charge for the incoming nuclei are used to calculate their classical gluon fields. Following the recent work by Chen et al. we calculate the energy momentum tensor of those fields at early times in the collision event-by-event. This can then be used for subsequent hydrodynamic evolution of the single events.

1. Introduction

Quark Gluon Plasma (QGP) is a state of matter created by high energy heavy ion collisions at the highest energies achieved at the Relativistic Heavy Ion Collider (RHIC) at Brookhaven National Laboratory and at the Large Hadron Collider (LHC) at CERN. The QGP state is believed to be close to thermal equilibrium soon after the collision, and is therefore studied by using viscous hydrodynamics to evolve the system until the point of kinetic freeze-out. Hydrodynamics requires an initial condition (for $\tau = \tau_{th}$), which can be motivated by describing the initial collision in terms of interactions between the colliding nuclei's gluon fields, which at high energies may be treated classically.

For ultrarelativistic collisions one particular approach for calculating these initial conditions is Color Glass Condensate (CGC). Ultrarelativistic nuclei are approximated to follow trajectories on the light cone. In the CGC approach, the gluon fields are treated classically, created by sources whose fluctuations are frozen in time. The framework to include corrections to the classical picture has been explored, but will not be addressed here (see, for example, [1]). In recent years, attempts have shifted from describing smooth, initial states representing average initial conditions to models which allow for fluctuations for a single event. A particularly successful implementation is the IP-Glasma event generator [2].

2. Background

Calculation of the gluon distribution at small values of Bjorken x, for large nuclei has been argued by McLerran and Venugopalan (MV) to be possible in systems with very high charge densities using weak coupling [3]. In such a case, the density of partons per unit area, μ, provides a scale much larger than the intrinsic QCD scale, where the strong coupling constant becomes small. This effect should be more pronounced in very large nuclei ($\mu \sim A^{1/3}$).

IOP Conf. Series: Journal of Physics: Conf. Series **832** (2017) 012018 doi:10.1088/1742-6596/832/1/012018

Figure 1. A sample of ρ for a single color and nucleus of radius $R = 3$ fm. The resolution scale is $\lambda = 0.2$ fm.

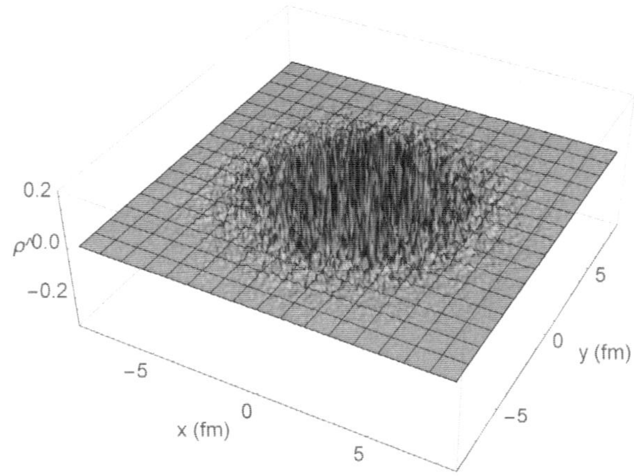

We follow analytical results presented by Guyangyao Chen et al. [4][5]. Here, we begin with presumed charge densities, ρ_1 and ρ_2, in the two nuclei giving the current

$$J_1^+(x) = \delta\left(x^-\right)\rho_1\left(\vec{x}_\perp\right), \qquad J_1^-(x) = 0, \tag{1}$$

for the particle traveling in the x^+ direction. The current for ρ_2 is set up analogously and there is no current in the transverse direction. To determine the gluon field in the forward light cone, we solve the Yang-Mills equations and continuity equation. This work uses a known recursive solution for the coefficients $A^\mu_{(n)}$ of an expansion of the gluon field A^μ in the forward light cone,

$$A^\mu\left(\tau, \vec{x}_\perp\right) = \sum_{n=0}^{\infty} \tau^n A^\mu_{(n)}\left(\vec{x}_\perp\right). \tag{2}$$

3. CGC Event Generator

While we simulate the collision on a grid, our work is analytic in character as much as possible. Whereas the grid parameters corresponding to the size of the system and the point spacing generally have effects on a numerical solution, we ensure that the physical infrared and ultraviolet scales in the problem are safe from these grid scales. The semi-analytic character of our event generator sets this work apart from the existing IP-Glasma event generator by Schenke et al. which is a numerical solver [2] [6]. However, we expect the results of both generators to be consistent, if the same average charge densities are used, up to times where our power series is convergent, roughly $\tau \sim \frac{1}{Q_s}$.

First, we must sample the charges in the nuclei. To describe the fluctuation of charges, we model the charge density as having Gaussian fluctuations with a strength μ and strong coupling g, or

$$P(\rho) = \frac{1}{\sqrt{2\pi g\mu}} e^{-\rho^2/2g\mu}, \tag{3}$$

where we assume the transverse density μ to be a Woods-Saxon integrated over the longitudinal coordinate, an assumption that can be generalized readily later to include, for example, the IP-Sat model. Figure 2 shows a typical result of charge sampling.

With the sampled ρ, we begin to solve for the gluon field. We initially solve in a covariant gauge where the gluon field is given by,

$$A^\mu_{cov} = \delta^{\mu\pm}\alpha. \tag{4}$$

IOP Conf. Series: Journal of Physics: Conf. Series **832** (2017) 012018 doi:10.1088/1742-6596/832/1/012018

The benefit is that in this gauge, the Yang-Mills equations reduce to a Poisson equation,

$$\Delta\alpha\left(x^{\mp},\vec{x}_{\perp}\right)=-\rho_{cov}\left(x^{\mp},\vec{x}_{\perp}\right). \tag{5}$$

A solution can be found with a regularized Green's function $G\left(x_{\perp}\right)=\frac{1}{2\pi}K_0\left(mx_{\perp}\right)$. Here, the effective gluon mass, m acts as an infrared cutoff for the system. We then convolute this with a Gaussian coarse-graining function to implement a finite resolution, λ,

$$G_C\left(\vec{x}_{\perp}\right)=\int\frac{1}{2\pi^2\lambda^2}K_0\left(m\vec{z}\right)e^{\left(\vec{x}_{\perp}-\vec{z}\right)^2/\lambda^2}dz. \tag{6}$$

This approach lets us control the infrared and ultraviolet scale of the problem independently from the grid we simulate on. We find the field, α, to be

$$\alpha\left(\vec{x}_{\perp}\right)=\sum_{S}G_C\left(\vec{x}_{\perp}-\vec{z}_S\right)\rho\left(\vec{z}_S\right), \tag{7}$$

where S indicates the source locations, $\vec{z}_S$.

While the solution for α is straightforward, it is not a simple gauge for calculating observables. For this goal, we transform our solution into the light cone gauge via the Wilson Line,

$$U\left(x_{\perp}\right)=\mathcal{P}\exp\left[-ig\int_{-\infty}^{x^{\pm}}\alpha\left(z^{\pm},x_{\perp}\right)dz^{\pm}\right]. \tag{8}$$

The integration over $z^{\pm}$ indicates a requirement to sample the charge distribution multiple times along the light cone for even a single nucleus. We find then our transformed gauge field is,

$$A^i\left(\vec{x}_{\perp}\right)=-\frac{1}{ig}U\left(\alpha,\vec{x}_{\perp}\right)\partial^iU^{\dagger}\left(\alpha,\vec{x}_{\perp}\right). \tag{9}$$

Here, we must calculate ∂^iU numerically, which we do using a finite difference scheme accurate to fourth order.

After the fields of the two nuclei are established we can now move on to calculate observables after the collision. We will start with the initial color-electric and color-magnetic fields. At $\tau=0$, we see that there are only longitudinal fields [5]

$$E_0=F^{+-}_{(0)}=ig\left[A_1^i,A_2^i\right],\qquad B_0=F^{21}_{(0)}=ig\epsilon^{ij}\left[A_1^i,A_2^i\right]. \tag{10}$$

Generally, the longitudinal components have non-zero terms for even orders, and transverse components have non-zero terms on odd powers of τ.

Now we move on to the stress-energy tensor, which we will use for our initial condition. We can solve order-by-order in τ. At $\tau=0$, only diagonal elements are non-zero,

$$T^{11}_{(0)}=T^{22}_{(0)}=\epsilon_0=-T^{33}_{(0)}, \tag{11}$$

Here, negative longitudinal pressure slows the participant remnants down, transferring energy into the glasma. The 1st order contributions give initial flow by contributing to the Poynting vector,

$$T^{0i}_{(1)}=\epsilon^{ij}\left(B_0E^j_{(1)}-E_0B^j_{(1)}\right)=\frac{1}{2}\alpha^i\cosh\eta+\frac{1}{2}\beta^i\sinh\eta, \tag{12}$$

$$T^{3i}_{(1)}=E_0E^i_{(1)}-B_0B^i_{(1)}=\frac{1}{2}\alpha^i\sinh\eta+\frac{1}{2}\beta^i\cosh\eta. \tag{13}$$

Figure 2. Energy density and initial flow at $\tau = 0$ at $\eta \gg 1$ in a head on collisions for nuclei radius $R = 3$ fm.

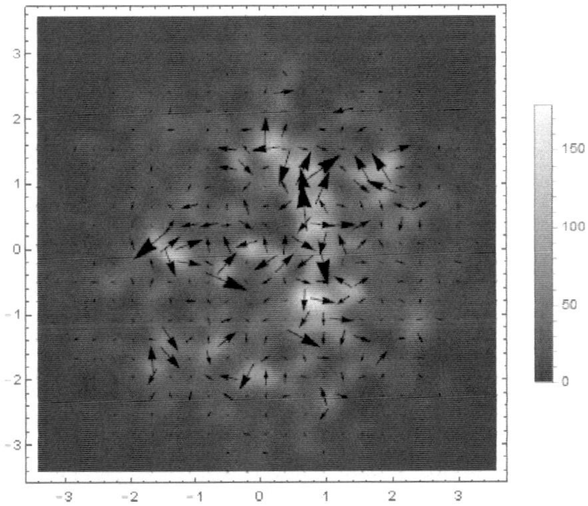

We have introduced the radial flow,

$$\alpha^i = -\nabla^i \epsilon_0, \tag{14}$$

which drives the transverse expansion, and the directed flow,

$$\beta^i = \epsilon^{ij} \left(\left[D^j, B_0\right] E_0 - \left[D^j, E_0\right] B_0 \right), \tag{15}$$

which originates from rapidity-odd transverse fields by an analogue to Gauss' Law. This analysis has been done to fourth order in work by G. Chen [5]. We have so far implemented up to second order in the event generator and plan to continue to higher orders (at least to fourth order). Figure 2 shows the initial energy density and flow field in the transverse plane for a typical event.

4. Conclusion

We have shown preliminary results from a CGC event generator currently under development. By utilizing analytic results, we will be enabled to investigate internal dynamics throughout the initial evolution of the system. A matching to viscous hydrodynamics, as in [6], will take us further to kinetic freeze-out. In terms of observables, we have shown how directed flow may be calculated, which is anticipated to allow for angular momentum calculated from first principles to be put into the initial conditions for the first time.

References

[1] Dusling K, Gelis F and Venugopalan R 2011, *Nucl. Phys. A* **872**, 161
[2] Schenke B, Tribedy P and Venugopalan R 2013 *Phys. Rev. Lett.* **108**, 25, 252301
[3] McLerran L and Venugopalan R 1994 *Phys. Rev.* D **49**, 2233
[4] Chen G and Fries R J 2013 *Phys. Lett.* B **723**, 417
[5] Chen G, Fries R J, Kapusta J I and Li Y 2015 *Phys. Rev.* C **92**, no. 6, 064912
[6] Gale C, Jeon S, Schenke B, Tribedy P and Venugopalan R 2012 *Phys. Rev. Lett.* **110**, 1, 012302

IOP Conf. Series: Journal of Physics: Conf. Series **832** (2017) 012019 doi:10.1088/1742-6596/832/1/012019

Multiplicity dependence of light flavour hadrons in small systems with the ALICE experiment

Nicolò Jacazio for the ALICE Collaboration

Department of Physics and Astronomy - Alma Mater Studiorum, Università di Bologna - Via Irnerio, 46 - 40126 Bologna, Italy

E-mail: nicolo.jacazio@bo.infn.it

Abstract. The large statistics of data collected at the high energies reached at the Large Hadron Collider have provided unprecedented opportunities to probe in more detail the mechanisms of particle production in small collision systems such as proton–proton (pp) and proton–lead (p–Pb) collisions. It is particularly interesting to perform such studies in high-multiplicity events, where, in the last years, features where found that are reminiscent of phenomena interpreted as signs of collective behaviour in lead–lead (Pb–Pb) collisions. These observations justify a comprehensive study of the production of identified particles to further investigate the dynamics in small collision systems. The ALICE detector, thanks to its excellent particle identification capabilities, allows the measurement of identified particles over a wide range of transverse momentum ($p_{\rm T}$). In these proceedings we report on the $p_{\rm T}$ distributions of π, K, p, K^0_S, K^*, Λ, Ξ and Ω measured as function of the charged-particle multiplicity density in pp collisions at $\sqrt{s} = 7$ TeV. We further report on the study of particle ratios in comparison to Monte Carlo models and in different collision systems. In particular, the production of hadrons containing strange quarks is also discussed as a function of the event multiplicity.

1. Introduction

High energy heavy-ion collisions, such as Pb–Pb collisions at the Large Hadron Collider, provide a unique way to study the deconfined state of matter known as Quark-Gluon Plasma (QGP). The onset of this strongly interacting phase is accompanied by some peculiar signatures in the final state. These include the collective flow of particles created in the collision and an increase in the relative production rates of strange and multi-strange hadrons with respect to minimum-bias pp and p–Pb collisions.

The transverse momentum ($p_{\rm T}$) distributions of identified particles are fundamental observables that can be used to study in detail the properties of the system created in the collision. Particle production measured in heavy-ion collisions is well described by models which include a hydrodynamic evolution [1]. A similar phenomenology to the Pb–Pb case has also been observed for high energy pp and p–Pb collisions when selecting events with a higher number of charged particles produced (the event multiplicity) with respect to the minimum-bias values. In particular some clear analogies have been found for the three systems when comparing long-range and near-side angular correlations [2, 3, 4, 5]. These observations motivate a comprehensive study of the identified particle production in order to investigate in more detail the origin of these features.

IOP Conf. Series: Journal of Physics: Conf. Series **832** (2017) 012019 doi:10.1088/1742-6596/832/1/012019

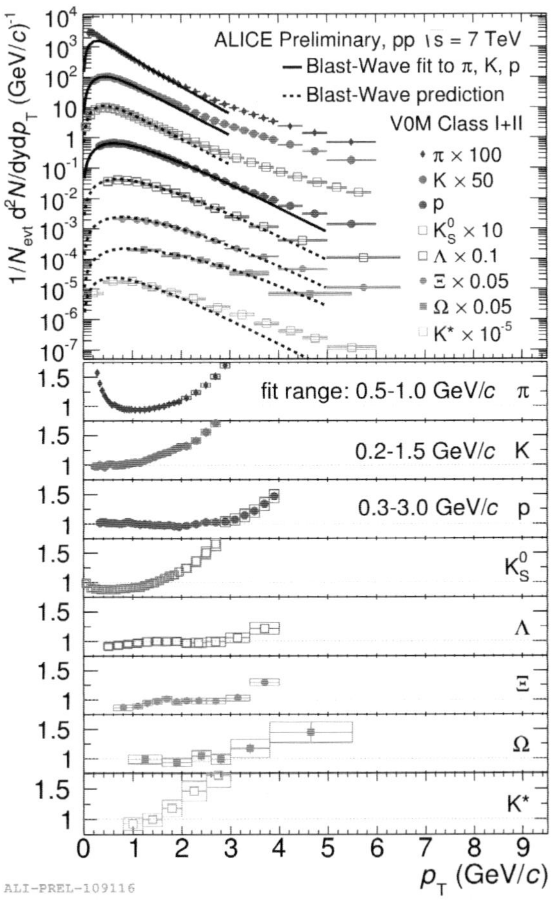

Figure 1. Spectra for π, K, p, K^0_S, K^*, Λ, Ξ and Ω obtained by merging the two highest multiplicity classes (I+II). A fit to the spectra of π, K and p with the Blast-Wave function [12] is shown in solid lines. The prediction for other particle species is reported in dashed lines. Different scaling factors are applied to the spectra in order to allow better visibility. In the lower panels the ratios between the measured spectra and the fit or the predicted function are shown.

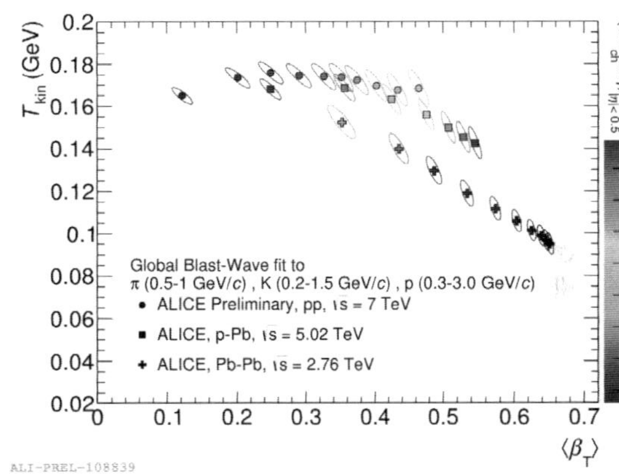

Figure 2. Blast-Wave function parameters ($\langle\beta_T\rangle$ and T_{kin}) obtained from the fit to the spectra of π, K and p for pp, p–Pb and Pb–Pb collisions for different multiplicities. A colour scale representing the corresponding multiplicity is shown to facilitate comparisons.

The ALICE experiment [6, 7] is particularly well suited for the study of identified particles thanks to the high detector granularity coupled with the excellent particle identification (PID) capabilities ensured over a wide range of transverse momentum by its different detectors. In particular, PID is performed in the region at mid-rapidity via the measurements of specific energy loss in the Inner Tracking System (ITS) and in the Time Projection Chamber (TPC). In addition the particle velocity is measured via the Time Of Flight (TOF) detector, while the High Momentum Particle IDentification (HMPID) detector identifies particles by measuring the angle of emission of Cherenkov light.

2. Particle spectra

We report on the transverse momentum spectra of π, K, p, K^0_S, K^*, Λ, Ξ and Ω measured as a function of multiplicity in pp collisions at $\sqrt{s} = 7$ TeV [8]. These measurements refers to primary particles, i.e. particles produced either directly in the collision or formed by non-weak particle decays. Weak decays and contributions from particle knock-out in the material were removed with the data driven approach described in [9, 10]. The systematic uncertainties were estimated by varying the PID techniques and the selection criteria used to select the track sample. The data sample, consisting of $\sim$ 100 M events collected in 2010 with a minimum-bias

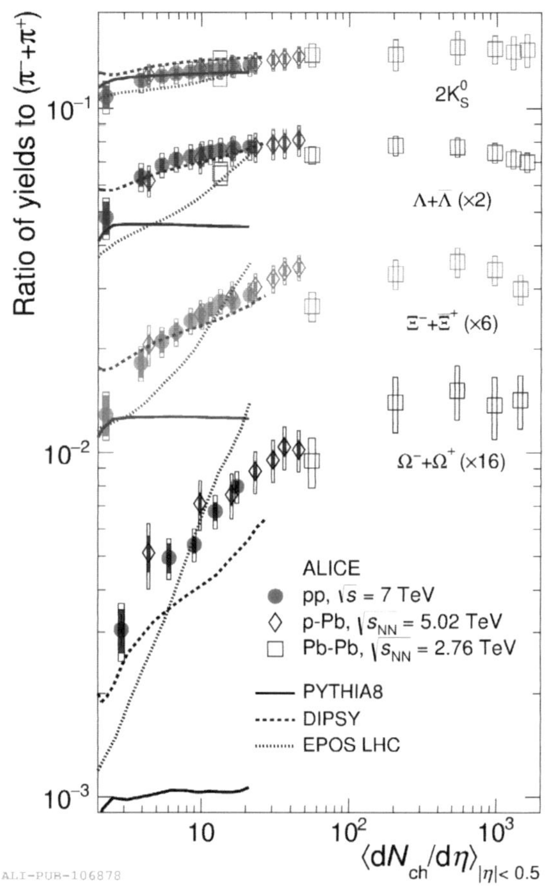

Figure 3. Ratios of K^0_S, Λ, Ξ and Ω yields to the one of π for pp, p–Pb and Pb–Pb collisions as a function of the charged multiplicity density compared to Monte Carlo models.

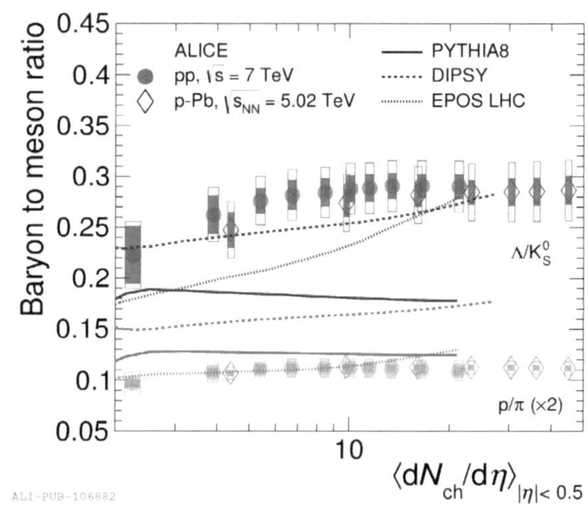

Figure 4. Λ/K^0_S and p/π ratios for pp and p–Pb as a function of multiplicity compared to Monte Carlo models.

trigger, was divided into ten classes (I-X), where class I (X) corresponds to the highest (lowest) multiplicity. The different classes were defined by measuring the total charge deposited in the V0 detectors (V0M amplitude), which consist of a set of two scintillator hodoscopes located in the pseudorapidity region $2.8 < \eta < 5.1$ (V0A) and $-3.7 < \eta < -1.7$ (V0C) and covering the full azimuth. The corresponding mean multiplicity density at mid-rapidity ($\langle dN_{ch}/d\eta \rangle$) was also measured for each multiplicity class as reported in [11]. In order to minimise spurious measurements originating from pile-up, events with more than one reconstructed vertex were rejected.

An estimate of the efficiency and acceptance corrections was performed by means of a full Monte Carlo simulation using events generated with PYTHIA6 Tune Perugia0 and propagating particles through a detailed description of the ALICE detector with the GEANT3 transport code.

The spectra of π, K, p, K^0_S, K^*, Λ, Ξ and Ω, obtained by combining the two highest multiplicity classes (I+II), are shown in Fig. 1. A mass-dependent shift of the spectral shapes towards higher momenta is observed when comparing different particle species, with the effect being more pronounced for particles with higher mass. This feature is similar to what is observed in Pb–Pb collisions, where the phenomenon is understood in terms of radial flow. The spectra for π, K and p are fitted simultaneously with the Blast-Wave model [12], while for the other species a prediction from the same model is computed by using the parameters obtained from the fit to π, K and p. From the ratio between the spectra and the Blast-Wave model, shown in the lower panels of Fig. 1, one can see that it offers a fairly good description of the data. Moreover,

IOP Conf. Series: Journal of Physics: Conf. Series **832** (2017) 012019 doi:10.1088/1742-6596/832/1/012019

when comparing the Blast-Wave fit parameters in pp and p–Pb as a function of multiplicity in Fig. 2, an almost continuous evolution is found.

From the ratios between the p_T-integrated yields of strange and multi-strange hadrons and the one of π, shown in Fig. 3 for the pp, p–Pb and Pb–Pb systems, a multiplicity-dependent increase in the relative yield for particles having non-zero net strangeness content is observed. Furthermore, the observed increase seems to be dependent on the number of constituent strange quarks, as the increase is larger for multi-strange hadrons. The comparison of the ratios with the predictions from Monte Carlo models shows that the increase is reproduced by the DIPSY event generator, while it is unable to reproduce the p/π ratio shown in Fig. 4. The EPOS model manages to describe only the magnitude of the effect and overestimates the rise while PYTHIA misses both trend and magnitude with increasing differences as the strangeness content increases.

The ratios between the p_T-integrated yields of particles with equal amount or no s quark, such as Λ/K^0_S and p/π respectively, are reported in Fig. 4 as a function of the charged particle multiplicity for pp and p–Pb events. The trends for both ratios are independent of the event multiplicity, thus suggesting that the increase in the production of strange particles is related to the strangeness content rather than due to a difference in the particle masses. Comparisons to Monte Carlo models show poor agreement, especially when trying to reproduce all observations simultaneously.

3. Conclusions

The ALICE Collaboration has presented the results on the production of identified light flavour hadrons as a function of the charged particle multiplicity in pp collisions at $\sqrt{s} = 7$ TeV. A mass-dependent hardening of the spectral shapes was observed for high multiplicity collisions. The Blast-Wave model describes the measured spectra of π, K and p well. This holds true also for other particle species whose spectra show a qualitative good agreement with the spectral shapes predicted by the Blast-Wave model. The fit parameters exhibit a continuous evolution with multiplicity in pp and p–Pb. An increase in the production, relative to π, of strange particles was found as a function of the multiplicity. This increase is more pronounced for particles with higher strangeness content. No significant increase is found when comparing the production of particles with the same number of constituent s quarks, indicating that this effect is related to the strangeness content rather than the particle masses. This observation is qualitatively reproduced by models such as DIPSY, while the same model is not in quantitative agreement with the analogous measurement for p/π.

In conclusion, the results presented in these proceedings show some clear similarities between the pp, p–Pb and Pb–Pb systems, possibly suggesting that the same particle production mechanisms are at play at high energies and particle densities, regardless of the colliding system.

References

[1] Abelev B et al. (ALICE Collaboration) 2013 *Phys. Rev.* C **88** 044910
[2] Khachatryan V et al. (CMS Collaboration) 2010 *JHEP* **1009** 091
[3] Aad G et al. (ATLAS Collaboration) 2016 *Phys. Rev. Lett.* **116**, 172301
[4] Abelev B et al. (ALICE Collaboration) 2013 *Physics Letters* B **719** 29 - 41
[5] Aad G et al. (ATLAS Collaboration) 2013 *Phys. Rev. Lett.* **110**, 182302
[6] Aamodt K et al. (ALICE Collaboration) 2008 *Journal of Instrumentation* **3** S08002
[7] Abelev B et al. (ALICE Collaboration) 2014 *Int. J. Mod. Phys.* A **29** 1430044
[8] Adam J et al. (ALICE Collaboration) 2016 arXiv:1606.07424
[9] Aamodt K et al. (ALICE Collaboration) 2011 *Eur. Phys. J.* C **71** 1594
[10] B. Abelev et al. (ALICE Collaboration) 2012 *Phys. Rev. Lett.* **109** 252301
[11] B. Abelev et al. (ALICE Collaboration) 2013 *Phys. Rev. Lett.* **110** 032301
[12] Schnedermann E, Sollfrank J, Heinz U W 1993 *Phys. Rev.* C **48** 2462

HQ16 IOP Publishing
IOP Conf. Series: Journal of Physics: Conf. Series **832** (2017) 012020 doi:10.1088/1742-6596/832/1/012020

Experimental overview on heavy-flavor production

Alessandro Grelli
Princetonplein 1, Oornestein laboratory, Utrecht, NL
E-mail: a.grelli@uu.nl

Abstract. Hadrons containing heavy-flavors are unique probes of the properties of the hot and dense QCD medium produced in heavy-ion collisions. Due to their large masses, heavy quarks are produced at the initial stage of the collision, almost exclusively via hard partonic scattering processes. Therefore, they are expected to experience the full collision history propagating through and interacting with the QCD medium. The parton energy loss, which is sensitive to the transport coefficients of the produced medium, can be studied experimentally by measuring the nuclear modification factor which accounts for the modification of the heavy- flavored hadron yield in Pb–Pb collisions with respect to pp collisions. In semi-central Pb–Pb collisions, the degree of thermalization of charm quarks in the QCD medium can be accessed via the measurement of the heavy flavor elliptic flow v_2 at low p_T. Furthermore, the measurement of heavy-flavors production in pp collisions allows testing the perturbative QCD calculations.
The PHENIX and STAR Collaborations at the Relativistic Heavy-Ion Collider and ALICE, CMS and ATLAS Collaborations at the Large Hadron Collider have measured the production of charmonium and bottonium states as well as open heavy flavor hadrons via their hadronic and semi-leptonic decays at mid-rapidity and in the semi-muonic decay channel at forward rapidity in pp, p-A and A–A collisions in an energy domain that ranges from $\sqrt{s} = 0.2$ TeV to $\sqrt{s} = 13$ TeV in pp collisions and from $\sqrt{s_{\rm NN}} = 0.2$ TeV to $\sqrt{s_{\rm NN}} = 5.02$ TeV in A–A collisions. In this contribution the latest experimental results will be reviewed.

1. Introduction

The lattice QCD calculations predict for the strongly-interacting matter, in conditions of high density and temperature such as the one produced in high-energy ultra relativistic heavy-ion collisions, quark de-confinement and the formation of the so called Quark-Gluon Plasma (QGP) [1]. Heavy-flavor hadrons are qualitatively and quantitatively unique probes for the investigation of the QGP. Open charm and beauty hadrons are expected to be sensitive to the energy density through the in-medium energy loss of their heavy quark constituents. In particular, charm and beauty quarks, due to the large masses, are produced at the early stage of the collision [2] and expected to loose less energy than light quarks and gluons while traversing the QGP due to dead-cone and color-charge effects [3, 4]. Furthermore, the in-medium thermal production is expected to be negligible [5]. In addition, if in-medium hadronization is the dominant mechanism of charm hadron formation at low-p_T then strange charm hadrons (i.e. D_s^+) should be enhanced relatively to non-strange charm hadrons (i.e D^0) [6]. The aforementioned properties make heavy quarks a clean and effective probe for the QGP properties.
In addition quarkonium states (i.e Charmonium and Bottonium) are expected to be sensitive to the initial temperature of the system via their dissociation due to color screening. At Large Hadron Collider (LHC) energy scale the recombination effects, that can be pictured as in-medium recombination and statistical recombination at the freeze-out, are expected to become

competitive with the dissociation producing a visible effect on the nuclear modification factor once compared with the same observable at lower collision energy (i.e. Relativistic Heavy-Ion Collider (RHIC)).
The nuclear modification factor R_{AA} of particles is well-established as a sensitive observable for the study of the interaction of hard partons with the medium. This factor is defined as the ratio of the p_{T}-differential yield measured in nucleus-nucleus (AA) collisions in a given centrality class to the yield calculated from the proton-proton cross-section scaled by the nuclear overlap function $< T_{\mathrm{AA}} >$ for that centrality class, obtained from Glauber model calculations of the collision geometry [7],

$$R_{\mathrm{AA}}(p_{\mathrm{T}}) = \frac{1}{< T_{\mathrm{AA}} >} \cdot \frac{\mathrm{d}N_{\mathrm{AA}}/\mathrm{d}p_{\mathrm{T}}}{\mathrm{d}\sigma_{\mathrm{pp}}/\mathrm{d}p_{\mathrm{T}}} \tag{1}$$

A strong suppression of the yield of charged particles was observed in A–A collisions at intermediate-high p_{T} relative to scaled pp collisions both at RHIC [8, 9] and LHC [10].
Further insight in the medium properties is provided by the measurement of anisotropy in the azimuthal distribution of particle momenta characterized by the Fourier coefficients $v_{\mathrm{n}} =< cos[n(\phi - \psi n)] >$, where n is the order of the harmonic, ϕ is the azimuthal angle of the particles momentum, and ψn is the azimuthal angle of the initial state symmetry plane for the n-th harmonic. At low-intermediate (< 6 $GeV/$c) transverse momentum a non zero v_2 reflects the collective expansion, while high transverse momentum a positive v_2 is expected as a consequence of the path-length dependence of the in-medium parton energy loss.
In this paper we review the results on heavy flavor production in pp collisions in a range of energies from $\sqrt{s} = 0.2$ TeV to $\sqrt{s} = 13$ TeV and in A–A collisions in a range of energies from $\sqrt{s_{\mathrm{NN}}} = 0.2$ TeV to $\sqrt{s_{\mathrm{NN}}} = 5.02$ TeV. The proton-proton results provide the baseline for the study of medium-induced effects in nucleus–nucleus collisions. Moreover they offer the possibility to test pQCD. Perturbative QCD predictions extracted using FONLL [25] (FixedOrder-plus-Next-to-Leading-Log) describe well the beauty production cross section measured in pp collisions at $\sqrt{s} = 1.96$ TeV by CDF at the Tevatron [26]. Additionally, the CMS experiment found that at the LHC energy scale the beauty production is well described by the theory [27]. In the case of charm, FONLL prediction agrees with data within errors. Nevertheless, the data points are systematically on the upper edge of the theory error band suggesting that charm production could be underestimated in the calculation as also seen at RHIC energies ($\sqrt{s}$= 0.2 and 0.5 TeV) [28].

2. Open heavy-flavor production in nucleon–nucleon collisions

At the RHIC and LHC the production of open heavy flavor hadrons has been measured via their hadronic and semi-leptonic decays at mid-rapidity in pp collisions in an energy domain that rages from $\sqrt{s} = 0.2$ TeV to $\sqrt{s} = 13$ TeV and in the semi-muonic decay channel at forward rapidity at $\sqrt{s} = 7$ TeV. The D meson production, as well as the total charm production cross-section, are discussed in section 2.1 while a review of heavy-flavor measurement via heavy-flavor hadron semi-leptonic decays can be found at [14, 15, 16, 17].

2.1. D meson production cross-section at central and forward rapidity

The study of open charm production via D meson hadronic decays is performed in a fairly similar way at RHIC and LHC. It is based either on displaced vertex reconstruction or on background simulation [18] (i.e event mixing, like-sign or rotational background). The displaced vertex reconstruction, is achieved thanks to the excellent tracking systems of the modern high-energy experiments that allow selection of decay vertices of open charm mesons ($c\tau \sim 120$ – 310 μm). In a second step an invariant mass analysis is performed. On the left panel of Fig. 1 the $D^0 \rightarrow K^-\pi^+$ production cross-section is shown in the transverse momentum region

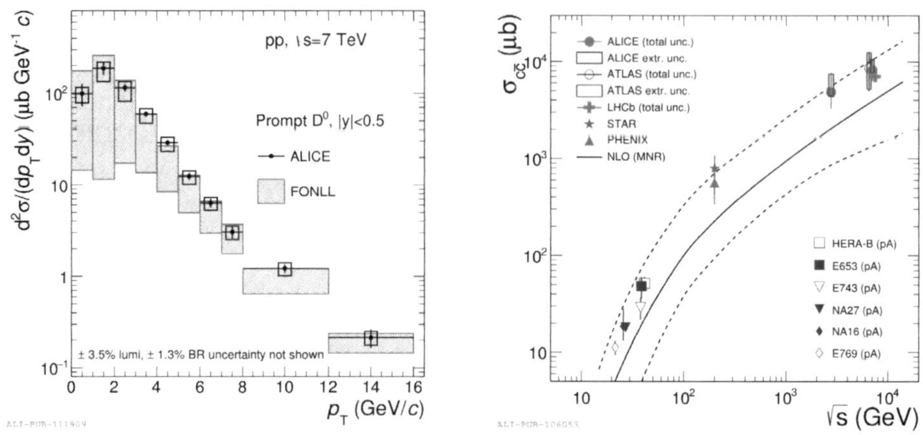

Figure 1. Left panel: D^0 p_{T}-differential production cross-section as measure by ALICE Collaboration super-imposed with FONLL pQCD calculation (solid red boxes). Right panel: total charm production cross-section versus $\sqrt{s}$ [18].

$0 < p_{\mathrm{T}} < 16\ GeV/c$ as measured by ALICE Collaboration in pp collisions at $\sqrt{s} =\ 7$ TeV at mid-rapidity ($|y| < 0.5$) [18]. The measured cross-section sits on the upper-side of the FONLL pQCD calculation error band. The result, qualitatively similar to the measurements performed at $\sqrt{s} = 0.2, 0.5$ TeV and $\sqrt{s} = 1.96$ TeV by STAR [28] and CDF [26] collaborations, respectively, suggests an underestimation of the charm production by the calculation. The right side of Fig. 1 shows the evolution of the total charm production cross-section versus the collision energy as measured by several collaborations [18]. The experimental results are compared with NLO MNR [19] calculations. The model is compatible with the measured data within uncertainties. It is noticeable the agreement of the ALICE, ATLAS [20] mid-rapidity measurements with the LHCb [21] forward rapidity measurement at $\sqrt{s} =\ 7$ TeV.

3. Open heavy-flavors in nucleus–nucleus collisions

Due to the QCD nature of parton energy loss discussed in the introduction, hierarchy in the R_{AA} is expected to be observed when comparing the mostly gluon-originated light-flavor hadrons to D and to B mesons [11, 12]: $R_{\mathrm{AA}}(\pi) < R_{\mathrm{AA}}(D) < R_{\mathrm{AA}}(B)$. Clearly there are caveats to be taken into account while interpreting the results, like the different parton p_{T} spectra that can modify this picture.

The production of D^0, D^{*+} and D^+ was measured [13, 12, 22, 23] in A–A collisions at mid-rapidity via the exclusive reconstruction of the decays $D^0 \to K^-\pi^+$, $D^{*+} \to D^0\pi^+$ and $D^+ \to K^-\pi^+\pi^+$.

ALICE measured the R_{AA} of prompt D mesons in 10% wide centrality bins from 0 to 80% and in the $8 < p_{\mathrm{T}} < 16\ GeV/c$ region [33]. The measurement was compared with the R_{AA} of non-prompt J/ψ (from B-mesons decays) measured by the CMS Collaboration [35]. As shown in the left panel of Fig. 2, the comparison of the two results, in a similar kinematic range, shows clear indication for a mass hierarchy in the suppression pattern being the non-prompt J/ψ less suppressed than D mesons, as expected from a larger energy loss of charm quark with respect to beauty quark. A comparison of the p_{T}-differential D meson R_{AA} measured at $\sqrt{s_{\mathrm{NN}}} = 0.2$ TeV by STAR Collaboration [34] and the one measured by ALICE Collaboration [24] at $\sqrt{s_{\mathrm{NN}}} = 2.76$ TeV hints a different trend at low-p_{T}. Namely the STAR result show an enhancement at about 1.5 GeV/c while in the ALICE case a suppression is found. The difference

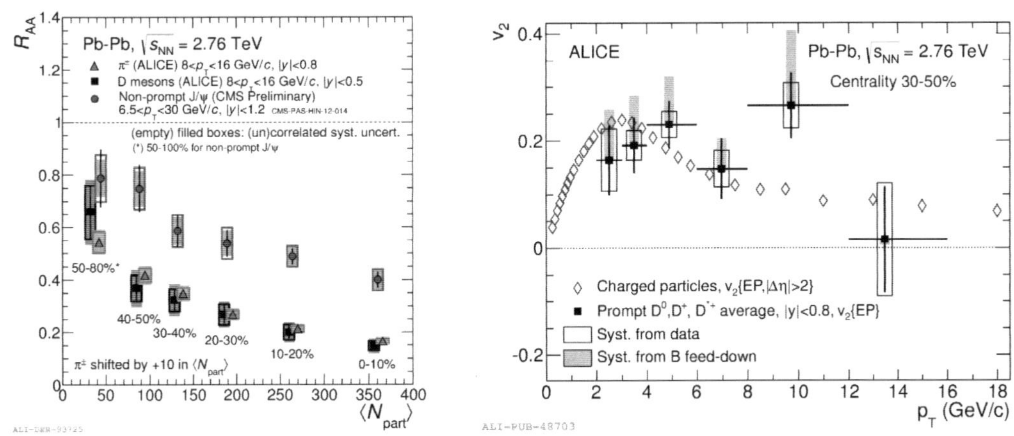

Figure 2. Left panel: D meson R_{AA} versus N_{part} measured by ALICE Collaboration [33] compared with the non-prompt J/ψ measured by CMS Collaboration [35]. Right panel: D meson elliptic flow compared with charged hadrons as measured by ALICE Collaboration [30].

may be related to initial state effects and elliptic flow but the relatively large uncertainties of the two measurements suggest the need for additional statistics before drawing any conclusion. In addition to the D mesons measurements, the R_{AA} of heavy-flavor hadron decay electrons was measured by ALICE, STAR and PHENIX [15, 16, 17]. Furthermore, ALICE measured the D-meson elliptic flow (v_2) as a function of transverse momentum in the Pb–Pb centrality interval $30 - 50\%$ [30]. As shown in the right panel of Fig. 2 the D meson v_2 shows a similar trend of the charged particle v_2, within the rather large systematic and statistical uncertainties. The measurement indicates, at 5σ confidence level, a positive v_2 for $2 < p_T < 6\ GeV/c$. More recent results from CMS (at LHC) [23] and STAR (at RHIC) [31] confirm the observation. The D meson results, supported by recent measurements in the semi-leptonic sector [32], suggests that charm quarks participate in the collective flow of the expanding medium.

4. Quarkonium measurements in nucleus-nucleus collisions

Quarkonium is considered one of the main probes of the investigation of QGP. Its production in A–A collisions is expected to be significantly suppressed with respect to the pp yield, scaled by the number of binary nucleon-nucleon collisions. The origin of such a suppression is thought to be the colour screening [36]. According to it the suppression should occur sequentially, following the binding energy of each meson. Strongly bound states, such as the $\Upsilon(1S)$ or the J/ψ, should melt at higher temperatures with respect to the more loosely bound states such as $\Upsilon(2S)$, or $\Upsilon(3S)$. The simple picture of color screening needs to be complemented by recombination effects at LHC energies.

On this regard, the centrality dependence of the inclusive J/ψ production at low-p_T in A–A collisions [38, 39, 40] has been measured by ALICE and PHENIX Collaborations. In Fig. 3 the inclusive J/ψ nuclear modification factor, as a function of the number of participant nucleons (N_{part}), is compared to the one measured by the PHENIX [38] in a similar kinematic range. It is evident a smaller J/ψ suppression at LHC energies, with respect to RHIC. The smaller suppression at the LHC is assumed to be due to the larger $c\bar{c}$ pair multiplicity which allows a larger recombination, resulting in a compensation of the suppression from color screening. The measured behavior is expected by the statistical model [29], where the J/ψ yield is completely determined by the chemical freeze–out conditions and by the abundance of $c\bar{c}$ pairs. A precise

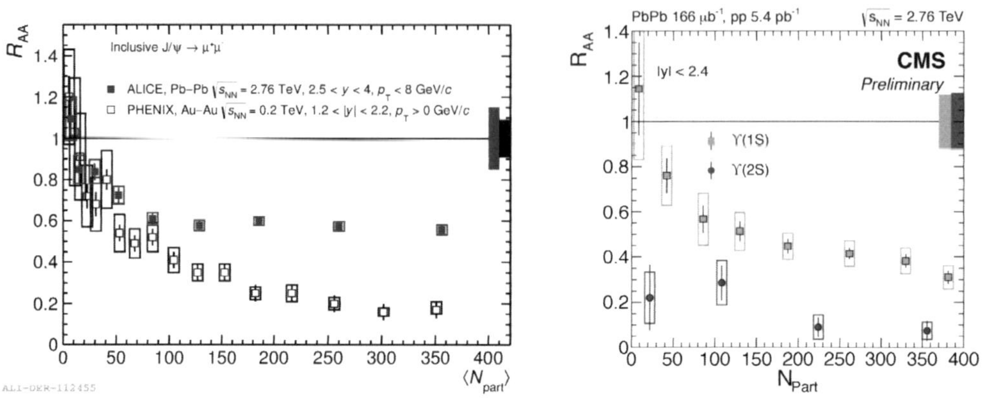

Figure 3. Left panel: ALICE [39] and PHENIX [38] inclusive low-p_{T} J/ψ nuclear modification factor versus the number of participant nucleons at forward rapidity. Right panel: R_{AA} ofΥ states as a function of centrality, as measured by CMS Collaboration [45].

comparison between the experimental result and the theory predictions will benefit from the improvements in the experimental measurement of the charm production cross-section. Due to kinamatical reasons the (re)combination process is expected to be dominant in central collisions and at low-p_{T}, becoming negligible as the $J/\psi p$ $_{\mathrm{T}}$ increases. This observation is confirmed by the mid-rapidity R_{AA} measured by CMS [41] and STAR [42] for high-p_{T} J/ψ. Further insight on charmonium production in Pb-Pb collisions can be achieved comparing the production yields of a higher mass resonance, as the $\psi(2S)$, to the J/ψ ones. The double ratio of the $\psi(2S)$ to J/ψ yields in Pb–Pb and in pp collisions as a function of centrality, can be found in CMS [43] and for ALICE [40, 44]. CMS observes values of the double ratio higher than one in the region $1.6 < |y| < 2.4$, $3 < p_T < 30$ $GeV/$c, implying a $\psi(2S)$ R_{AA} larger than the J/ψ one. Results obtained in the range $|y| < 1.6$ and $6.5 < p_T < 30$ $GeV/$c show a decreasing pattern towards most central collisions, i.e. the $\psi(2S)$ R_{AA} is, smaller than the corresponding J/ψ value, as expected by sequential suppression scenario. However, the large statistics and systematic uncertainties associated to the $\psi(2S)$ measurement preclude any strong conclusion.

Finally, the LHC energies allow the study of the resonances of the bottomonium family. CMS results [45], confirm the observation of sequential suppression for $\Upsilon(1S)$ ($R_{\mathrm{AA}} = 0.43 \pm 0.03 \pm 0.07$), $\Upsilon(2S)(R_{\mathrm{AA}} = 0.13 \pm 0.03 \pm 0.02)$ shown in Fig. 3 (right) and $\Upsilon(3S)(R_{\mathrm{AA}} = 0.14$ at 95% CL.). A comprehensive overview on quarkonium production can be found at [37].

5. Conclusions

In the last ten years the experimental investigation of heavy-flavor physics in heavy ion collisions had a boost with the start of the LHC operations at CERN and with the upgrade of the RHIC at BNL. A considerable amount of new measurements become available both in pp and A–A. In pp collisions the limit of ~ 0 p_{T} was reached, both at mid- and forward rapidity, allowing on one side a clear test of pQCD and on the other side the definition of a solid experimental reference for A–A studies. In A–A collisions the study of charm and beauty nuclear modification factors become available via hadronic and semi-leptonic decays of open heavy-flavor hadrons at mid-rapidity and in their semi-muonic decay channel at forward rapidity. In addition elliptic flow measurements proved that charm take part in the collective motion of the system. Quarkonium studies proved that at high collisional energy the simple picture of dissociation via color screening needs to be complemented with the concept of in-medium and statistical regeneration.

In this frame of new measurements and theoretical developments the run II of the LHC and the RHIC operations of the next few years will open the possibility of precision measurements in the charm sector. The run III of the LHC starting in 2019 will open the era of precision measurements in the beauty sector.

References

[1] D. J. Gross and F. Wilczek, Phys. Rev. Lett. 30, 1343, H.D. Politzer, Phys. Rev. Lett. 30, 1346 (1973).
[2] F.-M. Liu and S.-X. Liu, Phys. Rev. C89 no. 3, (2014) 034906, arXiv:1212.6587.
[3] R. Baier, Y. L. Dokshitzer, A. H. Mueller, S. Peigne, and D. Schiff, Nucl.Phys. B484 (1997) 265282, arXiv:hep-ph/9608322.
[4] M. Gyulassy and M. Plumer, Phys.Lett.B243 (1990) 432438
[5] P. Lvai and R. Vogt, Phys. Rev. C 56, 2707
[6] V. Greco, C. M. Ko, and R. Rapp, Phys. Lett. B595 (2004) 202208, arXiv:nucl-th/0312100. A. Andronic, P. Braun-Munzinger, K. Redlich, and J. Stachel,Phys. Lett. B571 (2003) 3644, arXiv:nucl-th/0303036.
[7] N. Armesto, C. A. Salgado and U. A. Wiedemann, Phys. Rev. D 69 (2004) 114003.
[8] S. S. Adler et al. [PHENIX Collaboration], Phys. Rev. C 69 (2004) 034910.
[9] J. Adams et al. [STAR Collaboration], Phys. Rev. Lett. 91 (2003) 172302.
[10] K. Aamodt et al. [ALICE Coll.], Phys.Lett. B696 (2011) 30-39
[11] A. Dainese, N. Arnesto, M. Cacciari, et al. AIP Conf. Proc., 842 (2006) pp. 74-76.
[12] B. Abelev et al. [ALICE Coll.], JHEP 1209 (2012) 112.
[13] B. Abelev et al. [ALICE Coll.], JHEP 1201 (2012) 128.
[14] Adare, A. [PHENIX Coll.], Physical Review C, 93 (2016). pp. 34904-29. ISSN 2469-9985
[15] A. Adare et al. [PHENIX Coll.], Phys.Rev. C84 (2011) 044905, arXiv:1005.1627.
[16] A. Adare et al. [PHENIX Coll.], Phys.Rev. C86 (2012) 024909, arXiv:1204.0754.
[17] B. Abelev et al. [STAR Coll.], Phys.Rev.Lett. 98 (2007) 192301, arXiv:nucl-ex/0607012.
[18] J. Adam et al. [ALICE Coll.], CERN-EP-2016-127 , arXiv:1605.07569.
[19] M. L. Mangano, P. Nason, and G. Ridolfi, Nucl. Phys. B373 (1992) 295345.
[20] G. Aad et al. [ATLAS Coll.], arXiv:1512.02913 [hep-ex].
[21] R. Aaij et al. [LHCb Coll.], Nucl. Phys. B871 (2013), arXiv:1302.2864. R. Aaij et al. [LHCb Coll.], JHEP 03 (2016) 159, arXiv:1510.01707.
[22] L. Adamczyk et al. [STAR Coll.], Phys.Rev.Lett. 113 no. 14, (2014) 142301, arXiv:1404.6185.
[23] CMS Collaboration, CMS PAS HIN-16-001 (2016). URL https://cds.cern.ch/record/2157844/
[24] J. Adam et al. [ALICE Coll.], JHEP 1603 (2016) 081
[25] M. Cacciari, M. Greco and P. Nason, JHEP 9805 (1998) 007; M. Cacciari, S. Frixione and P. Nason, JHEP 0103 (2001) 006.
[26] M. Cacciari, S. Frixione, M.L. Mangano, P.Nason and G.Ridolfi, JHEP 0407 (2004) 033.
[27] V. Khachatryan, et al. [CMS Coll.], Eur.Phys.J. C71, 2011, arXiv:1011.4193.
[28] L. Adamczyk et al.[STAR Coll.], Phys.Rev. D86 (2012) 072013
[29] A. Andronic, P. Braun-Munzinger, K. Redlich, J. Stachel, J. Phys. G38 (2011) 124081. arXiv:1106.6321.
[30] B. Abelev at al. [ALICE Coll.], PRL 111 (2013) 102301.
[31] Liang He [STAR Coll.], J.Phys.Conf.Ser. 736 (2016), 012007.
[32] J. Adam et al. [ALICE Coll.], JHEP 1609 (2016) 028.
[33] J. Adam et al. [ALICE Coll.], JHEP 1511 (2015) 205.
[34] L. Adamczyk et al. [STAR Coll.]. Phys. Rev. Lett. 113, 142301 (2014).
[35] CMS Collaboration, CMS-PAS-HIN-12-014 (2013).
[36] T. Matsui and H. Satz, Phys. Lett. B178 (1986) 416
[37] R. Arnaldi, arXiv:1604.03139
[38] A. Adare, et al. [PHENIX Coll.], Phys. Rev. C84 (2011) 054912.
[39] B. B. Abelev, et al. [ALICE Coll.], Phys. Lett. B734 (2014) 314327. arXiv:1311.0214.
[40] J. Adam, et al. [ALICE Coll.], arXiv:1506.08804.
[41] CMS Collaboration: CMS-PAS-HIN-12-014 (2014). URL https://cdsweb.cern.ch/record/1472735
[42] L. Adamczyk, et al. [STAR Coll.], Phys. Lett.B722 (2013) 5562. arXiv:1208.2736.
[43] V. Khachatryan, et al. [CMS Coll.], Phys. Rev. Lett. 113 (2014) 262301. arXiv:1410.1804.
[44] P. Braun-Munzinger, J. Stachel, Phys. Lett. B490 (2000) 196202. arXiv:nucl-th/0007059.
[45] CMS Collaboration: Nuclear modification of Υ states in Pb–Pb (CMS-PAS-HIN-15-001). URL https://cds.cern.ch/record/2030083

IOP Conf. Series: Journal of Physics: Conf. Series **832** (2017) 012021 doi:10.1088/1742-6596/832/1/012021

J/ψ azimuthal anisotropy in Au+Au collisions at $\sqrt{s_{NN}}$ = 200 GeV

Chensheng Zhou(for the STAR Collaboration)

Shanghai Institute of Applied Physics, Chinese Academy of Sciences, Shanghai 201800, China and

Brookhaven National Laboratory, Upton, NY 11973, USA

Abstract.

The study of J/ψ azimuthal anisotropy allows for a disentanglement of various production mechanisms and possibly an access to charm quark azimuthal anisotropy. J/ψ meson produced from direct pQCD processes have little azimuthal anisotropy due to the lack of collectivity or azimuthal preference of initial emitting, while J/ψ meson produced from recombination of charm quarks in the medium are expected to inherit the azimuthal anisotropy of the constituent charm quarks. We present measurements of J/ψ azimuthal anisotropy, with J/ψ reconstructed via the di-electron channel, as a function of transverse momentum in Au+Au collisions at $\sqrt{s_{NN}}$ = 200 GeV. This analysis is carried out with data taken by the STAR experiment during RHIC operation in year 2011. Combined with the published results using 2010 data, the updated results provide further support to the conclusion that J/ψ production above 2 GeV/c is unlikely to be dominated by regeneration of fully thermalized charm quarks.

1. Introduction

Mounting evidences [1–6] have supported that a quark-gluon plasma (QGP) has been produced in ultra-relativistic heavy-ion collisions at the Relativistic Heavy-Ion Collider (RHIC) at Brookhaven National Laboratory, as well as at the Large Hadron Collider at CERN. Yet the property of QGP is still under intensive investigations. Out of many probes that are used to study QGP, charm quark is a unique one [7, 8]. Charm quarks are produced in early steps of the heavy-ion collisions and experience the entire evolution of the QGP. Charm quarks have large masses, which make them a genuine hard probe even at low p_T. Sizable collective motion of charm quarks is an indication of early thermalization, measurement of which is useful for understanding the dynamics responsible for fast thermalization at RHIC. The measurement of J/ψ azimuthal anisotropy, especially its second Fourier coefficient v_2 (elliptic flow) [9], can be used to infer the relative contributions from different production sources [10, 11], as they each give its own characteristic pattern in terms of v_2 magnitudes and p_T dependence. In that sense, J/ψ v_2 is a unique probe for testing production mechanisms and useful for constraining models.

There are many models trying to describe the experimental results of J/ψ v_2. The one that is based on the coalescence of fully thermalized charm quarks [7] gives large v_2, and its increase with p_T starts at relatively large p_T. In the model that considers the direct pQCD production [10], as heavy quarks are expected to thermalize much more slowly than light quarks, the phase space distribution of massive J/ψ is governed by transport rather than kinetic equation, which produces limited v_2 due to the leakage effect. In a recent work, the J/ψ

v_2 is studied under the influence of external magnetic field [12]. With the presence of strong magnetic field, generated early and short-lived, the Lorentz force and the harmonic potential break the rotational symmetry in the coordinate space which eventually introduces additional J/ψ asymmetry at relatively large p_T ($\sim$ 10 GeV/c).

The results for Au+Au collisions at $\sqrt{s_{NN}}$ = 200 GeV recorded at RHIC in 2010 (Run10) have been published [13]. Since the publication, more data have been taken in 2011 (Run11). Our analysis utilized the Run11 dataset, which was then combined with Run10 results, in order to reduce the statistical uncertainty and hopefully achieve a stronger conclusion.

2. J/ψ reconstruction

In this analysis, the J/ψ particles were reconstructed via the $J/\psi \to e^+e^-$ channel, which has a branching ratio of 5.9%. Two daughter tracks of the J/ψ particles were required to have more than 20 space points in the Time Projection Chamber (TPC), and their distances of closest approach (DCA) less than 1 cm from the primary vertex. Low-momentum electrons and positrons can be separated from hadrons by applying a cut on the inverse velocity ($0.97 < 1/\beta < 1.03$), which was calculated from the time of flight measured by the Time-of-Flight detector, and the path length measured by the TPC. At larger momenta ($p > 1.5$ GeV/c), a cut on the momentum-to-energy ratio ($0.3 < p/E < 1.5$) was required to select electrons and positrons, where the energy was measured by the Barrel Electromagnetic Calorimeter. More than 15 TPC space points were required to calculate the energy loss ($\langle dE/dx \rangle$) inside the TPC. The electrons and positrons were further identified by their specific $\langle dE/dx \rangle$ with an asymmetric cut around the expected value, because the lower side has more hadron contamination. The cut also varies according to whether the candidate track passes the $1/\beta$ and/or p/E cut for efficiency optimization. With this combination of $1/\beta$, p/E and $\langle dE/dx \rangle$ cuts, we can identify electrons and positrons in a wide momentum range. The measured J/ψ particles cover the rapidity range $-1 < y < 1$, favoring J/ψ particles near $y = 0$ because of detection efficiency variation due to acceptance and decay kinematics.

3. Measuring J/ψ v_2

In our analysis, we obtained the J/ψ v_2 using event plane method [9]. Firstly, we divided the measurements of J/ψ yields into 10 bins in $(\phi - \Psi)$, ranging from 0 to π. ϕ is the azimuthal angle of J/ψ momentum in cylindrical coordinate system, and Ψ is the event plane angle obtained from the azimuthal distribution of TPC tracks. Two bins at supplementary angles were combined into one. For example, the bin at 0-0.1π and the bin at 0.9π-π were combined. Different event plane resolutions (R) were used for different centrality bins, as listed in Table 1. Entries in the invariant mass distribution of electron and positron pairs were weighted by 1/R according to the centrality [14].

Table 1. Event plane resolution (R) for different centralities.

Centrility(%)	0-5	5-10	10-20	20-30	30-40	40-50	50-60	60-70	70-80
R	0.360	0.478	0.613	0.727	0.798	0.360	0.793	0.695	0.555

Then the J/ψ yield within each combined $(\phi - \Psi)$ bin was obtained by fitting the weighted invariant mass distribution with a Crystal Ball function [15] as signal, sitting on top of a second-order polynomial function as background. The Crystal Ball function (see equation 1) consists

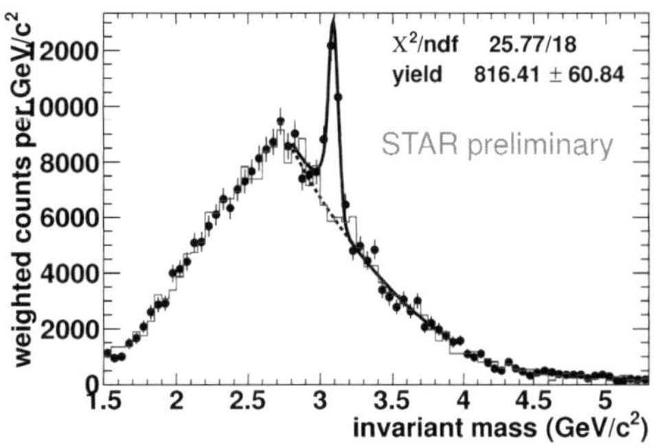

Figure 1. An example of obtaining J/ψ yield. ($0 < p_T < 2$ GeV/c, centrality 10-40% and $0.1\pi < \phi - \Psi < 0.2\pi$).

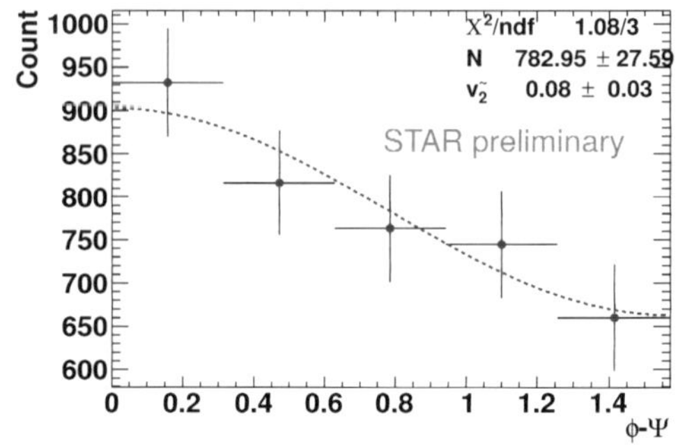

Figure 2. An example of obtaining J/ψ v_2 ($0 < p_T < 2$ GeV/c, centrality 10-40%).

of a Gaussian core and a power-law tail at low mass to account for the energy loss of electrons and J/ψ radiative decays.

$$f(x;\alpha,n,\bar{x},\sigma) = N \cdot \begin{cases} \exp\left(-\frac{(x-\bar{x})^2}{2\sigma^2}\right), & \text{for } \frac{x-\bar{x}}{\sigma} > -\alpha \\ \left(\frac{n}{|\alpha|}\right)^n \cdot \exp\left(-\frac{|\alpha|^2}{2}\right) \cdot \left(\frac{n}{|\alpha|} - |\alpha| - \frac{x-\bar{x}}{\sigma}\right)^{-n}, & \text{for } \frac{x-\bar{x}}{\sigma} \le -\alpha \end{cases} \quad (1)$$

Fig. 1 is an example of e^+e^- invariant mass distribution and the corresponding fit for J/ψ peak in a selected bin. With yield obtained for each $(\phi - \Psi)$ bin, we extracted v_2 by fitting the J/ψ yield as a function of $(\phi - \Psi)$ with the function $y = N(1 + v_2 \cos[2(\phi - \Psi)])$, as shown in Fig. 2. Finally the observed v_2 was scaled by 1/R and $0.1\pi/\sin(0.1\pi)$ to correct for the event plane resolution and bin width effect, where 0.1π was the bin width of our $(\phi - \Psi)$ bins.

4. Combining Run10 and Run11 results

The same track selection and fitting method were used for both Run10 and Run11 analyses. The consistency of these two results has been checked by performing a so-called Z-test [16], as showed in Fig. 3. For each pair of points at a given p_T, their consistency can be quantified by $Z_i = (a_i - b_i)/\sqrt{s_{a_i}^2 + s_{b_i}^2}$, where a_i and b_i are the J/ψ v_2 extracted from Run10 and Run11 datasets, and s_{a_i} and s_{b_i} are the corresponding uncertainties. The value of each Z_i is small (< 2), therefore the difference between each pair is below 2σ. For two consistent datasets, the sum of Z_i should follow a normal distribution with the mean equal to 0 and the σ equal to $\sqrt{N}$, where N is the number of pairs. In our analysis, $N = 4$ and the $\sum Z_i$ is also below 2σ. Therefore we can safely combine the results of two runs and take their difference as a conservative estimate of the systematic uncertainty.

Fig. 4 shows the combined result of Run10 and Run11 analyses together with model calculations. It is noticeable that for p_T above 2 GeV/c, especially the second bin, our measured v_2 is consistent with zero. Based on the calculations of v_2 for J/ψ from the coalescence of thermalized charm quarks [7] and the possible non-flow contribution to the measured v_2, we can infer that in case the charm quarks are fully thermalized, the contribution of regeneration to J/ψ production above 2 GeV/c is unlikely to be the dominant source.

5. Summary

J/ψ v_2 from Run11 data has been obtained and combined with previous Run10 results. The combined result for Au+Au collisions at $\sqrt{s_{NN}} = 200$ GeV is presented as a function of J/ψ

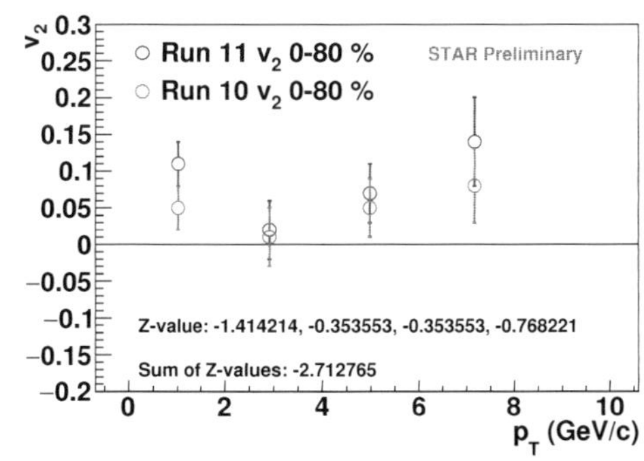

Figure 3. The Z-test of Run10 and Run11 results.

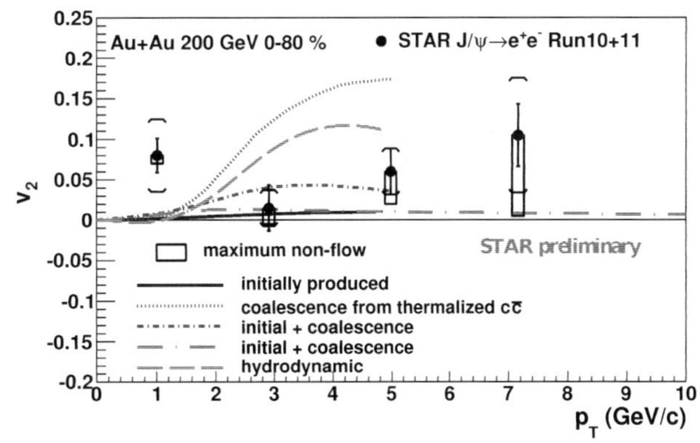

Figure 4. The combined result of Run10 and Run11.

transverse momentum. For p_T above 2 GeV/c, the measured v_2 is consistent with zero within statistical and systematic uncertainties. Comparing to model calculations, the measured v_2 disfavors the scenario that the dominant contribution to the observed yields at this kinematic region is the regeneration of fully thermalized charm quarks. Future work is needed to finalize the systematic uncertainties by comparing different fitting methods, studying trigger bias, etc.

Acknowledgments

This work was supported in part by National Natural Science Foundation of China under contract Nos. 11220101005 and 11421505, the Major State Basic Research Development Program in China under Contract No. 2014CB845401.

References

[1] Arsene I *et al.* (BRAHMS) 2005 *Nucl. Phys.* **A757** 1–27 (*Preprint* nucl-ex/0410020)
[2] Back B B *et al.* 2005 *Nucl. Phys.* **A757** 28–101 (*Preprint* nucl-ex/0410022)
[3] Adams J *et al.* (STAR) 2005 *Nucl. Phys.* **A757** 102–183 (*Preprint* nucl-ex/0501009)
[4] Adcox K *et al.* (PHENIX) 2005 *Nucl. Phys.* **A757** 184–283 (*Preprint* nucl-ex/0410003)
[5] Chatrchyan S *et al.* (CMS) 2012 *JHEP* **05** 063 (*Preprint* 1201.5069)
[6] Abelev B *et al.* (ALICE) 2012 *Phys. Rev. Lett.* **109** 072301 (*Preprint* 1202.1383)
[7] Greco V, Ko C M and Rapp R 2004 *Phys. Lett.* **B595** 202–208 (*Preprint* nucl-th/0312100)
[8] Ravagli L and Rapp R 2007 *Phys. Lett.* **B655** 126–131 (*Preprint* 0705.0021)
[9] Poskanzer A M and Voloshin S A 1998 *Phys. Rev.* **C58** 1671–1678 (*Preprint* nucl-ex/9805001)
[10] Yan L, Zhuang P and Xu N 2006 *Phys. Rev. Lett.* **97** 232301 (*Preprint* nucl-th/0608010)
[11] Liu Y, Xu N and Zhuang P 2010 *Nucl. Phys.* **A834** 317C–319C (*Preprint* 0910.0959)
[12] Guo X, Shi S, Xu N, Xu Z and Zhuang P 2015 *Phys. Lett.* **B751** 215–219 (*Preprint* 1502.04407)
[13] Adamczyk L *et al.* (STAR) 2013 *Phys. Rev. Lett.* **111** 052301 (*Preprint* 1212.3304)
[14] Masui H, Schmah A and Poskanzer A M 2016 *Nucl. Instrum. Meth.* **A833** 181–185 (*Preprint* 1212.3650)
[15] Gaiser J 1982 *Charmonium Spectroscopy From Radiative Decays of the J/ψ and ψ′* Ph.D. thesis SLAC
[16] Sprinthall R C 2011 *Basic statistical analysis* 9th ed (Pearson Education)

IOP Conf. Series: Journal of Physics: Conf. Series **832** (2017) 012022 doi:10.1088/1742-6596/832/1/012022

Heavy quark dynamics within a Boltzmann transport model: radiative vs collisional energy loss

Gabriele Coci[1,2], Francesco Scardina[2] and Vincenzo Greco[1,2]

[1] Department of Physics and Astronomy, University of Catania, Via S. Sofia 64, ZIP-95125 Catania (Italy)

[2] Laboratori Nazionali del Sud, INFN-LNS, Via S. Sofia 62, ZIP-95125 Catania (Italy)

E-mail: coci@lns.infn.it

Abstract. We discuss the propagation of heavy quarks (charm and bottom) through the QGP by means of a relativistic Boltzmann transport approach including both collisional and radiative energy loss mechanisms. In particular we investigate the impact of induced gluon radiation by dynamical QCD medium implementing in our transport model a formula for the emitted gluon spectrum calculated in a higher-twist scheme. We notice that in the region of high transverse momentum ($p_T > 10$ GeV) radiative processes play an essential role giving a dominant contribution to the generation of R_{AA} and v_2 at momentum values for which the energy loss by collisions is in the perturbative regime.

1. Introduction

Heavy quarks (HQs) represent an ideal probe for investiganting the properties of hot and dense QCD matter created at ultra-Relativistic Heavy Ion Collisions (uRHICs). They are created at early stages of collisions and due to their large masses they have a large thermalization time with respect to the bulk medium. Thus they conserve memory of propagation during the entire space-time evolution of the QGP. As they diffuse through the medium, HQs suffer collisions and radiative effects as a kind of gluon bremsstrahlung. Both processes are responsible for the substantial quenching of high-p_T HQs and also for the coupling of their dynamics with the bulk collective flow. The two key observables in HQ sector are the nuclear suppression factor $R_{AA}(p_T)$, i.e. the ratio between the spectra of heavy flavor hadrons and their decay products measured in nucleus-nucleus collisions with the same spectra in proton-proton ones, and the elliptic flow $v_2(p_T)$ which is a measure of the developed anisotropies in the angular distributions in momentum space. Both observables have been measured at RHIC and LHC facilities [1] [2] and the challenge of each theoretical framework is to give a consistent explanation of the surprising low R_{AA} and high v_2 observed in the HQ sector, hence solving the so called $R_{AA} - v_2$.

2. Our transport model for heavy quarks in QGP

We describe the HQ propagation inside the QGP by means of relativistic Boltzmann equation which is written as follows:

$$p^\mu \partial_\mu f_{HQ}(x,p) = C[f_{HQ}](x,p) \quad (1)$$

f_{HQ} is the HQ phase-space distribution function and $C[f_{HQ}]$ is the relativistic Boltzmann-like collision integral which contains all the interaction mechanisms between HQs and light

IOP Conf. Series: Journal of Physics: Conf. Series **832** (2017) 012022 doi:10.1088/1742-6596/832/1/012022

partons of the bulk, which for simplicity we assume to be formed by gluons. We consider only elastic two-body collisions whose total cross section is derived from the scattering matrices of $g + HQ \rightarrow g + HQ$ processes calculated at pQCD leading-order [3]. A realistic Debye screening mass $m_D^2 = 4\pi\alpha_s T^2$ for regulating the t-channel gluon propagator and the pQCD running of the coupling constant $\alpha_s(T)$ with bulk temperature T are also included. The integrated distribution function for the medium components contained in $C[f_{HQ}]$ is determined in analogous manner by solving the related Boltzmann-Vlasov equation [4]. The Boltzmann equation is solved numerically dividing the coordinate space in a 3D lattice and the test particle method is used to sample the distribution function in each cell. The collision integral is solved by means of the stochastic algorithm for evaluating the collision probability [5] [6].

The goal of this work is to include inside this framework a model which could account for the energy loss of HQs by inelastic processes. With this aim we implemented a mechanism for medium-induced gluon radiation calculated in multiple-scattering approximation and pQCD factorization scheme which was already included in other models where HQ dynamics was described by Langevin [7] or Linearized Boltzmann equation [8]. The differential spectrum of emitted gluons is derived from the higher-twist calculation for the energy loss of fast partons [9] [10] and heavy quarks [11] and it is given by the following expression

$$\frac{dN_g}{dx dk_\perp^2 dt} = \frac{2\alpha_s C_A P(x)}{\pi k_\perp^4}\hat{q}\left(\frac{k_\perp^2}{k_\perp^2 + x^2 M^2}\right)^4 \sin^2\left(\frac{t}{2\tau_f}\right) \tag{2}$$

where x and $k_\perp$ are respectively the fraction of HQ energy E carried away by the emitted gluon and the transverse gluon momentum with respect to the propagating direction.

$P(x)$ is the gluon splitting function derived from medium-modified DGLAP equations [12], while the $\hat{q}$ represents the HQ transport coefficient depending on E and T. In this initial work we simply relate the $\hat{q}$ with the momentum-space diffusion coefficient, i.e. $\hat{q} = 4\kappa$, which has a better significance if Fokker-Planck approach for HQ dynamics is applied. In further study a more precise treatment of $\hat{q}$ within our Boltzmann transport model will be considered referring also to the Jet Collaboration efforts done in extracting such jet quenching parameter [13]. Destructive interference known as LPM effect [14] which occurs if the time between two successive collisions is smaller than the formation time $\tau_f = 2Ex(1-x)/(k_\perp^2 + x^2 M^2)$ and suppression of radiation at small angle of emission ("dead-cone" effect [15]) due to non-negligible HQ mass M are both included as visible from Eq. (2).

The differential spectrum is numerically integrated in x and $k_\perp^2$ and the result multiplied by the timestep of evolution Δt in order to get the single-emission radiation probability $P_{rad}(t, T, E)$. A physical lower cut-off of the gluon energy equivalent to m_D is considered in order to avoid the divergence at $x \rightarrow 0$. In the same manner we did for the collision probability, we use $P_{rad}(t, T, E)$ to sample the number of test particles which radiate at each timestep and we distribute the four-momentum components of the emitted gluon according to the differential spectrum Eq. (2). By such a fully reconstruction of the final state we are able to relax the path-length approximation for HQ propagation and include its diffusion during the radiation mechanism, which we predict has a remarkable effect on v_2 formation (see Fig. 4) since it couples the HQ dynamics at high p_T more strongly to the collective flow of the bulk, keeping a similar suppression pattern.

3. Results at RHIC

We have carried out simulations at RHIC (Au+Au at $\sqrt{s} = 200$ AGeV, $b = 7.5$ fm) events. The initial conditions in the coordinate space are given by the standard geometrical Glauber model, while for initial distributions in momentum space we employ a Boltzmann-Jüttner function for bulk gluons and a power law function calculated within pQCD NLO [16] for charm quarks.

Our purpose in this study was to investigate the impact of various energy loss scenarios on R_{AA} and v_2 observables, therefore we have not included the hadronization mechanisms yet.

IOP Conf. Series: Journal of Physics: Conf. Series **832** (2017) 012022 doi:10.1088/1742-6596/832/1/012022

It is well know that pQCD alone is not able to reproduce the quenching in HQ sector. The problem is usually cleared by introducing a constant k factor to enhance cross section for HQ scatterings. The physical reason for introducing such factor is the idea to include non-perturbative effects inside the HQ dynamics which become dominant at low p_T [17]. However in the region of high p_T where radiative energy loss is much higher than collisional one, such non-perturbative effects should be negligible. Referring to [8] we model this effect by multiplying both collisional and radiative probabilities with a p_T-dependent k-enhancement which for simplicity is assumed to have a gaussian profile: $k(p_T) = \left[1 + (k_0 - 1)\exp\left(-\frac{p_T^2}{2\sigma_p^2}\right)\right]$.
We set the width parameter σ_p in order to reach the condition $k(p_T) \to 1$ at $p_T > 10$ GeV with a good approximation. The amplitude k_0 which corresponds to the maximum enhancement in the limit $p_T \to 0$ is fixed instead by fitting the R_{AA} with a non-physical scenario where only collisions occur. The R_{AA} suppression in all studied scenarios is shown in Fig. 1. In Fig. 2 the energy loss distributions of charm quarks as function of their initial p_T is plotted.

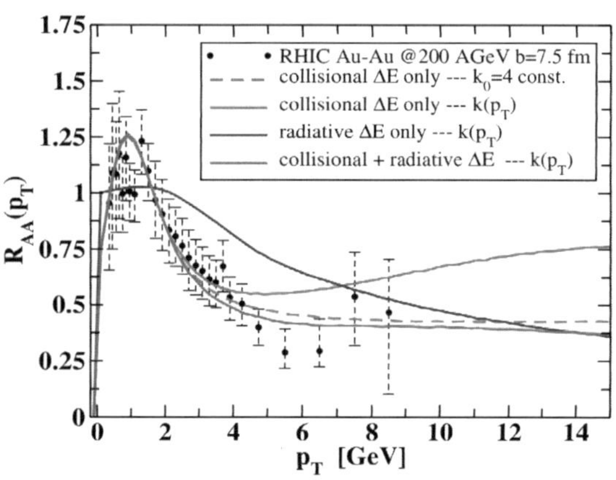

Figure 1. Measured R_{AA} at RHIC for single-non photonic electrons compared to the suppression of charm quarks. The red dashed line is the non-physical collisional scenario with constant k_0=4, while the solid lines are respectively collisional only (red), radiative only (blue) and collisional+radiative scenarios (green) with k_0=4 and σ_p=5 GeV.

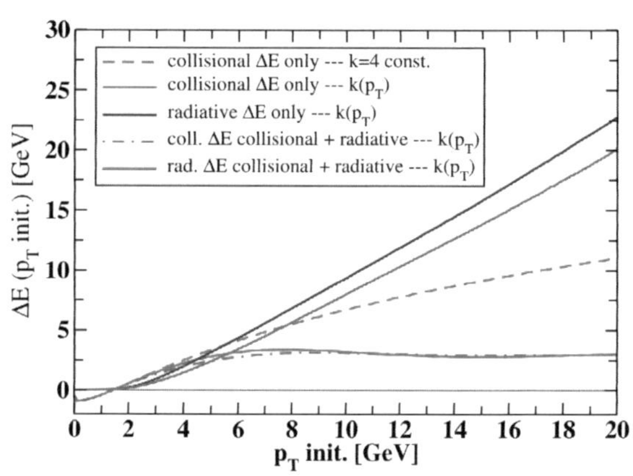

Figure 2. Collisional and radiative energy loss distributions for charm quarks as function of initial p_T. Colours and parameters are the identical to the R_{AA} plot in Fig. 1. The green dashed and green solid lines are respectively the collisional and radiative contributions in case both energy loss processes are considered.

Once the R_{AA} is fixed at least in the region of experimental points by tuning k_0 and σ_p, we can look at the impact on the elliptic flow as it is visible from Fig. 4. The production of v_2 at $p_T > 10$ GeV is mostly determined by the radiative mechanism. As said before, we notice that in our approach the formation of v_2 is naturally beyond the simple path-length approximation; in fact the HQs can both radiate or absorb energy and can deviate from their trajectory due to gluon emission. The discrepancy with experimental data at $p_T < 2 - 3$ GeV is known to be reduced by means of a T-dependent k-factor in the collisional energy loss [18].

4. Conclusions

In this work we studied the HQ propagation in QGP by means of Boltzmann equation and we implemented both a radiative and collisional energy loss mechanism. For the low p_T region we modelled non-perturbative effects using a $k(p_T)$ enhancement which disappears once the

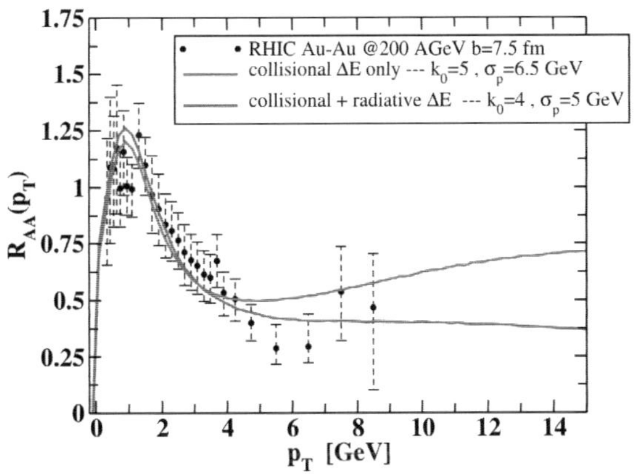

Figure 3. Suppression factor R_{AA} obtained at RHIC energies fitted by pure collisional quenching (red line) and collisional+radiative energy loss scenario (green line). Values of k_0 and σ_p parameters are reported in the legend.

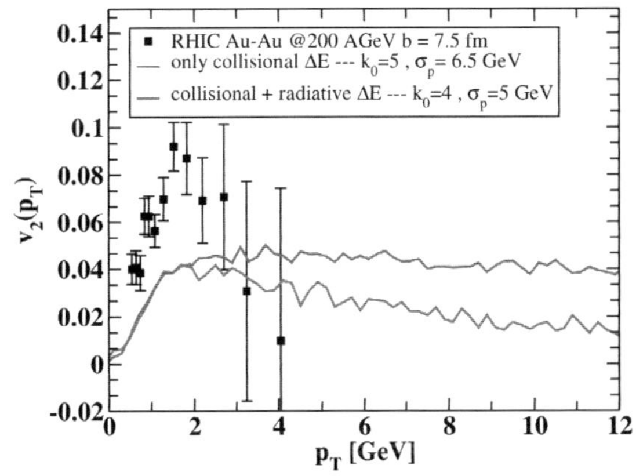

Figure 4. Comparison between elliptic flow v_2 measured at RHIC energies produced in a scenario where only collisions are considered (red line) against a more complete case where gluon radiation is also included (green line).

momentum scale is sufficiently higher to have a perturbative regime ($p_T \approx 10$ GeV). In future, we will concentrate on the possibility of implementing a multiple-emission algorithm in order to increase the role of coherent effects during radiative processes and also include the emitted gluons within the bulk dynamics and the subsequent rescattering. We will focus also on accounting non-perturbative effects with a different T-dependent behaviour of the interaction strenght in order to study the v_2 formation at low p_T. The goal is to include the HQ transport model in a more complete description of uRHICs where also initial-stage dynamics and hadronization mechanisms are accounted. The aim is not only to solve the $R_{AA} - v_2$ puzzle but also to investigate more differential observables, such as azimuthal correlations, which are going to be measured both at RHIC and LHC energies hopefully.

References

[1] PHENIX Collaboration, A. Adare *et al.*, *Phys. Rev.* C **84**, 044905 (2011), arXiv:1005.1627
[2] ALICE Collaboration, B. Abelev *et al.*, *J. High Energy Phys.* 09, 112 (2012).
[3] B. Combridge, *Nucl. Phys.* B vol. **151**, pp. 429-456 (1979).
[4] M. Ruggieri, F. Scardina, S. Plumari and V. Greco, *Phys. Rev.* C **89**, 054914 (2014).
[5] Z. Xu, C. Greiner, *Phys. Rev.* C **71**, 064901 (2005).
[6] F. Scardina, M. Colonna, S. Plumari and V. Greco, *Phys. Lett.* B **724**, pp. 296-300 (2013).
[7] S. Cao, G.-Y. Qin, S. Bass and B. Müller, *Nucl. Phys.* A **904-5**, pp. 653c-656c (2014).
[8] S. Cao, T. Luo, G.-Y. Qin and X.-N. Wang, *Phys. Rev.* C **94**, 014909 (2016).
[9] X.-N. Wang, X.-F. Guo, *Nucl. Phys.* A **696**, 788 (2001).
[10] A. Majumder, *Phys. Rev.* D **85**, 014023 (2012).
[11] B.-W. Zhang, E. Wang and X.-N. Wang, *Nucl. Phys.* A **757**, pp. 493-524 (2005).
[12] W.-T. Deng and X.-N. Wang, *Phys. Rev.* C **81** 024902 (2010).
[13] Jet Collaboration, K.M. Burke *et al.*, *Phys. Rev.* C 90, 014909 (2014).
[14] Landau and Pomeranchuk, *Dokl. Akad. Nauk Ser. Fiz* **92** 535 (1953); Migdal, *Phys. Rev.* **103**, 1811 (1956).
[15] Y.L. Dokshitzer and D.E. Kharzeev, *Phys. Lett.* B **519** 199 (2001).
[16] M. Cacciari, P. Nason and R. Vogt, *Phys. Rev. Lett.* **95**, 122001 (2005).
[17] H. van Hees, M. Mannarelli, V. Greco and R. Rapp, *Phys. Rev. Lett.* 100, 192301 (2008).
[18] S.K. Das, F. Scardina, S. Plumari and V. Greco, *Phys. Lett.* B **747**, 260-264 (2015).

IOP Conf. Series: Journal of Physics: Conf. Series **832** (2017) 012023 doi:10.1088/1742-6596/832/1/012023

Open-charm measurements in pp and Pb–Pb collisions at central rapidity with ALICE

Julien Hamon, for the ALICE collaboration
Université de Strasbourg, CNRS-IN2P3, IPHC UMR 7178, F-67000 Strasbourg, France
E-mail: julien.hamon@iphc.cnrs.fr

Abstract. An overview of recent D^+, D^0, D^{*+} and D_s^+ measurements performed by ALICE at central rapidity in proton-proton collisions at $\sqrt{s} = 7$ and 8 TeV, as well as lead-lead collisions at $\sqrt{s_{NN}} = 2.76$ TeV, is presented. An emphasis is put on the discussion of theoretical predictions with respect to the LHC Run I data.

1. Introduction

Heavy quarks (charm and beauty) are unique probes to investigate the properties of the hot and dense matter created in ultra-relativistic heavy-ion collisions, known as the Quark-Gluon Plasma. Due to their large masses, heavy quarks are mainly produced at the early stage of the collision in hard parton scatterings, before the plasma formation, and thus participate in all the subsequent stages of the collision. In their propagation through the medium, they interact with other partons and lose energy via elastic (collisional) and inelastic (gluon radiation) processes, commonly treated in the perturbative approach of Quantum Chromo-Dynamics (QCD). These processes are expected to depend on the parton in-medium path length, colour-charge and mass [1, 2]. This results in a characteristic ordering of the energy loss of partons: $\Delta E(\mathrm{g}) > \Delta E(\mathrm{u,d,s}) > \Delta E(\mathrm{c}) > \Delta E(\mathrm{b})$.

Experimentally, heavy-flavour hadrons are the front door to access medium properties probed by heavy quarks. During the LHC Run I, ALICE [3] has tackled the measurements of charm and beauty hadrons in several rapidity ranges. At central rapidity ($|y_{\mathrm{lab}}| < 0.5$), open-charm mesons have been studied in various colliding systems and energies. The tracking and particle identification abilities of the ALICE central barrel detectors have allowed for measuring D^+, D^0, D^{*+} and D_s^+ over a wide transverse momentum (p_T) range and for different collision centralities.

2. pp collisions at $\sqrt{s} = 7$ and 8 TeV

Proton-proton (pp) collisions provide a suitable environment for testing several predictions of QCD. The large mass of charm quark allows for perturbative calculations of the differential charm production down to low p_T, as well as its total production cross section. At LHC energies, the study of small systems also permits to examine the gluon and heavy-quark distributions inside the nucleon. Besides, pp collisions represent a crucial baseline for understanding the phenomena observed in heavy-ion collisions, by acting as a reference where final state effects are negligible.

In this context, the p_T-differential production cross section of prompt D^0 has been recently measured down to $p_T = 0$ GeV/c in pp collisions at $\sqrt{s} = 7$ TeV by ALICE [4], with the help of an analysis method which does not exploit the D-meson decay (displaced) vertex. Several pQCD models based on a factorisation theorem (FONLL [5], GM-VFNS [6], k_{T}-factorisation [7]) are in agreement with this measurement within the current theoretical precisions. A similar agreement is achieved for all other D-meson species. Furthermore, the total charm cross section at $\sqrt{s} = 7$ TeV has been updated and is well reproduced by pQCD calculations at next-to-leading order [8], within quite large theoretical uncertainties though, as presented in figure 1 (left).

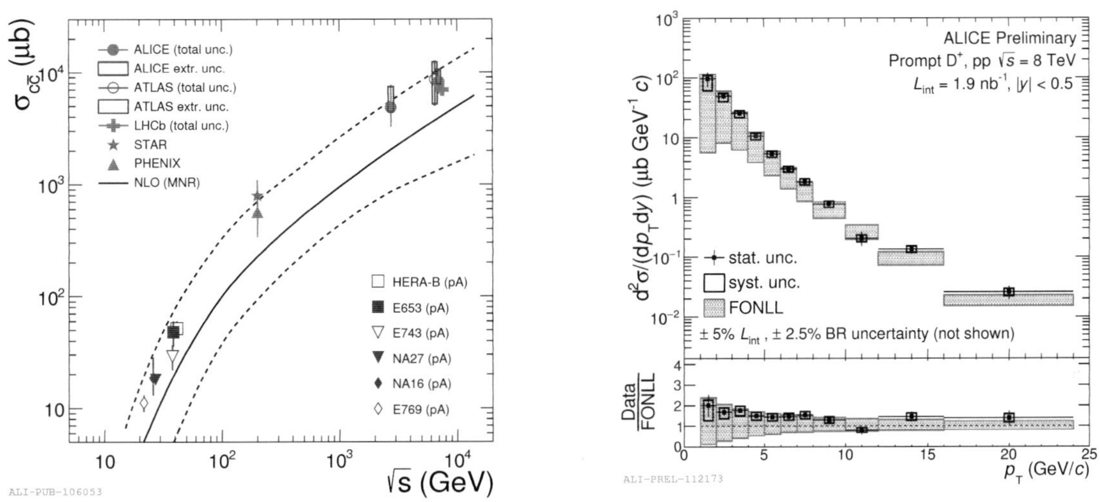

Figure 1. Left: Total charm production cross section in nucleon-nucleon collisions as a function of $\sqrt{s}$ [4]. Right: Preliminary p_{T}-differential production cross section of prompt D^{+} in pp collisions at $\sqrt{s} = 8$ TeV, measured by ALICE.

Currently, the production cross sections of prompt D^{+} and D^{*+} as a function of p_{T} are being studied by ALICE in pp collisions at $\sqrt{s} = 8$ TeV. The D^{+} preliminary results are shown in figure 1 (right). As already noticed at lower energies, the FONLL pQCD framework describes the data reasonably well within uncertainties, although the data still sit at the upper edge of the calculations. As expected, the production cross section obtained at 8 TeV is slightly higher than the one at 7 TeV.

3. Pb–Pb collisions at $\sqrt{s_{\mathrm{NN}}} = 2.76$ TeV

The parton energy loss, expected for heavy quarks in nucleus-nucleus collisions, leads to a reduction of the average parton momentum, which can be experimentally studied – to some extent – through the measurement of the nuclear modification factor of open-charm hadrons:

$$R_{\mathrm{AA}}(p_{\mathrm{T}}) = \frac{1}{\langle T_{\mathrm{AA}} \rangle} \cdot \frac{\mathrm{d}N_{\mathrm{AA}}/\mathrm{d}p_{\mathrm{T}}}{\mathrm{d}\sigma_{\mathrm{pp}}/\mathrm{d}p_{\mathrm{T}}} \quad (1)$$

where $\langle T_{\mathrm{AA}} \rangle$ is the average nuclear overlap function, proportional to the average number of nucleon-nucleon binary collisions, $\mathrm{d}N_{\mathrm{AA}}/\mathrm{d}p_{\mathrm{T}}$ and $\mathrm{d}\sigma_{\mathrm{pp}}/\mathrm{d}p_{\mathrm{T}}$ the transverse momentum differential yield and cross section in nucleus-nucleus and proton-proton collisions, respectively.

ALICE has extensively studied the nuclear modification factor of D mesons in lead-lead (Pb–Pb) collisions at $\sqrt{s_{\mathrm{NN}}} = 2.76$ TeV [9–11]. The observed stronger high-p_{T} suppression going from semi-central collisions ($R_{\mathrm{AA}}(p_{\mathrm{T}} \sim 6\ \mathrm{GeV}/c) \geq 0.4$) to central ones ($R_{\mathrm{AA}}(p_{\mathrm{T}} \sim 6\ \mathrm{GeV}/c) \leq 0.2$) is interpreted as an increase of the medium density, size and lifetime leading to larger energy loss of charm quarks. Additionally, the nuclear modification factor of D mesons has been measured as a function of $\langle N_{\rm part}\rangle$, the average number of nucleons participating in the Pb–Pb collisions [12]. The results are shown in figure 2 (left), where they are compared to those obtained by ALICE for charged pions [13] and by CMS for J/Ψ coming from beauty decays [14], in a similar kinematic range (high $p_{\rm T}$ and central rapidity). As can be seen, $R_{\rm AA}({\rm J}/\Psi \leftarrow {\rm B})$ is found to be significantly larger than $R_{\rm AA}({\rm D}^+, {\rm D}^0, {\rm D}^{*+})$ in central collisions, which is consistent with the parton energy loss hierarchy. Such an observation is in qualitative agreement with several theoretical predictions involving a mass dependence of parton energy loss processes (Djordjevic *et al.* [15], MC@sHQ+EPOS2 [16] and TAMU elastic [17]). On the other hand, the data show a quite similar $R_{\rm AA}$ for high-$p_{\rm T}$ pions and D mesons. This might be explained as due to the interplay between the colour-charge dependence of parton energy loss, and the different $p_{\rm T}$ and fragmentation distributions for gluons and charm quarks. At low $p_{\rm T}$, any direct interpretation is complicated since, in contrast to heavy-flavour hadrons, light-flavour hadron yields do not scale with the number of nucleon-nucleon binary collisions but rather with the number of participant nucleons. Moreover, radial flow and cold nuclear matter effects affect differently pions and D mesons.

The first measurement of the ${\rm D_s^+}$ nuclear modification factor has been performed by ALICE in Pb–Pb collisions at $\sqrt{s_{\rm NN}} = 2.76$ TeV [11] and is presented in figure 2 (right). Although $R_{\rm AA}({\rm D_s^+})$ is compatible with the average $R_{\rm AA}$ of non-strange D mesons within uncertainties, the current precision does not allow for drawing a firm conclusion on a possible signature of recombination mechanism working for charm quarks. An increase of the relative fraction of strange versus non-strange D mesons is indeed expected by the TAMU model, at low and intermediate $p_{\rm T}$ [18].

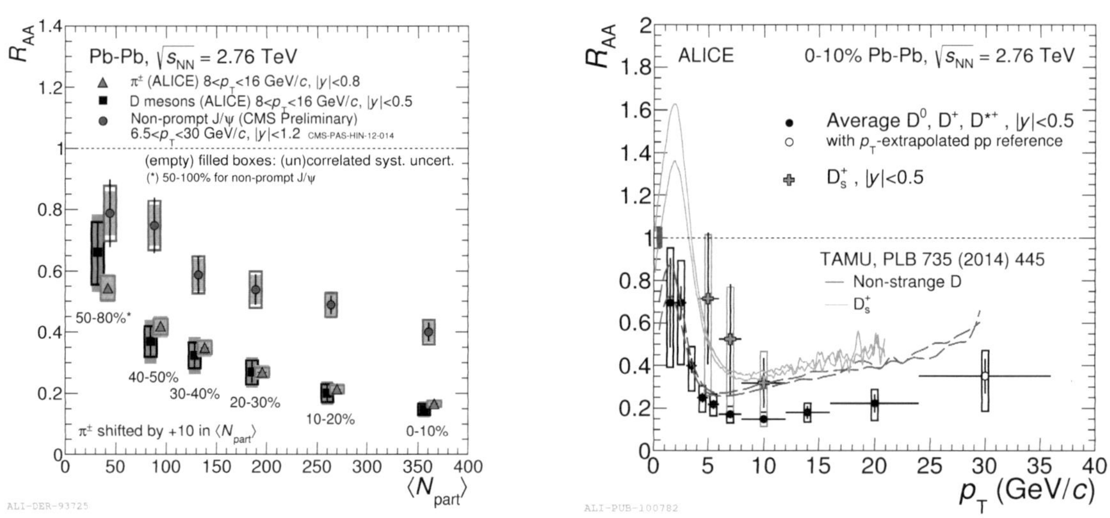

Figure 2. Nuclear modification factor of non-strange D mesons as a function of $\langle N_{\rm part}\rangle$ [12], compared to that of charged pions [9] and non-prompt J/Ψ [14] (left); and as function of $p_{\rm T}$ compared to that of ${\rm D_s^+}$ for central Pb–Pb collisions at $\sqrt{s_{\rm NN}} = 2.76$ TeV [10, 11] (right).

The diversity of open-charm measurements carried out by ALICE offers a valuable basis to constrain models. The results encompass both nuclear modification factor and elliptic flow [19] of different D-meson species as a function of $p_{\rm T}$ and the collision centrality. An attempt to describe simultaneously all measured observables, for charged pions and D mesons, within various theoretical models has been made in a recent ALICE paper [10]. While all models rely on a colour-charge and mass dependence of energy loss, the interaction mechanisms are not yet clearly accessed. As a general rule, gluon radiation seems to be needed to describe high-$p_{\rm T}$ $R_{\rm AA}$(D), whereas collisional interactions appear to be important for the understanding

IOP Conf. Series: Journal of Physics: Conf. Series **832** (2017) 012023 doi:10.1088/1742-6596/832/1/012023

of low-p_{T} R_{AA} and v_2 measurements. Although the role of the recombination – which should help describing low-p_{T} v_2 – is open to debate, it is generally accepted that an expanding medium gives rise to the observed non-zero v_2. As shown in [10], models fairly describing heavy-flavour hadron data do not necessarily manage to simultaneously reproduce pion data. In particular, only few models are able to describe both hadron species but only at high p_{T}: Djordjevic *et al.* [15], CUJET3.0 [20] and Vitev *et al.* [21]. It is worth noting that the latest includes in-medium meson formation and dissociation as an energy loss mechanism, on top of collisional interactions, instead of the commonly used radiative process.

4. Summary

Charm quark is an efficient tool to probe QCD in its perturbative regime and to give insights into the various stages of a heavy-ion collision, from its early stage to the interaction processes occurring in the hot and dense medium until hadronization. By exploiting LHC Run I, ALICE started providing constraints on theoretical models aiming at the description of charm physics. Energy loss processes can be studied with the diversity of measured observables. However, a simultaneous description of all open-charm data stays challenging. More sophisticated models together with more precise measurements, coming from the ongoing LHC Run II data taking, will be further needed to draw a clearer picture of the interaction mechanisms taking place throughout heavy-ion collisions.

References

[1] Baier R, Dokshitzer Y L, Mueller A H, Peigne S and Schiff D 1997 *Nucl. Phys.* D **483** 291
[2] Dokshitzer Y L and Kharzeev D E 2001 *Phys. Lett.* B **519** 199–206
[3] ALICE Collaboration 2008 *JINST* **3** S08002
[4] ALICE Collaboration 2016 (*Preprint* nucl-ex/1605.07569)
[5] Cacciari M, Frixione S, Houdeau N, L M M, Nason P and Ridolfi G 2012 *JHEP* **10** 137
[6] Kniehl B A, Kramer G, Schienbein I and Spiesberger H 2012 *Eur. Phys. J.* C **72** 2082
[7] Maciula R and Szczurek A 2013 *Phys. Rev.* D **87** 094022
[8] Mangano M L, Nason P and Ridolfi G 1992 *Nuclear Physics* B **373** 295 – 345
[9] ALICE Collaboration 2012 *JHEP* **09** 112
[10] ALICE Collaboration 2016 *JHEP* **03** 081
[11] ALICE Collaboration 2016 *JHEP* **03** 082
[12] ALICE Collaboration 2015 *JHEP* **11** 205
[13] ALICE Collaboration 2014 *Phys. Lett.* B **736** 196–207
[14] CMS Collaboration 2012 CMS–PAS–HIN–12–014
[15] Djordjevic M, Djordjevic M and Blagojevic B 2014 *Phys. Lett.* B **737** 298–302
[16] Nahrgang M, Aichelin J, Gossiaux P B and Werner K 2014 *Phys. Rev.* C **89** 014905
[17] He M, Fries R J and Rapp R 2013 *Nucl. Phys.* A **910-911** 409–412
[18] He M, Fries R J and Rapp R 2014 *Phys. Lett.* B **735**
[19] ALICE Collaboration 2014 *Phys. Rev.* C **90** 034904
[20] Xu J, Liao J and Gyulassy M 2015 *Chin. Phys. Lett.* **32** 092501
[21] Sharma R, Vitev I and Zhang B 2009 *Phys. Rev.* C **80** 054902

IOP Conf. Series: Journal of Physics: Conf. Series **832** (2017) 012024 doi:10.1088/1742-6596/832/1/012024

Separation of the Charm and Beauty Production in p–Pb and Pb–Pb Collisions with ALICE

Martin Völkl for the ALICE Collaboration

Physikalisches Institut der Universität Heidelberg, Heidelberg, Germany

E-mail: voelkl@physi.uni-heidelberg.de

Abstract. Measurements of heavy (charm and beauty) quarks provide useful insights into the properties of the quark–gluon plasma. The good particle identification capabilities of ALICE make a measurement based on the electrons from semi-leptonic decays of heavy-flavour hadrons feasible. This approach makes use of the large branching ratios ($\approx 10 - 20\%$) of heavy–flavour hadrons into electrons. Separation of the contribution from beauty-hadron decay electrons was achieved by utilizing the large decay length of the associated hadrons, making the investigation of beauty quarks in the medium possible. By comparing measurements in p–Pb and Pb–Pb collisions, it is possible to disentangle effects of cold nuclear matter from those of the hot and dense medium. The results show a strong change in the transverse momentum distribution of electrons from beauty-hadron decays in central Pb–Pb collisions with respect to the proton–proton case. No significant change from proton–proton collisions can be seen in the p–Pb case, suggesting that the modification is a final state effect.

1. Introduction

The interaction of heavy quarks with the medium produced in an ultrarelativistic heavy-ion collision can yield insight into the properties of the quark–gluon plasma. The use of heavy quarks as probes has several advantages [1], among them the defined moment of creation in the initial hard scatterings of the collision and the clear association of the heavy quark in the medium and the hadron it forms. In the interaction with the medium, heavy quarks with large momenta may loose energy to the medium, while those with low momenta can participate in the collective motion of the medium. Measurements of these processes give an insight into the properties of the medium. To assess the strength of the interaction, it is useful to compare the transverse momentum distribution of the resulting hadrons in central Pb–Pb collisions with that in pp collisions. In addition, it is also useful to compare to a measurement in p–Pb collisions to gain an insight into initial state effects due to the use of nuclei as projectiles, such as the modification of the parton distribution function. A useful property of heavy-flavour hadrons is the typically large branching ratio of decays with an electron in the final state. As a result most electrons at mid-rapidity and with a momentum of a few GeV/c come from heavy-flavour sources. With the excellent particle identification capabilities of the ALICE detector, an approach based on the measurement of electrons becomes feasible. The comparatively large decay length of beauty hadrons ($c\tau \approx 500\,\mu$m) can be utilized to separate the contribution from beauty from that of charm and other background sources. The measurements can broadly be split into two main steps: Identifying a clean sample of electrons from all measured particles and secondly identifying the contribution from beauty-hadron decays.

2. Electron Identification

The Pb–Pb data set discussed here was recorded by the ALICE detector setup in 2010 [2, 3], while the p–Pb data set was recorded in 2013. Electrons were identified using the signal of the Time Projection Chamber (TPC) and Time-of-Flight detector (TOF). The time of flight was required to be compatible with the expectation for an electron within 3σ. The TPC measures the energy deposit of particles in the gas ($\mathrm{d}E/\mathrm{d}x$). The signal of the main hadronic background – charged pions – lies below the electron signal, suggesting an asymmetric selection range which was set to be $-0.5 < n_\sigma^{\mathrm{TPC}} < 3$, where n_σ^{TPC} is the deviation of the signal to an electron hypothesis in units of the detector resolution.

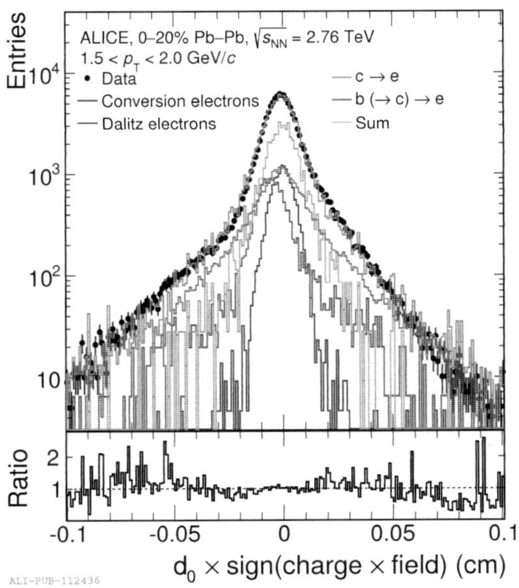

Figure 1. Impact parameter distributions in Pb–Pb collisions [4].

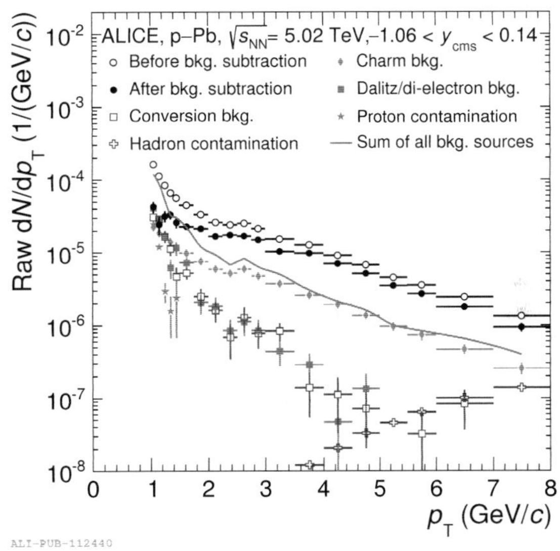

Figure 2. Background contributions in the p–Pb electron sample with applied impact parameter requirement [4].

3. Signal Extraction

The signal electrons can come from the decay of the beauty hadrons directly or from the decay of an intermediate charm state. A differentiating property of the beauty–hadron decays is the fairly large decay length of the beauty hadrons. From the single electron tracks it is not possible to deduce the decay length directly. An indirect way is the use of the track impact parameter. The impact parameter is obtained by tracing the measured track back towards the reconstructed interaction vertex. If the electron is produced at a secondary vertex, then this will typically not lead to the primary vertex. The impact parameter is the minimum distance to the primary vertex in the transverse plane. In the definition used for the analysis, it has a positive or negative sign depending on whether the interaction vertex is inside or outside of the circle the track describes in the transverse plane. Electrons from beauty-hadron decays will typically have a larger (absolute) impact parameter value than electrons from the background sources.

To discuss the different electron sources, it is useful to group them by the shape of their impact parameter distribution. The sources can be separated into the following four categories, which are also shown in Fig. 1:

Electrons from beauty-hadron decays. Due to the large decay length ($c\tau \approx 500\ \mu$m) of beauty hadrons this distribution is the widest among the different electron sources.

IOP Conf. Series: Journal of Physics: Conf. Series **832** (2017) 012024 doi:10.1088/1742-6596/832/1/012024

Electrons from charm-hadron decays. The smaller decay length of the charm hadrons ($c\tau \approx 100 - 300$ μm) leads to a narrower distribution as compared to electrons from beauty-hadron decays.

Electrons from photon conversions in the detector material. Due to the small opening angle of photon conversions, a non-zero impact parameter appears only due to the influence of the magnetic field. With the definition given above, this leads to an asymmetric impact parameter distribution with a maximum at negative values.

All remaining electron sources. This contribution is dominated by electrons from the decay of light hadrons with a small contribution from the decay of particles with strangeness. Many of these electrons come from Dalitz decays of π^0 and they will thus be referred to as Dalitz electrons in the following.

The approach of the analysis for p–Pb collisions is to require a minimum (absolute) impact parameter for the measured electrons. This strongly reduces the background. The remaining background can then be subtracted. After a correction for the associated efficiency, this leads to an estimate for the number of electrons from beauty-hadron decays. The estimate for the background contributions comes from the information of other measurements by ALICE, which are corrected with the corresponding efficiency. The resulting contributions are shown in Fig. 2.

For the Pb–Pb analysis, the information about the different contributions is obtained by fitting the full impact parameter distribution as shown in Fig. 1. For this purpose, estimates for the distributions are added with a free amplitude parameter to reproduce the full distribution. The information about these distributions comes from Monte Carlo simulations of the event properties and the detector setup. These have finite statistics, which has to be taken into account in the fitting procedure [5].

In both cases, the accurate reproduction of the impact parameter distributions by the simulations has to be carefully assessed and corrected where necessary [4].

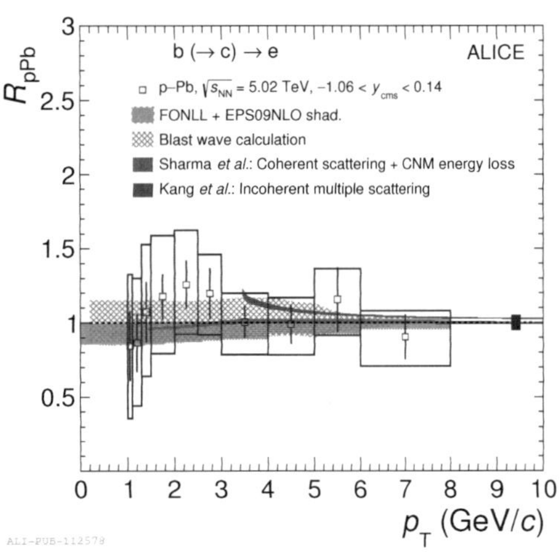

Figure 3. Measured R_{pPb} compared with models [4].

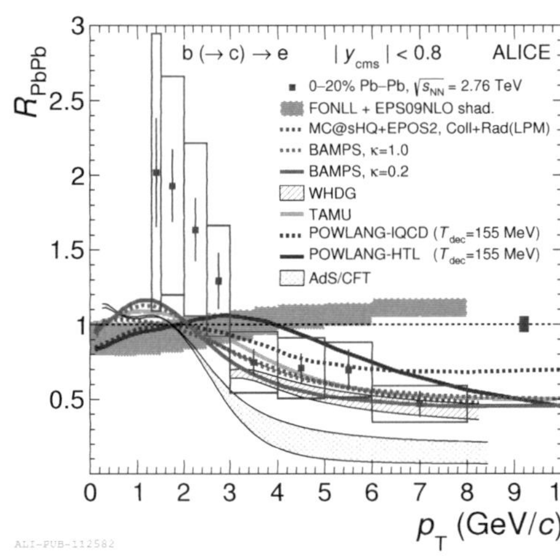

Figure 4. Measured R_{PbPb} in comparison with models [4].

4. Results

The comparison of the results was done using the nuclear modification factor, which compares the p_{T}-distribution of the electrons per nuclear collision to the p_{T}–distribution of electrons in

IOP Conf. Series: Journal of Physics: Conf. Series **832** (2017) 012024 doi:10.1088/1742-6596/832/1/012024

an equivalent number of proton–proton collisions [6]:

$$R_{\mathrm{AA}}(p_{\mathrm{T}}) = \frac{1}{\langle N_{\mathrm{coll}}\rangle}\frac{\mathrm{d}N_{\mathrm{AA}}/\mathrm{d}p_{\mathrm{T}}}{\mathrm{d}N_{\mathrm{pp}}/\mathrm{d}p_{\mathrm{T}}} = \frac{1}{\langle T_{\mathrm{AA}}\rangle}\frac{\mathrm{d}N_{\mathrm{AA}}/\mathrm{d}p_{\mathrm{T}}}{\mathrm{d}\sigma_{\mathrm{pp}}/\mathrm{d}p_{\mathrm{T}}},$$

where N_{coll} is the expected number of nucleon-nucleon collisions and T_{AA} is the nuclear overlap function. The resulting ratio should be unity if a heavy-ion collision is equivalent to a superposition of independent nucleon-nucleon collisions.

In p–Pb collisions (Fig. 3) the nuclear modification factor is consistent with unity and also in agreement with the expectations from theoretical models. This means, that no large deviation from unity due to such effects is expected in the R_{PbPb} .

The R_{PbPb} shown in Fig. 4 is below unity above $p_{\mathrm{T}} = 3\,\mathrm{GeV/c}$ and rises towards lower momenta. Given that cold nuclear matter effects are expected to be small, this suggests that fast quarks in the medium are slowed down by the interaction with the medium. Most of the models based on in-medium energy loss show a similar trend and also reasonable agreement with each other [4]. The AdS/CFT-inspired model [7] underestimates the nuclear modification factor, while the POWLANG model [8] based on hard thermal loops gives a higher estimate. In conclusion, modification of the p_{T}-distribution of electrons from heavy-flavour hadron decays at mid-rapidity is caused by the interaction with the hot and dense medium. The magnitude of the effects can be reproduced by several model calculations.

References

[1] Prino F and Rapp R 2016 (*Preprint* arXiv: 1603.00529)
[2] Aamodt K *et al.* (ALICE Collaboration) 2008 *JINST* **3** S08002
[3] Abelev B B *et al.* (ALICE Collaboration) 2014 *Int. J. Mod. Phys.* **A29** 1430044 (*Preprint* arXiv: 1402.4476)
[4] Adam J *et al.* (ALICE Collaboration) (*Preprint* arXiv: 1609.03898)
[5] Barlow R and Beeston C 1993 *Comp. Phys. Comm.* **77** 219
[6] Abelev B *et al.* (ALICE Collaboration) 2013 *Phys. Lett.* **B721** 13–23 (*Preprint* arxiv: 1208.1902)
[7] Horowitz W A 2015 *Phys. Rev.* **D91** 085019 (*Preprint* 1501.04693)
[8] Alberico W M, Beraudo A, De Pace A, Molinari A, Monteno M, Nardi M and Prino F 2011 *Eur. Phys. J.* **C71** 1666 (*Preprint* 1101.6008)

IOP Conf. Series: Journal of Physics: Conf. Series **832** (2017) 012025 doi:10.1088/1742-6596/832/1/012025

Latest results on Υ production in heavy-ion collisions from the STAR experiment

R Vértesi (for the STAR Collaboration)

Nuclear Physics Institute of the Academy of Sciences of the Czech Republic, Řež 130, 25068 Husinec-Řež, Czech Republic

Wigner Research Centre for Physics of the Hungarian Academy of Sciences RMI, H-1525 Budapest 114, P.O.Box 49, Hungary

E-mail: vertesi.robert@wigner.mta.hu

Abstract. We report on the latest measurements of the production of Υ mesons in heavy-ion collisions from the STAR experiment at RHIC. New measurements of the nuclear modification factors of the Υ(1S+2S+3S) and Υ(1S) states in U+U collisions at $\sqrt{s_{\rm NN}} = 193$ GeV are presented as a function of the number of participants ($N_{\rm part}$) in the collisions. In addition, the suppression of Υ(1S) and Υ(2S+2S) is presented versus the quark-antiquark binding energy. Preliminary results on Υ suppression in Au+Au collisions at $\sqrt{s_{\rm NN}} = 200$ GeV, reconstructed via the dimuon channel, are also reported.

1. Introduction

There is extensive experimental evidence, eg. in Refs. [1, 2], that a hot and strongly interacting medium is formed in high-energy heavy-ion collisions. It is expected that color screening of the quark-antiquark binding potential within such a medium would render quarkonium formation less likely [3], an effect colloquially referred to as "melting" of the bound state. Since quarkonium states with stronger binding energies would melt at higher temperatures than the more weakly bound states, measuring the yields of different quarkonium states may reveal a sequential suppression pattern, thus serving as a thermometer for the medium [4]. Although the suppression of charmonia was anticipated as a key signature of the quark-gluon plasma formation [3], later studies revealed that several concurrent effects, such as recombination of uncorrelated $c\bar{c}$ pairs in the medium, cold nuclear matter (CNM) effects and feed-down from higher-mass states, complicate the interpretation of the results [5, 6, 7]. Bottomonia, on the other hand, are less affected by recombination at top RHIC energies, and CNM effects are also expected to be moderate at mid-rapidity. Thus the measurements of the $\Upsilon(n{\rm S})$ states provide a cleaner probe of the strongly interacting medium.

Recent results from STAR [8] show that in central Au+Au collisions at $\sqrt{s_{\rm NN}} = 200$ GeV, the Υ(1S+2S+3S) production is suppressed to an extent that cannot be explained by CNM effects only. In the same study the yields of the excited states are consistent with a complete suppression. Since the energy density in central U+U collisions is estimated to be about 20% higher on the average than that in central Au+Au collisions at $\sqrt{s_{\rm NN}} = 200$ GeV [9], studies of quarkonium production in U+U collisions can provide further tests of the sequential suppression hypothesis. Recent installation of the Muon Telescope Detector (MTD) [10], on the other hand, opened the door for precision measurements of quarkonia at STAR via the dimuon decay channel.

IOP Conf. Series: Journal of Physics: Conf. Series **832** (2017) 012025 doi:10.1088/1742-6596/832/1/012025

2. First measurement of Υ production in U+U collisions

The STAR detector complex [11] provides clean identification of electrons at mid-rapidity ($|\eta| < 1$) in the full azimuth angle. STAR collected data in U+U collisions at $\sqrt{s_{\rm NN}} = 193$ GeV triggered by high-energy electrons (or positrons) with an integrated luminosity of 263.4 μb^{-1}. The analysis of this data, briefly recapitulated here, is described in details in Ref. [12]. The tracks of the electron candidates are reconstructed in the Time Projection Chamber (TPC), and then matched with energy deposits in the Barrel Electromagnetic Calorimeter (BEMC). Electron identification is carried out in the TPC based on fractional energy loss, and in the BEMC based on the shape of the electromagnetic shower. Electron candidates have to have their energy-to-momentum ratio close to unity. Identified electron and positron candidates are then paired in order to reconstruct the invariant mass of the Υ candidates. The combinatorial background is then reconstructed using like-sign electron (positron) pairs. Since Drell-Yan processes and open $b\bar{b}$ production yield correlated backround in the signal region, combinatorial background cannot be simply subtracted to get the signal. Instead, a simultaneous fit is applied on the unlike-sign and like-sign invariant mass data. The former is constructed as a sum of the three Υ (nS) peaks, the correlated and the combinatorial background. A priori knowledge on the shapes of the signal and background shapes were obtained from simulations [8, 12] in order to reduce the number of free parameters in the fit. Figure 1 shows the invariant mass distributions together with the fitted functions on the combinatorial and correlated backgrounds as well as the signal peaks.

We use the nuclear modification factor to quantify suppression, defined as $R_{\rm AA} = \frac{\sigma_{pp}^{inel}}{\sigma_{AA}^{inel}} \frac{1}{\langle N_{\rm coll}\rangle} \frac{d\sigma_{\Upsilon}^{AA}/dy}{d\sigma_{\Upsilon}^{pp}/dy}$, where $\sigma_{AA(pp)}^{inel}$ is the total inelastic cross-section of the U+U (p+p) collisions, and $d\sigma_{\Upsilon}^{AA(pp)}/dy$ denotes the Υ production cross-section in U+U (p+p) collisions.

Figure 2 shows the nuclear modification factor of Υ(1S+2S+3S) and Υ(1S), respectively, for U+U collisions at $\sqrt{s_{\rm NN}} = 193$ GeV and Au+Au collisions at $\sqrt{s_{\rm NN}} = 200$ GeV. The new U+U data consolidate observations made previously in Au+Au data: considering 0-10% centrality Au+Au and U+U points together, both the Υ(1S+2S+3S) and Υ(1S) points are significantly suppressed, similar in extent to data from the LHC [14] but this suppression is not complete [12]. Theory comparisons [15, 16, 17] show that data favor scenarios where the $q\bar{q}$ pairs are strongly bound, over the weakly bound ones. Figure 3 shows the nuclear modification factor of Υ states measured in U+U collisions at $\sqrt{s_{\rm NN}} = 193$ GeV and Au+Au collisions at $\sqrt{s_{\rm NN}} = 200$ GeV, compared to high-$p_{\rm T}$ J/ψ in Au+Au collisions at $\sqrt{s_{\rm NN}} = 200$ GeV [18]. The emerging picture supports sequential melting of the quarkonium states with different binding energies. While the U+U data show a non-significant presence of the excited states, it is still consistent with the upper limit established previously in the Au+Au measurement.

3. Production of Υ in Au+Au collisions with the MTD

Measurements in the dielectron channel suffer from the electron bremsstrahlung tail that causes the higher mass states to give contribution to the peak region of lower mass states. The advantage of $\Upsilon \rightarrow \mu^+\mu^-$ measurements is that muons are less affected by bremsstrahlung, thus it is less of a challenge to separate the different Υ states with comparable statistics. The Muon Telescope Detector (MTD), completed in 2014, is designed for precision measurements of quarkonia production via the dimuon channel. It is installed outside the solenoidal magnet of STAR, covering 45% of the azimuth angle within the pseudorapidity range $|\eta| < 0.5$. Its Multi-gap Resistive Plate Chamber (MRPC) technology provides means to trigger on muons and identify them. Figure 4 shows the invariant mass spectrum of Υ candidates reconstructed via the dimuon channel. Here, as in the dielectron channel, the raw yields of the Υ states are obtained from a simultaneous fit on the like-sign and unlike-sign muon pairs. The individual Υ(nS) states are modelled with Gaussian distributions, each centered at the PDG value of the given state. Figure 5 shows the excited-to-ground state ratio Υ(2S+3S)/Υ(1S) via the dimuon

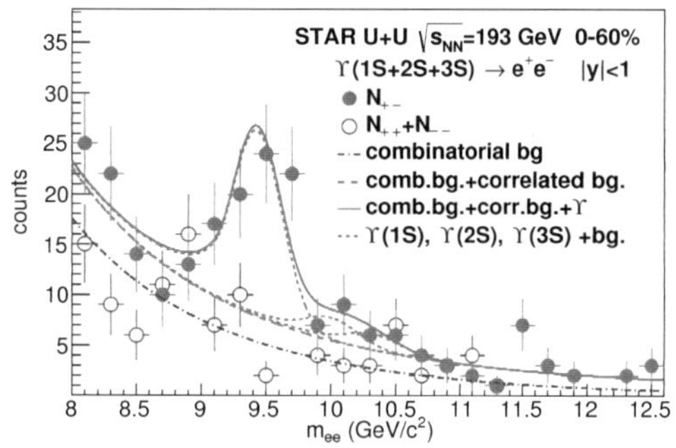

Figure 1. Invariant mass distribution of Υ candidates and the combinatorial background in the 0-60% centrality class of U+U collisions at $\sqrt{s_{\mathrm{NN}}} = 193$ GeV [12].

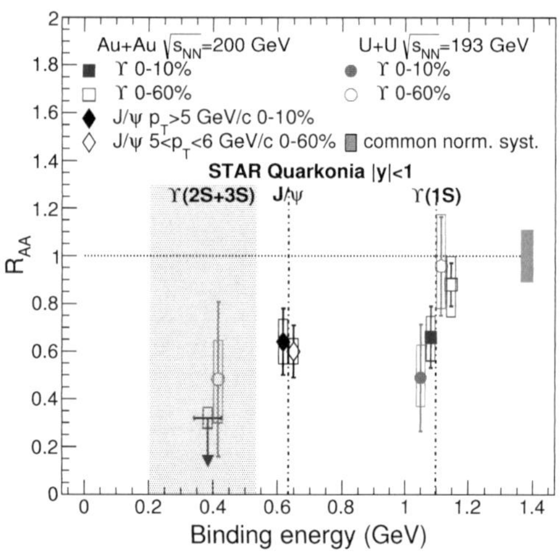

Figure 3. R_{AA} versus binding energy for Υ (1S) and Υ(2S+3S) in U+U collisions at $\sqrt{s_{\mathrm{NN}}} = 193$ GeV [12], compared to the Υ and high-p_{T} J/ψ results in Au+Au collisions at $\sqrt{s_{\mathrm{NN}}} = 200$ GeV [8, 18]. The 95% upper confidence bound is shown for Υ(2S+3S) measured in Au+Au collisions.

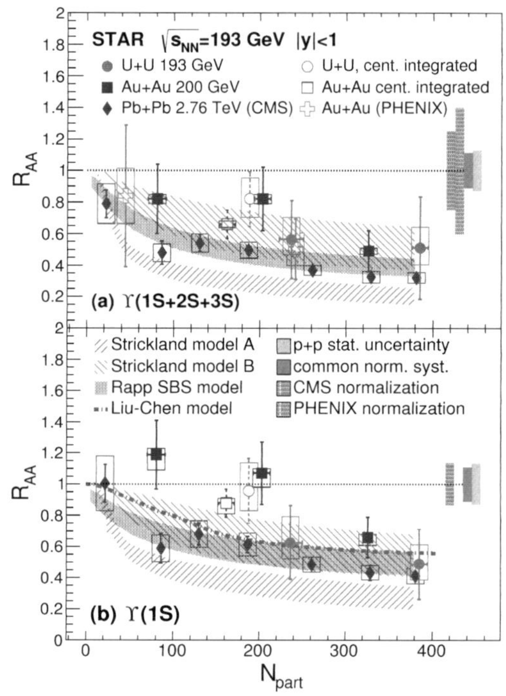

Figure 2. R_{AA} of (a) Υ(1S+2S+3S) and (b) Υ(1S) as a function of N_{part} in U+U and Au+Au collisions from STAR (solid circles and squares, respectively) [8, 12], compared to PHENIX [13] (crosses) and CMS [14] data (diamonds) as well as several theoretical calculations [15, 16, 17].

channel in Au+Au collisions at $\sqrt{s_{\mathrm{NN}}} = 200$ GeV. The result is compared to worldwide data in p+p collisions, as well as measuremens from CMS in Pb+Pb collisions at $\sqrt{s_{\mathrm{NN}}} = 2.76$ TeV [19]. The result suggests that the excited states may not be as strongly suppressed at RHIC as at the LHC, although the errors are large.

4. Summary and outlook

Measurements of Υ(1S+2S+3S) and Υ(1S) production in U+U collisions at $\sqrt{s_{\mathrm{NN}}} = 193$ GeV consolidate the trends observed in Au+Au collisions and extend them toward higher N_{part}

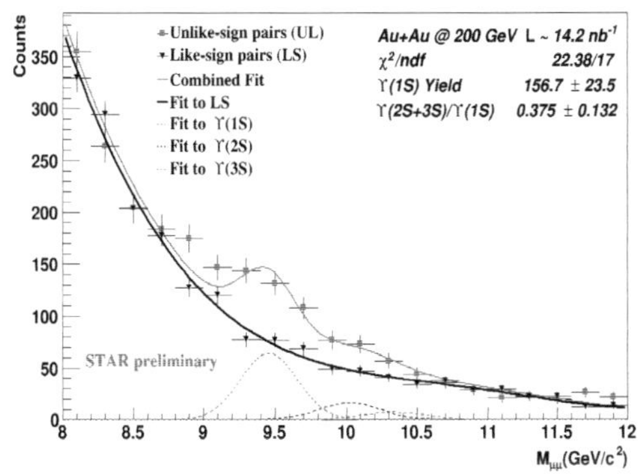

Figure 4. Invariant mass distribution of $\Upsilon \rightarrow \mu^+\mu^-$ candidates and combinatorial background in the 0-80% centrality class of Au+Au collisions at $\sqrt{s_{\rm NN}} = 200$ GeV.

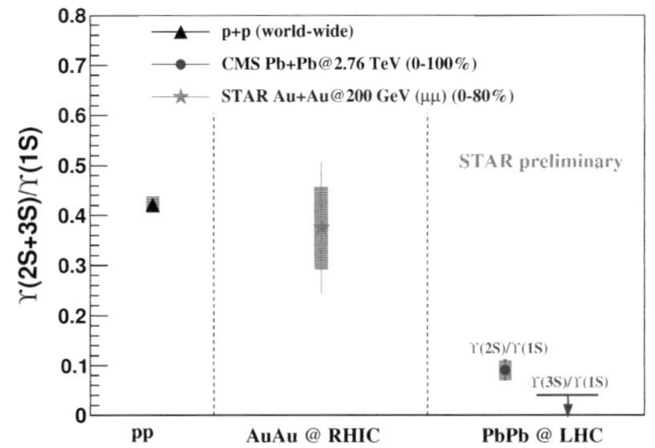

Figure 5. Excited-to-ground state ratio of the Υ mesons via the dimuon channel, compared to CMS measurements [19].

values. We observe a significant suppression of the Υ (1S+2S+3S) signal in heavy-ion collisions at RHIC top energies. The suppression of the Υ(1S) state in central heavy-ion collisions is also confirmed by the new U+U data. New preliminary measurements of Υ production via the dimuon decay channel show an indication that the excited Υ(2S+3S) states are not completely suppressed in Au+Au collisions at $\sqrt{s_{\rm NN}} = 200$ GeV, and hint that the dissociation of the excited Υ(2S+3S) states is less prominent at RHIC than at LHC energies. There are ongoing analyses of the full 2014–2016 dataset of Au+Au collisions in both the dimuon and dielectron channels. Once completed, these will provide more precise information about the extent of excited state suppression. In parallel to that, analysis of p+A data recorded in 2015 will confirm or exclude subtantial CNM effects at mid-rapidity.

This work has been supported by the Hungarian NKFIH/OTKA NK 106119 and K 120660 grants, the János Bolyai scholarship of the Hungarian Academy of Sciences, as well as the grant 13-20841S of the Czech Science Foundation (GAČR).

References

[1] Adams J *et al.* (STAR Collaboration) 2005 *Nucl. Phys. A* **757** 102
[2] Adare A *et al.* (PHENIX Collaboration) 2010 *Phys. Rev. Lett.* **104** 132301
[3] Matsui T and Satz H 1986 *Phys. Lett. B* **178** 416
[4] Mócsy Á and Petreczky P 2007 *Phys. Rev. Lett.* **99** 211602
[5] Grandchamp L, Rapp R and Brown G E 2004 *J. Phys. G* **30**, S1355
[6] Vogt R, Nelson R E and Frawley A D 2012 *PoS ConfinementX* **203**
[7] Vértesi R 2016 *Int. J. Mod. Phys. A* vol. **31** nos. 28-29 1645036
[8] Adamczyk L *et al.* (STAR Collaboration) 2014 *Phys. Lett. B* **735** 127; 2015 *Phys. Lett. B* **743** 537
[9] Kikoła D, Odyniec G and Vogt R 2011 *Phys. Rev. C* **84** 054907
[10] Ruan L *et al.* 2009 *J. Phys. G* **36** 095001
[11] Ackermann K H *et al.* (STAR Collaboration) 2003 *Nucl. Instrum. Meth. A* **499** 624
[12] Adamczyk L *et al.* (STAR Collaboration) 2016 *Preprint* arXiv:1608.06487 (Submitted to *Phys. Rev. C*)
[13] Adare A *et al.* (PHENIX Collaboration) 2015 *Phys. Rev. C* **91** 024913
[14] Khachatryan V *et al.* (CMS Collaboration) 2011 *Phys. Rev. D* **83** 112004
[15] Emerick A, Zhao X and Rapp R 2012 *Eur. Phys. J. A* **48** 72
[16] Strickland M and Bazow D 2012 *Nucl. Phys. A* **879** 25
[17] Liu Y, Chen B, Xu N and Zhuang P, 2011 *Phys. Lett. B* **697** 32
[18] Adamczyk L *et al.* (STAR Collaboration) 2013 *Phys. Lett. B* **722** 55
[19] Chatrchyan S *et al.* (CMS Collaboration) 2012 *Phys. Rev. Lett.* **109** 222301

IOP Conf. Series: Journal of Physics: Conf. Series **832** (2017) 012026 doi:10.1088/1742-6596/832/1/012026

J/ψ production in p+p at $\sqrt{s}$ = 500 GeV collisions and Au+Au collisions at $\sqrt{s_{NN}}$ = 200 GeV at the STAR experiment

Qian Yang (for the STAR Collaboration)
University of Science and Technology of China, Hefei, Anhui 230026, China
Brookhaven National Laboratory, Upton, NY 11973, USA
E-mail: ayqlyl@mail.ustc.edu.cn

Abstract. In this paper, we present measurements of J/ψ cross-section and yield dependence on charged-particle multiplicity in p+p collisions at $\sqrt{s}$ = 500 GeV at mid-rapidity in the transverse momentum range of 0-20 GeV/c. Measurements of J/ψ nuclear modification factors in Au+Au collisions at $\sqrt{s_{NN}}$ = 200 GeV up to 14 GeV/c via the di-muon channel with the full data sample taken during RHIC 2014 run is also presented.

1. Introduction

The study of J/ψ meson production has been a long-term interest both in elementary collisions and heavy-ion collisions. In p+p collisions, the study of J/ψ production provides test ground for calculations based on perturbative Quantum Chromodynamics (pQCD) and the evolution of produced $c\bar{c}$ pair. However, none of the existing theoretical models can fully describe J/ψ production. J/ψ measurements at a new beam energy of $\sqrt{s}$ = 500 GeV in p+p collisions will provide additional insights into its production mechanism. On the other hand, the production of J/ψ has also been extensively used to probe the Quark-Gluon Plasma (QGP) created in heavy-ion collisions [1]. The suppression of J/ψ in a deconfined medium due to Debye color screening of the charm quark potential was proposed as a signature of the QGP formation. Interpretation of the J/ψ suppression is, however, still a challenge due to the contributions from the regenerated J/ψ by the coalesced uncorrelated $c\bar{c}$ pairs in the medium and the cold nuclear matter effects. Precise measurements of the nuclear modification factor (R_{AA}) for J/ψ over a broad kinematic range in Au+Au collisions can help better understand the feature of Debye color screening. At STAR, the newly installed Muon Telescope Detector (MTD), which provides both the di-muon trigger and the muon identification capability at mid-rapidity, opens the door to measuring J/ψ via the di-muon decay channel for the first time at STAR. The STAR experiment recorded data corresponding to integrated luminosities of 28.3 pb^{-1} and 22 pb^{-1} for p+p collisions at $\sqrt{s}$ = 500 GeV in the RHIC 2013 and 2011 run, respectively, by using the MTD di-muon trigger and the Barrel Electromagnetic Calorimeter (BEMC) trigger, and 14.2 nb^{-1} for Au+Au collisions at $\sqrt{s_{NN}}$ = 200 GeV in the RHIC 2014 run via the di-muon trigger. In this proceedings, we report 1) the measurements of J/ψ cross-section and yield dependence on event multiplicity in p+p collisions at $\sqrt{s}$ = 500 GeV; and 2) the measurements of J/ψ R_{AA} in Au+Au collisions at $\sqrt{s_{NN}}$ = 200 GeV.

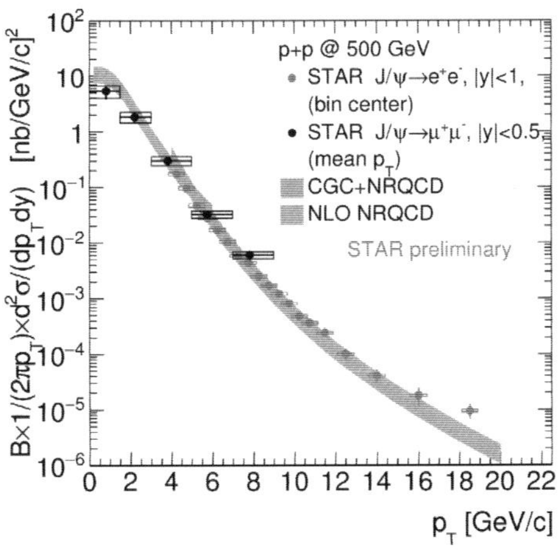

Figure 1. J/ψ cross section scaled by the branching ratio B as a function of p_T in the di-muon decay channel (black cicle) and in the di-electron decay channel (red circle).

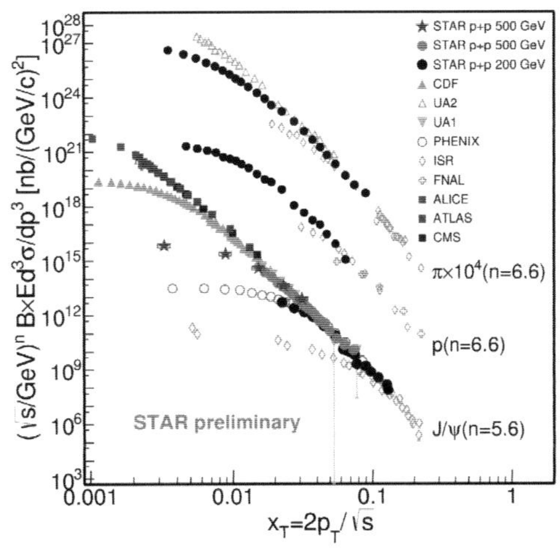

Figure 2. x_T scaling of J/ψ cross section scaled by the branching ratio B in the di-muon decay channel (blue star) and in the di-electron decay channel (red circle).

2. J/ψ measurements in p+p collisions at $\sqrt{s} = 500$ GeV

2.1. J/ψ transverse momentum spectrum and x_T scaling

Figure 1 shows the inclusive cross section of J/ψ in p+p collisions at $\sqrt{s} = 500$ GeV measured via the di-electron and di-muon channels for the transverse momentum (p_T) range of $0 < p_T < 20$ GeV/c. The measurement via the di-muon decay channel extends p_T reach down to 0. The results in these decay channels are consistent in the overlapping range of $4 < p_T < 9$ GeV/c. The experimental results can be well described by CGC+NRQCD calculations at low p_T [3] and NLO NRQCD calculations at high p_T [4]. Figure 2 shows the $x_T = 2p_T/\sqrt{s}$ scaling of J/ψ cross section [5], where the J/ψ cross section in p+p collisions at $\sqrt{s} = 500$ GeV follows a common trend as a function of x_T at high p_T. The breaking of the x_T scaling at low p_T can be attributed to soft processes.

2.2. J/ψ yield versus event activity

At STAR, we use the number of tracks matched to the Time-of-Flight (TOF) detector in the pseudorapidity interval $|\eta| < 1$ to characterize event activity. The relative yields of J/ψ as a function of event activity from both the di-electron and di-muon channels are presented in Fig. 3. The high-p_T J/ψ results ($p_T > 4$ GeV/c) are from the di-electron channel using BEMC-triggered data. The low-p_T J/ψ results ($p_T > 0$ GeV/c) are from the di-muon channel using MTD-triggered data. A similar stronger-than-linear trend is observed both at LHC [6] and RHIC. Comparisons of STAR measurements to different model calculations are shown in Fig. 4 for different kinematic ranges. PYHTIA8 with a default setting, including the Multiple-Parton Interaction (MPI) contributions to particle production, can describe the rising trend and p_T dependence in data. Percolation model [7] also qualitatively reproduces the trend in data.

3. J/ψ measurement in Au+Au collisions at $\sqrt{s_{NN}} = 200$ GeV

Figure 5 shows the J/ψ invariant yield as a function of p_T for different centralities in Au+Au collisions at $\sqrt{s_{NN}} = 200$ GeV within $|y| < 0.5$. The measurements from the di-muon channel are consistent with the published di-electron results [8, 9] at the same collision energy within

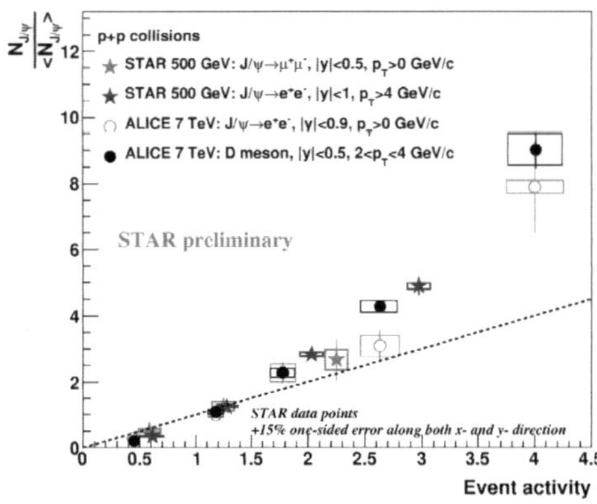

Figure 3. Relative J/ψ yield as a function of event activity in the di-muon decay channel (blue star) and in the di-electron decay channel (red star). Measurements for open (hidden) charm mesons from LHC are shown as open (filled) circles .

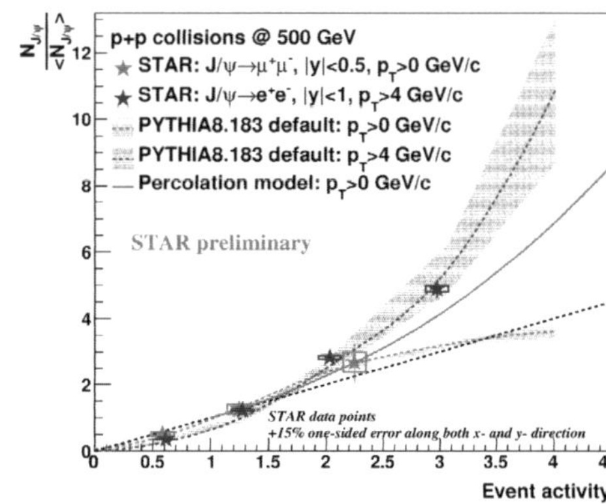

Figure 4. Relative J/ψ yield as a function of event activity. Different model calculations are presented: PYTHIA8 and the percolation model. The dash shaded bands represent the statistical errors of the PYTHIA calculations.

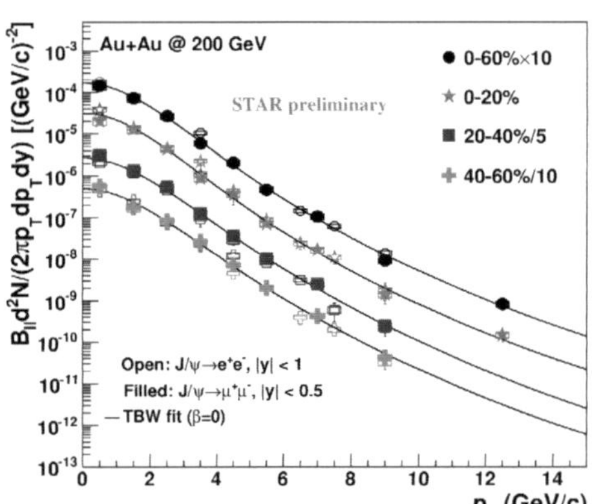

Figure 5. J/ψ invariant yield as a function of p_T for different centralities in di-moun channel (closed) and di-electron channel (open).

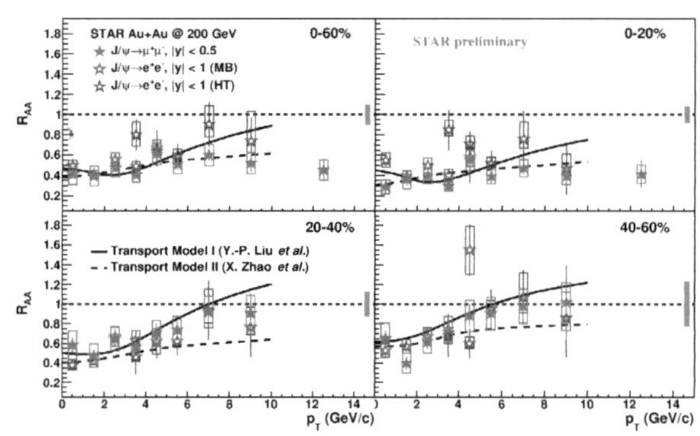

Figure 6. J/ψ R_{AA} as a function of p_T for different centralities in di-muon channel (closed) and di-electron channel (open). Transport Model predictions are shown as solid and dashed lines.

$|y| < 1$. Figure 6 shows the J/ψ R_{AA} in different centrality bins for Au+Au collisions. A strong suppression at low p_T range in all centrality bins is observed. This suppression could be due to the effects of dissociation and cold nuclear matter (CNM). The rising trend as a function of p_T can be a result of formation-time effects and more feed-down contribution from B hadrons at high p_T. Figures 7 and 8 show the centrality dependence of J/ψ R_{AA} at low-p_T and high-p_T ranges, separately. The smaller suppression at the LHC [10, 11] in central collisions at low p_T indicates larger regeneration contribution due to higher charm cross section, while the larger suppression of J/ψ at LHC at high p_T indicates larger dissociation rate due to higher temperature of the medium or shadowing effects. Transport models from Tsinghua [12, 13] and Texas A&M University [14, 15], which include both dissociation and regeneration effects, are also shown. At low-p_T range, both models can describe the centrality dependence at RHIC, but tends to underestimate R_{AA} at LHC. At high-p_T range, there is tension between models and data.

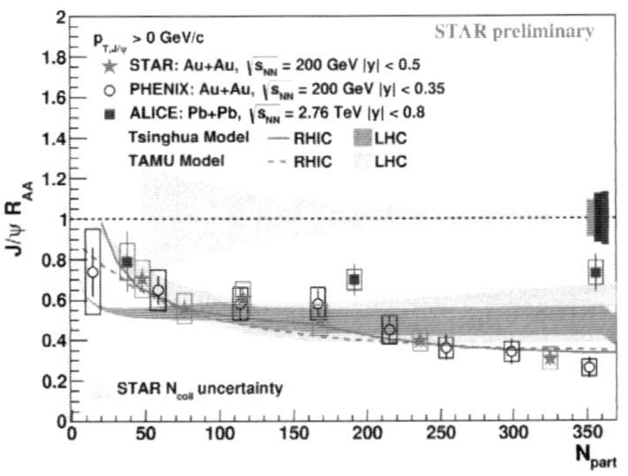

Figure 7. J/ψ R_{AA} as a function of N_{part} for $p_T > 0$ in di-muon channel (red star). Measurements from PHINEX and ALICE are shown as open circles and blue squares. Transport Models predictions are shown as lines and bands for RHIC and LHC energes.

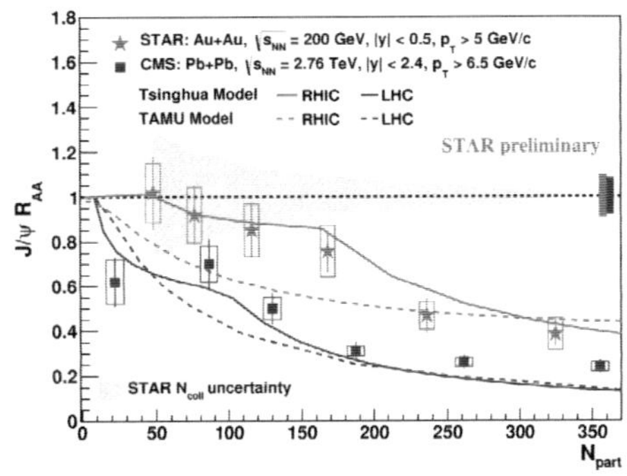

Figure 8. J/ψ R_{AA} as a function of N_{part} for $p_T > 5$ GeV/c in di-muon channel (red star). Measurements from CMS are shown as blue squares. Transport Model predictions are show as solid and dashed lines.

4. Summary

We present J/ψ production in p+p collisions at $\sqrt{s} = 500$ GeV and Au+Au collisions at $\sqrt{s_{NN}} = 200$ GV at RHIC. In p+p collisions, the inclusive J/ψ spectrum is measured down to $p_T = 0$ and the spectra can be well described by CGC + NRQCD and NLO NRQCD calculations. The relative J/ψ yield grows rapidly as the charged-particle multiplicity increases and a stronger-than-linear rise at higher multiplicities for $p_T > 4$ GeV/c is observed. PYTHIA8 and the percolation model can qualitatively describe the observed trend in data. In Au+Au collisions, we observe clear J/ψ suppression at $p_T < 5$ GeV/c at all centralities and J/ψ R_{AA} can be qualitatively described by transport models including dissociation and regeneration effects.

References

[1] T. Matsui and H. Satz, Phys. Lett. B 178 416-422 (1986)
[2] S. Chatrchyan et. al. (CMS collaboration), JHEP 09 (2010) 091
[3] Yan-Qing Ma and Raju Venugopalan, Phys. Rev. Lett. 113 192301 (2015)
[4] H. Shao et. al., JHEP 05 103 (2015)
[5] L. Adamczyk et. al. (STAR Collaboration), Phys. Rev. C 80 041902 (2009)
[6] B. Abelev et. al. (ALICE collaboration), JHEP 09 (2015) 148
[7] E.G. Ferreiro and C. Pajares, Phys. Rev. C 86 (2012) 034903
[8] L. Adamczyk et. al. (STAR Collaboration), Phys. Let. B 722 55-62 (2013)
[9] L. Adamczyk et. al. (STAR Collaboration), Phys. Rev. C 90 024906 (2014)
[10] B. Abelev et. al. (ALICE Collaboration), Phys. Let. B 734 314-327 (2014)
[11] S. Chatrchyan et. al. (CMS Collaboration), JHEP 05 063 (2012)
[12] Y. Liu et. al., Phys. Let. B 678 72-76 (2009)
[13] K. Zhou et. al., Phys. Rev. C 89 054911 (2014)
[14] X. Zhao and R. Rapp, Phys. Rev. C 82 064905 (2010)
[15] X. Zhao and R. Rapp, Nucl. Phys. A 859 114-125 (2011)

IOP Conf. Series: Journal of Physics: Conf. Series **832** (2017) 012027 doi:10.1088/1742-6596/832/1/012027

Charmonium production in Pb-Pb collisions at $\sqrt{s_{\rm NN}} = 2.76$ TeV and $\sqrt{s_{\rm NN}} = 5.02$ TeV measured with ALICE at the LHC

Benjamin Audurier for the ALICE Collaboration
Centre National de la Recherche Scientifique, 3 rue Michel-Ange, 75794 Paris cedex 16, France
E-mail: benjamin.audurier@subatech.in2p3.fr

Abstract. We report on the results of the charmonium production in Pb-Pb collisions at $\sqrt{s_{\rm NN}} = 2.76$ and 5.02 TeV measured with the ALICE detector at the LHC. In particular, we focus on the new measurements obtained at $\sqrt{s_{\rm NN}} = 5.02$ TeV for the J/ψ at forward rapidity in the dimuon decay channel and their comparison with previous measurements at lower energy and model calculations.

1. Introduction

Quantum chromodynamics predicts the existence of a deconfined phase of quarks and gluons at high temperature and density where chiral symmetry is restored [1]. This state of matter is known as the Quark Gluon Plasma (QGP) [2], and can be formed in high-energy heavy-ion collisions. Its characterization is one of the goals of ultra-relativistic heavy-ion collision studies. The charmonium states (for instance J/ψ and ψ(2S)) are one of the probes studied to investigate the properties of the QGP. Indeed, the presence of a deconfined medium should modify the charmonium production yield, due to the color screening of the charm quark anti-quark potential and interaction with the surrounding medium [3]. Such a suppression was already observed in heavy-ion collisions at SPS and RHIC energies. However, at LHC energies, this suppression is presumably counterbalanced by the so-called regeneration mechanism [4], i.e. a recombination mechanism of c and $\bar{c}$ quarks, occurring during the deconfined phase or at the hadronization. This process is favoured by the increase of the charm cross section at LHC energies.

Two observables are frequently used to quantify the effect of the deconfined state of matter on charmonia. The first one, the so-called *Nuclear Modification Factor* $R_{\rm AA}$, is defined for a given centrality i as

$$R^{i}_{AA} = \frac{dY^{PbPb,i}/dp_T dy}{< N_{coll} > dY^{\rm pp}/dp_T dy} \tag{1}$$

where $dY^{PbPb,i}/dp_T dy$ ($dY^{\rm pp}/dp_T dy$) is the differential yield in Pb-Pb (pp) interactions and $\langle N_{\rm coll}\rangle$ is the mean number of binary collisions. Any deviation from unity of the $R_{\rm AA}$ can be interpreted as a suppression ($R_{\rm AA} < 1$) or an enhancement ($R_{\rm AA} > 1$) of the charmonium production yield in Pb-Pb collisions compared to binary scaled pp collisions.

The second observable, the *elliptic flow* v_2 is the second coefficient of the Fourier expansion of the azimuthal particles distributions and is defined as:

$$v^{i}_{2}(p_T, y) = \langle \cos[2(\varphi - \Psi_{RP})]\rangle^{i} \tag{2}$$

where ϕ is the particle azimuthal angle and Ψ_{RP} is the reaction plane angle defined by the beam axis and the impact parameter direction. The elliptic flow is related to the anisotropy of the colliding system. If charmonia are created mainly via the regeneration mechanism, the v_2 measurement can be a stringent test for the degree of thermalization of the charm quarks in the QGP. Those two observables are measured by ALICE for the inclusive J/ψ production.

2. The ALICE Detector

ALICE measures the J/ψ production both at forward ($2.5 < y < 4$) and mid-rapidity ($|y| < 0.9$) through the dimuon and dielectron decay channel respectively, down to zero transverse momentum. Depending on the rapidity range, central barrel detectors (mid-rapidity) or muon arm detectors (forward rapidity) are involved. A review of the ALICE detectors can be found in [5].

3. Results

The first data campaign at $\sqrt{s_{\rm NN}} = 2.76$ TeV (2011) shows a clear J/ψ suppression both at forward and mid-rapidity [6]. This suppression was already measured by PHENIX [7, 8] at $\sqrt{s_{\rm NN}} = 0.2$ TeV (Fig.1), but the ALICE measurements show a weaker centrality dependence and smaller suppression for central events. This behavior is expected in a regeneration scenario, as more J/ψ should be produced in central with respect to peripheral collisions.

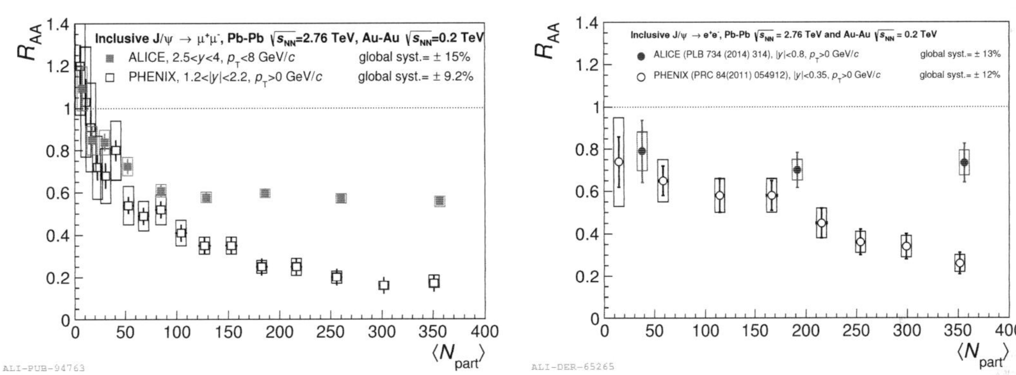

Figure 1: Results for the $R_{\rm AA}$ versus the average number of participant $\langle N_{part}\rangle$ in the dimuon (left panel) and dielectron (right panel) channel. The red points are ALICE results in Pb-Pb at $\sqrt{s_{\rm NN}} = 2.76$ TeV, the open points are PHENIX results in Au-Au at $\sqrt{s_{\rm NN}} = 0.2$ TeV. The error bars represent the statistical uncertainties, the boxes around the data points the uncorrelated systematic uncertainties [6].

In addition, the measurement of the J/ψ flow in semi-central collisions at forward rapidity shows a hint of a positive v_2 [9], in agreement within uncertainties with transport models including regeneration mechanism (Fig.2).

Finally, an unexpected excess of J/ψ was measured by ALICE at very low $p_{\rm T}$ [10] in the most peripheral collisions (Fig.3). A possible explanation for this excess is the coherent photoproduction mechanism for Pb-Pb collisions with $b < 2R$, where b and R are the impact parameter and the nuclei radius respectively.

ALICE recently performed measurements in Pb-Pb collisions at $\sqrt{s_{\rm NN}} = 5.02$ TeV. The detailed analysis can be found in [11]. ALICE has collected $\sim 225\ \mu b^{-1}$ integrated luminosity at $\sqrt{s_{\rm NN}} = 5.02$ TeV, representing 7 times more statistics with respect to 2011 data campaign. The fully integrated $R_{\rm AA}$ for inclusive J/ψ production at forward rapidity is $R_{\rm AA} = 0.69 \pm 0.01(\text{stat}) \pm 0.07(\text{syst})$. Comparing this value with the corresponding one at $\sqrt{s_{\rm NN}} = 2.76$ TeV, $R_{\rm AA} = 0.58 \pm 0.01\ (\text{stat}) \pm 0.09\ (\text{syst})$ [6], an increase of $R_{\rm AA}$ by $\sim$19% is found, with a 1.0σ significance. Results versus $\langle N_{\rm part}\rangle$ are shown in Fig.4. While a significant reduction of the systematic uncertainties has been performed, results are compatible within uncertainties, showing an increasing suppression with centrality up to $\langle N_{\rm part}\rangle \sim 100$, followed by a roughly

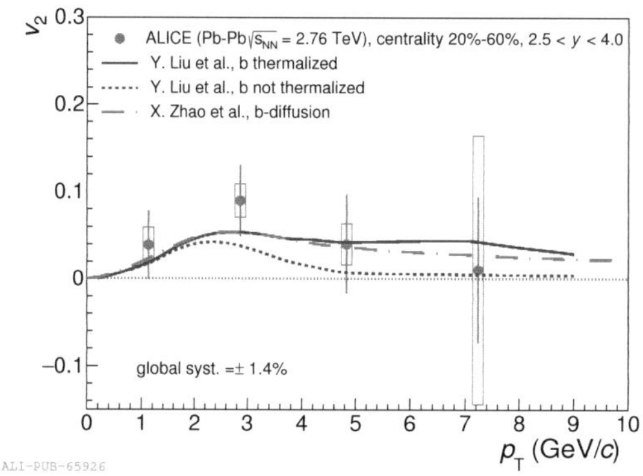

Figure 2: J/ψ elliptic flow as a function of p_T in the 20−60% centrality range compared to transport models. The error bars represent the statistical uncertainties, the boxes around the data points the uncorrelated systematic uncertainties [9].

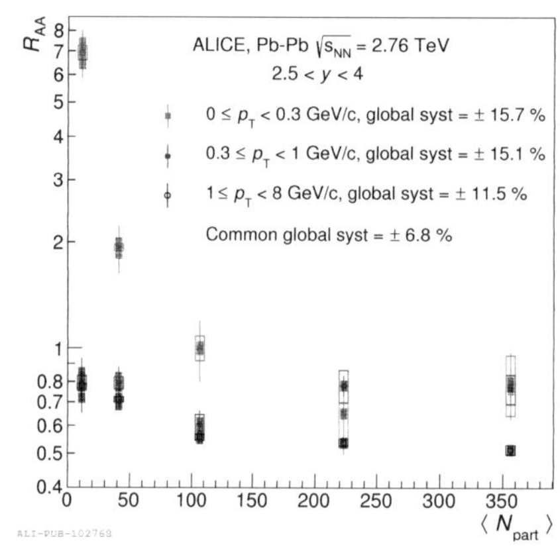

Figure 3: The R_{AA} versus $\langle N_{part} \rangle$ for several p_T ranges [10].

constant R_{AA} value. A $p_T > 0.3$ GeV/c selection has been applied to the data, removing $\sim 80\%$ of the photoproduced J/ψ mentioned earlier, this production source being a contamination for these hadronic observables. Data are well described by various models, even if they have rather large uncertainties due to the choice of the corresponding input parameters, $\mathrm{d}\sigma_{c\bar{c}}/dy$ in particular (see [11] for details). Making assumptions regarding the non-prompt component of the inclusive J/ψ nuclear modification factor, the R_{AA} of prompt J/ψ would be about 10% higher if $R_{\mathrm{AA}}^{\text{non-prompt}} = 0$ and about 3-4% (1%) smaller if $R_{\mathrm{AA}}^{\text{non-prompt}} = 1$ for central (peripheral) collisions.

The p_T dependence of the R_{AA} in central events is shown in Fig.5. The p_T range has been extended up to 12 GeV/c with respect to $\sqrt{s_{\mathrm{NN}}} = 2.76$ TeV data due to the higher p_T reach for pp measurements at $\sqrt{s_{\mathrm{NN}}} = 5.02$ TeV. Similarly to what has been observed studying the centrality dependence of the J/ψ R_{AA} , the p_T dependent suppression factor for central collisions increases systematically by about 20% with energy with respect to 2011 results. In this case, the region $p_T < 0.3$ GeV/c was not excluded, the contribution of J/ψ photoproduction being negligible with respect to the hadronic one for central events [10]. Making the same kind of assumption concerning the non-prompt component, the prompt J/ψ R_{AA} is expected to be 7% larger (1% smaller) for $p_T < 1$ GeV/c and 17% larger (20% smaller) for $6 < p_T < 8$ GeV/c when the beauty contribution is fully (not) suppressed.

4. Conclusions

ALICE measures the inclusive J/ψ R_{AA} both at $\sqrt{s_{\mathrm{NN}}} = 2.76$ and 5.02 TeV and the J/ψ elliptic flow at $\sqrt{s_{\mathrm{NN}}} = 2.76$ TeV. Results at $\sqrt{s_{\mathrm{NN}}} = 2.76$ TeV show significant differences with previous results obtained at RHIC at $\sqrt{s_{\rm NN}}$ = 0.2 TeV and are in good agreement within uncertainties with theoretical models including both a suppression and a regeneration mechanism affecting the probe. The inclusive J/ψ $R_{\rm AA}$ obtained at $\sqrt{s_{\rm NN}}$ = 5.02 TeV is compatible within uncertainties with the $\sqrt{s_{\rm NN}}$ = 2.76 TeV measurement while systematic uncertainties has been greatly reduced. The same trend is found both for the $\langle N_{\rm part}\rangle$ and $p_{\rm T}$ dependences of the $R_{\rm AA}$ between $\sqrt{s_{\rm NN}}$ = 2.76 and 5.02 TeV. Data and theoretical models are compatible within uncertainties and support a picture of competing J/ψ suppression and regeneration in the QGP.

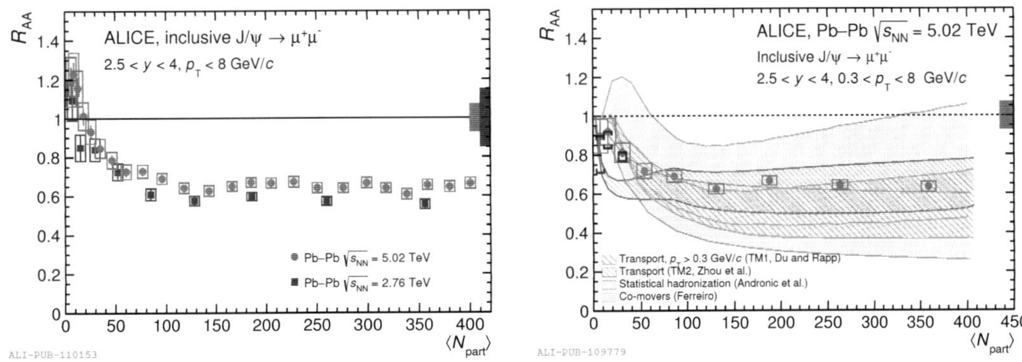

Figure 4: Results for the inclusive J/ψ $R_{\rm AA}$ versus $\langle N_{\rm part}\rangle$ at forward rapidity for $0 < p_{\rm T} < 8$ GeV/c (left panel) and for $0.3 < p_{\rm T} < 8$ GeV/c (right panel). The error bars represent the statistical uncertainties, the boxes around the data points the uncorrelated systematic uncertainties, while the global uncertainties are shown as a filled box around $R_{\rm AA} = 1$ [11].

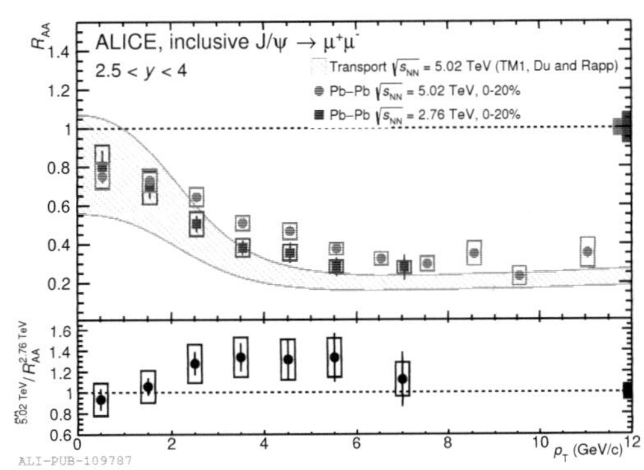

Figure 5: Inclusive J/ψ $R_{\rm AA}$ versus $p_{\rm T}$ at $\sqrt{s_{\rm NN}}$ = 5.02 TeV, compared to the corresponding result at $\sqrt{s_{\rm NN}}$ = 2.76 and to the prediction of a transport model (see text for reference). The error bars represent statistical uncertainties, the boxes around the points uncorrelated systematic uncertainties, while global uncertainties are shown as a filled box around $R_{\rm AA} = 1$ on the right [11].

References

[1] Karsh F 2002 *Lect. Notes Phys.* **583** 209-249
[2] Shuryak E V 1980 *Phys. Rept.* **61** 71-158
[3] Matsui and Statz 1986 *Phys. Lett.* **B178** 416
[4] Braun-Munzinger P and Stachel J 2000 *Phys. Lett.* **B490** 196-202
[5] The ALICE Collaboration 2008 *JINST* **3** S08002
[6] The ALICE Collaboration 2016 *JINST* **05** 179 arxiv:1506.08804
[7] The PHENIX Collaboration 2007 *Phys. Rev. Lett.* **98** 232301 arxiv:nucl-ex/0611020
[8] The PHENIX Collaboration 2011 *Phys. Rev. Lett.* **C84** 054912 arxiv:1103.6269
[9] The ALICE Collaboration 2013 *Phys. Rev. Lett.* **111** 162301
[10] The ALICE Collaboration 2016 *Phys. Rev. Lett.* **116** 222301
[11] The ALICE Collaboration 2016 *arXiv:1606.08197*

IOP Conf. Series: Journal of Physics: Conf. Series **832** (2017) 012028 doi:10.1088/1742-6596/832/1/012028

Measurements of open charm hadrons at the STAR experiment

M Simko[1,2] for the STAR Collaboration

[1] Nuclear Physics Institute of the Czech Academy of Sciences, Na Truhlarce 39/64, 18086 Praha, Czech Republic;

[2] Faculty of Nuclear Sciences and Physical Engineering, Czech Technical University in Prague, Brehova 7, 11519 Praha, Czech Republic.

E-mail: simko@ujf.cas.cz

Abstract. Because of their large masses, charm quarks are predominantly produced in the early stages of the heavy-ion collisions via hard scatterings. Therefore, they experience the entire evolution of the Quark-Gluon Plasma created in such collisions. Compared to light quarks, charm quarks thermalize more slowly. Therefore, the open charm hadrons present a unique probe to the properties of the hot and dense nuclear matter by measuring their energy loss and degree of thermalization in the medium. Furthermore, with the combined measurements of D^0 and D_s mesons, we can study multiple modes of coalescence of charm quarks with light quarks in heavy-ion collisions. Heavy Flavor Tracker at the STAR experiment enables full topological reconstruction of open charm hadrons which greatly improves measurements of D^0 mesons and opens the door to reconstructing the D_s mesons for the first time at RHIC. In this paper, we present the nuclear modification factor and azimuthal anisotropy for the D^0 and D_s mesons as well as the ratio of D_s/D^0 in Au+Au collisions at the center-of-mass energy $\sqrt{s_{NN}} = 200\,\text{GeV}$.

1. Introduction

Charm quarks are considered to be an excellent probe to study the strongly coupled quark-gluon plasma (sQGP) as they are produced predominantly in hard scatterings during the early stages of heavy-ion collisions due to their large masses which are mostly unaffected by the medium. Thus, they experience the full evolution of the system. Analogous to the Brownian motion, charm quarks are sensitive to the transport properties of the sQGP, e.g. $2\pi T D_s$, where T is the temperature of the system and D_s the spatial diffusion coefficient for the c-quark [1].

The inclusive production of the c-quark has been measured at STAR [2] and exhibits scaling with the number of binary nucleon–nucleon collisions ($N_{\rm coll}$) in Au+Au collisions. The D^0 meson, being the lightest hadron that contains a charm quark, provides, therefore, an excellent calibrated probe to the behavior of the medium. Recent measurements of the D^0 meson at Relativistic Heavy-Ion Collider (RHIC) and the Large Hadron Collider (LHC) [3–5] show suppression of yields at high transverse momenta ($p_{\rm T}$) and suggest a non-zero v_2 at intermediate to high $p_{\rm T}$. Measurements of better precision are, however, needed to provide more stringent constraints on model calculations.

The D_s meson, which contains a strange quark, provides an additional handle on the hadronization process of charm quarks. Recent calculations [6] suggest an enhancement of the D_s meson yield compared to the D^0 meson because of the process of quark coalescence.

2. The STAR experiment and open charm hadron reconstruction

The Solenoidal Tracker at RHIC (STAR) is a large-acceptance multi-purpose detector that covers the full azimuth and pseudorapidity range of $|\eta| < 1$ [7]. The main tracking detector of STAR is the Time Projection Chamber, which also provides the $\mathrm{d}E/\mathrm{d}x$ information for particle identification (PID). The Time-Of-Flight detector is also used to improve the PID capabilities.

Since the beginning of 2014, a new detector, Heavy Flavor Tracker (HFT), has been installed at STAR. The HFT consists of 4 layers of silicon detectors: one layer of double strip, one layer of silicon pad, and finally two layers of PiXeL (PXL) detectors using the state-of-the-art slimmed-down Monolithic Active Pixel Sensor technology for the first time in a collider experiment. It provides excellent distance of closest approach resolution down to $\sim$ 30 µm at high-p_{T} while maintaining a very small material budget as the first layer of PXL has a radiation length of $\sim 0.4\,\%\, X_0$.

The excellent track pointing resolution provided by the HFT allows for a direct topological reconstruction of the secondary vertices of open charm meson decays via hadronic channels, i.e. $\mathrm{D}^0 \to \pi^{\pm} + \mathrm{K}^{\mp}$, $\mathrm{D}_s^{\pm} \to \pi^{\pm} + \phi(1020) \to \pi^{\pm} + \mathrm{K}^{\pm} + \mathrm{K}^{\mp}$. This greatly reduces the combinatorial background for these measurements. In the case of the $\mathrm{D_s}$ meson, the decay channel via $\phi(1220)$ is used to place an additional constraint on the $\mathrm{K}^{\pm} + \mathrm{K}^{\mp}$ invariant mass and, therefore, to reduce the background even further. STAR has recorded $\sim$ 3.2 B minimum-bias events in Au+Au collisions at $\sqrt{s_{\mathrm{NN}}} = 200$ GeV in the years 2014 and 2016. Results from $\sim$ 780 M of these events are shown in this paper.

3. Results

3.1. D^0 measurements

Although the inclusive charm quark production scales with N_{coll} [2], the D^0 spectrum shape is significantly modified in Au+Au collisions. Fig. 1a shows the D^0 nuclear modification factor R_{AA} as a function of p_{T} in the most central (0–10 %) collisions, where R_{AA} is the ratio between the yield in Au+Au collisions and that in p+p collisions scaled by N_{coll}. The new results (black full circles) obtained with the HFT are consistent with the published R_{AA} from 1.1 B minimum-bias events taken in the years 2010 and 2011 without the HFT (red empty circles) [3]. For the new results, a much better precision is achieved despite the less statistics used. The D^0 production is significantly suppressed at high-p_{T} which indicates strong interactions between charm quarks and the medium in this kinematic region. In the intermediate p_{T} range (~ 0.7–$2\,\mathrm{GeV}/c$), data show an enhancement which can be described by models including coalescence of charm quarks.

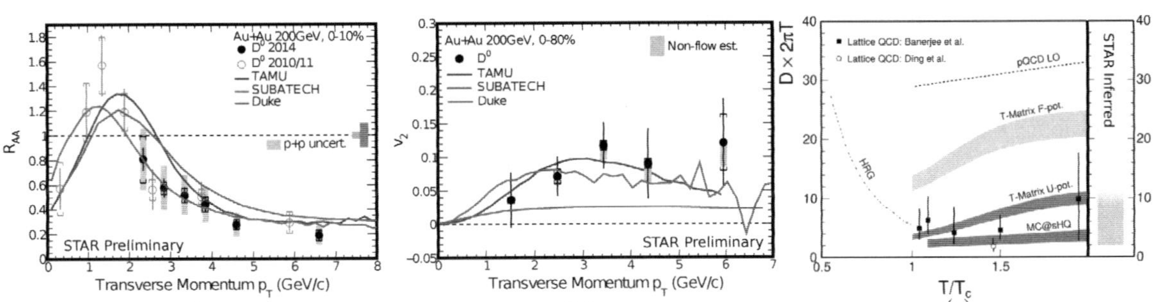

(a) (b) (c)

Figure 1. (a) D^0 R_{AA} as a function of p_{T} for 0–10 % central Au+Au collisions. The gray bands are systematic uncertainties from the p+p baseline and the light and dark green vertical bands around unity are uncertainties related to the N_{coll} in Au+Au collisions and the global normalization in the p+p collisions, respectively; **(b)** D^0 v_2 as a function of p_{T} for Au+Au collisions at $\sqrt{s_{\mathrm{NN}}} = 200\,\mathrm{GeV}$; **(c)** Charm-quark diffusion coefficient used in various models compared to that inferred from the STAR data.

The HFT also enables the measurement of the D^0 v_2 for the first time at RHIC, as shown in Fig. 1b. The vertical bars (brackets) indicate the statistical (systematic) uncertainties while the gray bands represent the estimated non-flow contribution inferred from D meson–hadron correlations in p+p collisions. The data show that the v_2 is significantly larger than 0 above 2 GeV/c.

Several models, which use different approaches to determine the charm quark diffusion coefficient, are compared to the measurements of R_{AA} and v_2. The group from TAMU [8] (blue) employs a non-perturbative T-matrix approach with the assumption that two body interactions can be described by a potential, which is a function of the transferred 4-momentum. This model predicts the charm quark diffusion coefficient multiplied by temperature as $3 \leq 2\pi T D_s \lesssim 11$. The SUBATECH group [9] (green) uses a pQCD approach with the Hard Thermal Loop approximation for soft collisions. In this approach, the diffusion coefficient is within $2 \leq 2\pi T D_s \leq 4$. The model by the Duke university group [11] uses $2\pi T D_s$ as a free parameter. The red curves shown in Figs. 1a and 1b use the value $2\pi T D_s = 7$ which is fixed to match the D^0 R_{AA} measured at the LHC. The Duke model can describe the shape of R_{AA} well, however it systematically underestimates the v_2. The other two models are consistent with both R_{AA} and v_2 data. Fig. 1c shows the $2\pi T D_s$ values obtained from different models [8, 12–17] compared to the range inferred from the STAR data which is drawn as the yellow band. The STAR inferred range of $2 \leq 2\pi T D_s \lesssim 12$ is consistent with the lattice QCD calculations [16, 17] shown as the black points.

3.2. D_s measurements

Thanks to the HFT, the D_s meson, consisting of a charm quark and a strange quark, is measured for the first time at RHIC. Such measurements are expected to shed more light on the mechanism of the charm quark coalescence.

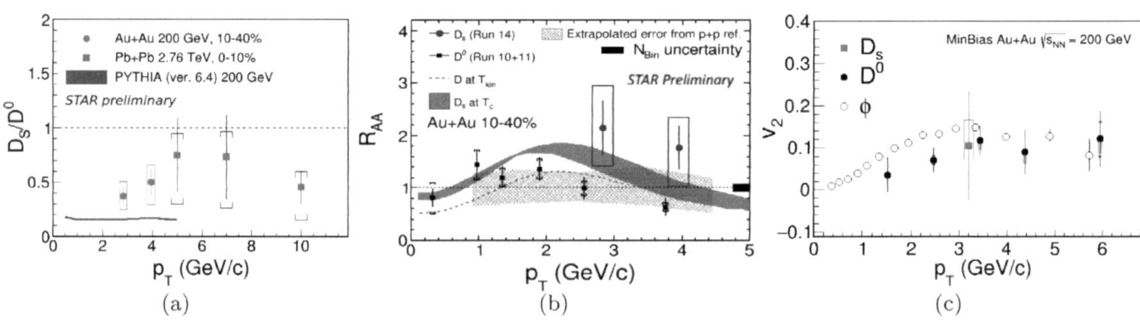

Figure 2. **(a)** The ratio between D_s and D^0 yield in Au+Au collisions at $\sqrt{s_{NN}} = 200$ GeV for 10–40 % centrality; **(b)** D_s R_{AA} in the 10–40 % centrality class; **(c)** D_s v_2 in minimum-bias Au+Au collisions at $\sqrt{s_{NN}} = 200$ GeV.

In Fig. 2a, the yield ratio of produced D_s to D^0 is shown as a function of p_T in 10–40 % central Au+Au collisions at $\sqrt{s_{NN}} = 200$ GeV (red circles). The D^0 spectrum is obtained from the published STAR data [3]. The measured D_s/D^0 ratios are compared to similar measurements at the LHC for minimum-bias Pb+Pb collisions [18] (red squares). Both of the measurements are consistent with each other within uncertainties. To compare our measurement to the D_s/D^0 in p+p collisions, PYTHIA 6.4 [19] (purple curve) is used. The STAR measurement is slightly enhanced compared to the p+p ratio; however, the enhancement is statistically insignificant.

In Fig. 2b, the nuclear modification factor for D_s is shown as blue points. The p+p baseline is obtained from the measured total charm cross-section by STAR [2] multiplied by the $c \to D_s$ fragmentation factor obtained from the measurements at HERA [20, 21]. The uncertainty of the baseline is indicated by the green hashed band and the uncertainty on N_{coll} is plotted as the

IOP Conf. Series: Journal of Physics: Conf. Series **832** (2017) 012028 doi:10.1088/1742-6596/832/1/012028

black rectangle. The D_s R_{AA} is compared to the calculation done by the TAMU group [6] and the published D^0 R_{AA} by STAR [3]. Again, we observe a hint of D_s enhancement compared to D^0, which can be described by the TAMU model within uncertainties; however, more data is needed to draw firmer conclusions.

Fig. 2c shows the first result of D_s v_2 at RHIC as the red filled square. The data slightly prefer a non-zero v_2, albeit not significantly. The D_s v_2 is compared to the v_2 of D^0 and ϕ meson [22] which, like the D_s, contains strange quarks. The D_s v_2 is consistent with both measurements within uncertainties.

4. Summary and outlook

We report the first measurements of the open charm hadrons using the state-of-the-art vertex detector HFT. D^0 v_2 is measured for the first time at RHIC. These data are significantly above zero and favor models with charm diffusion. Moreover, the D^0 R_{AA} is measured with a much improved precision, compared to the previous measurements without the HFT. Comparing both R_{AA} and v_2 measurements to different models, STAR is able to infer the value of the charm spatial diffusion coefficient D_s multiplied by the temperature T to be $2 \leq 2\pi T D_s \lesssim 12$, which is consistent with the lattice QCD calculations.

The first measurement of the D_s meson at RHIC is enabled by the HFT. The ratio of the production yield of D_s/D^0 is compatible with a simmilar LHC measurement and indicates an enhancement of the D_s mesons in Au+Au collisions compared to the p+p collisions. The R_{AA} of the D_s is measured and is consistent with model calculations. Moreover, the D_s v_2 is also measured for the first time at RHIC.

In the years 2014 and 2016, $\sim$ 4 times more minimum-bias Au+Au events were recorded compared to the shown results. All of these measurements will benefit greatly from the increased statistics. In addition, there have been significant improvements in terms of HFT performance in 2016 as well as offline software, which greatly enhance the tracking efficiency. In 2015, STAR took reference p+p and p+Au collisions at $\sqrt{s_{NN}} = 200\,\text{GeV}$, which will provide a baseline of better precision and help to quantify the cold nuclear matter effects.

This work has been supported by the grant LG15001 of the Ministry of Education of the Czech Republic.

References

[1] Abreu L M, Cabrera D, Llanes-Estrada F J and Torres-Rincon J M 2011 *Annals of Physics* **326** 2737–2772
[2] Adamczyk L *et al.* (STAR Collaboration) 2012 *Phys. Rev. D* **86** 072013
[3] Adamczyk L *et al.* (STAR Collaboration) 2014 *Phys. Rev. Lett.* **113** 142301
[4] Adam J *et al.* (ALICE Collaboration) 2016 *Journal of High Energy Physics* **2016** 1–43
[5] Abelev B *et al.* (ALICE Collaboration) 2013 *Phys. Rev. Lett.* **111** 102301
[6] He M, Fries R J and Rapp R 2013 *Phys. Rev. Lett.* **110** 112301
[7] Ackermann K H *et al.* (STAR Collaboration) 2003 *Nucl. Inst. Meth. A* **499** 624–632
[8] He M, Fries R J and Rapp R 2012 *Phys. Rev. C* **86** 014903
[9] Nahrgang M, Aichelin J, Bass S, Gossiaux P B and Werner K 2015 *Phys. Rev. C* **91** 014904
[10] Berrehrah H, Bratkovskaya E, Cassing W, Gossiaux P B and Aichelin J 2015 *Phys. Rev. C* **91** 054902
[11] Cao S, Qin G Y and Bass S A 2013 *Phys. Rev. C* **88** 044907
[12] Andronic A *et al.* 2016 *Eur. Phys. J. C* **76** 107
[13] Kaczmarek O 2014 *Nuclear Physics A* **931** 633–637
[14] Riek F and Rapp R 2010 *Phys. Rev. C* **82** 035201
[15] Svetitsky B 1988 *Phys. Rev. D* **37** 2484–2491
[16] Ding H T, Francis A, Kaczmarek O, Karsch F, Satz H and Soeldner W 2012 *Phys. Rev. D* **86** 014509
[17] Banerjee D, Datta S, Gavai R and Majumdar P 2012 *Phys. Rev. D* **85** 014510
[18] Adam J *et al.* (ALICE Collaboration) 2016 *Journal of High Energy Physics* **2016** 1–33
[19] Sjstrand T, Mrenna S and Skands P 2006 *Journal of High Energy Physics* **2006** 026
[20] The H1 Collaboration 2005 *Eur. Phys. J. C* **38** 447–459
[21] The ZEUS Collaboration 2005 *Eur. Phys. J. C* **44** 351–366
[22] Adamczyk L *et al.* (STAR Collaboration) 2016 *Phys. Rev. Lett.* **116** 062301

IOP Conf. Series: Journal of Physics: Conf. Series **832** (2017) 012029 doi:10.1088/1742-6596/832/1/012029

Measurement of J/ψ production in pp collisions at LHC energies with ALICE

S G Weber on behalf of the ALICE Collaboration
Research Division, GSI Helmholtzzentrum für Schwerionenforschung, Darmstadt, Germany
E-mail: S.Weber@gsi.de

Abstract. An overview of ALICE results on the measurement of J/ψ production in pp collisions at $\sqrt{s} = 7\,\mathrm{TeV}$ collected during the LHC Run-1 period is presented, as well as first results at forward rapidity from pp collisions at $\sqrt{s} = 13\,\mathrm{TeV}$ collected during the LHC Run-2 period. In particular, the measurement of J/ψ production as a function of transverse momentum and charged-particle multiplicity are discussed and compared to theoretical model calculations.

1. Introduction

1.1. Motivation

The role of charmonium production as an excellent probe for the deconfined medium in ultra-relativistic heavy-ion collisions has been proposed a long time ago [1] and is universally accepted in the field. It is understood that a high quality baseline from proton-proton collisions is crucial for the quantification of hot medium effects taking place in nucleus-nucleus collisions, as well as cold nuclear matter effects in proton-nucleus collisions.

Besides that, hadronic charmonium production in proton-proton collisions is already a complex, intrinsically multi-scale process. In a simplified way it can be thought of as a two-stage process: in a first step charm-anticharm quark pairs are produced in elementary parton-parton interactions. This high energy scale process is well understood in the context of perturbative Quantum ChromoDynamics (pQCD). The subsequent binding into charmonium states happens at lower energy scales and is thus a non-perturbative process. There are mainly three theoretical frameworks that aim to describe hadronic charmonium production:

- Color Evaporation Model (CEM) [2]: It is based on quark-hadron duality. The cross section of a given charmonium state is proportional to the cross section of constituting heavy quark pair, independent of energy, transverse momentum or rapidity.
- Color Singlet Model (CSM) [3]: pQCD is used to model the production of on-shell charm-anticharm quark pairs with the same quantum numbers as the charmonium states into which they will evolve, hence only color-singlet states are taken into account.
- Non-Relativistic QCD (NRQCD) [4]: Also contributions from color-octet states are considered. The transition into a color-singlet state is treated non-perturbatively as an expansion in powers of the relative velocity between the two heavy quarks. It is parametrized using universal long-range matrix elements, obtained from fits to measured data sets.

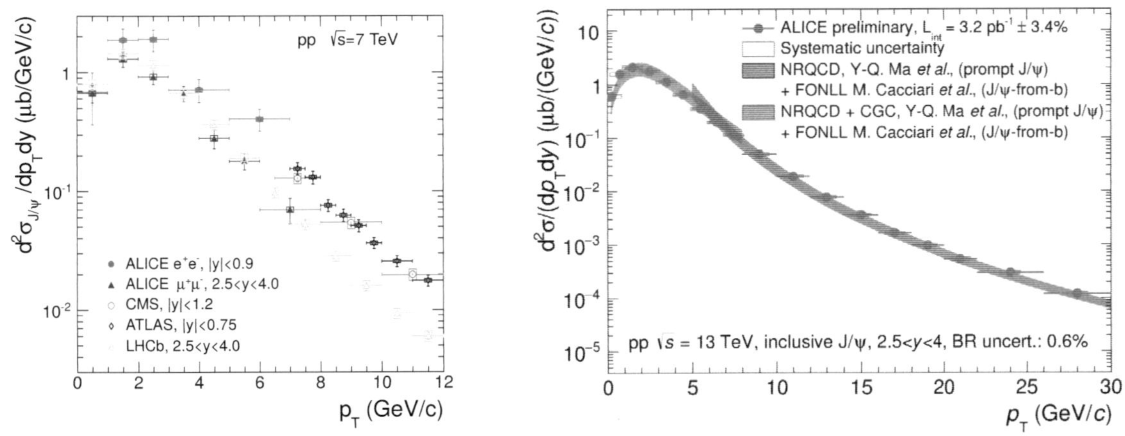

Figure 1. Inclusive J/ψ cross section in pp collisions as a function of transverse momentum. Left: Results at $\sqrt{s} = 7\,\text{TeV}$ at forward (blue triangles) and mid-rapidity (red squares). For comparison, results from other LHC experiments at the same collision energy are also shown. [5, 6, and references therein]. Right: Results at $\sqrt{s} = 13\,\text{TeV}$ at forward rapidity with model predictions from the NRQCD model [7], NRQCD and CGC [8] for the prompt contribution and FONLL [9] for the non-prompt contribution.

These models make predictions on the transverse momentum and energy dependence of hadronic J/ψ production as well as the polarization.

Another observable is the J/ψ yield as a function of the underlying event activity. It provides insight into the interplay between the hard and soft mechanisms relevant for charmonium production, and into the importance of multiple interactions [10]. It might also be useful to address new proposed phenomena, such as the onset of collective effects in high multiplicity proton-proton collisions. In the analysis presented here, the event activity is quantified by the charged particle multiplicity at mid-rapidity.

1.2. Experimental setup and analysis methods

The ALICE detector is capable of reconstructing J/ψ particles in two kinematic regions in two different decay channels, i.e. at mid-rapidity ($|y| < 0.9$) in the dielectron decay channel and at forward rapidity ($2.5 < y < 4.0$) in the dimuon decay channel. At mid-rapidity the Inner Tracking System is used for vertexing and tracking, and the Time Projection Chamber as the main tracking and Particle IDentification (PID) device. At forward rapidity, the dedicated muon arm is used for triggering, tracking and PID. The charged particle density is calculated using the number of tracklets reconstructed from hits in the two innermost layers of the Inner Tracking System. More details on the ALICE detector can be found in [11], details on the analyses in [5] and [12].

When comparing experimental results with model predictions one should keep in mind the different sources of J/ψ production: In an inclusive measurement about 60 % of the J/ψ cross section is directly produced, 20 % to 30 % is feed-down from heavier charmonium states, and 10 % to 20 % comes from the weak decay of B-mesons. The first two sources are commonly referred to as prompt production, the last as non-prompt. Models often make predictions for direct or prompt J/ψ production, therefore for a quantitative comparison to data it is important to take all the components into account.

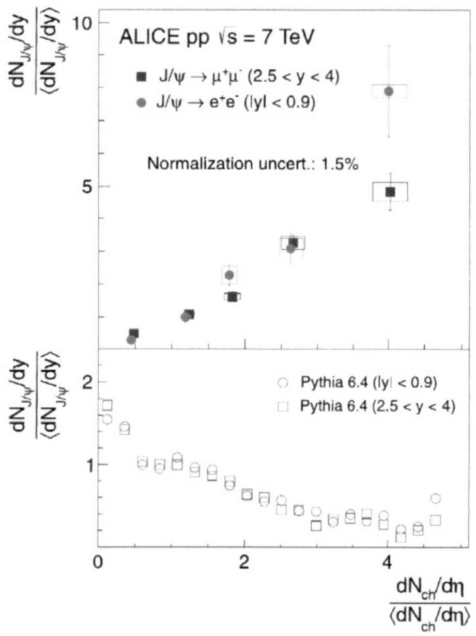

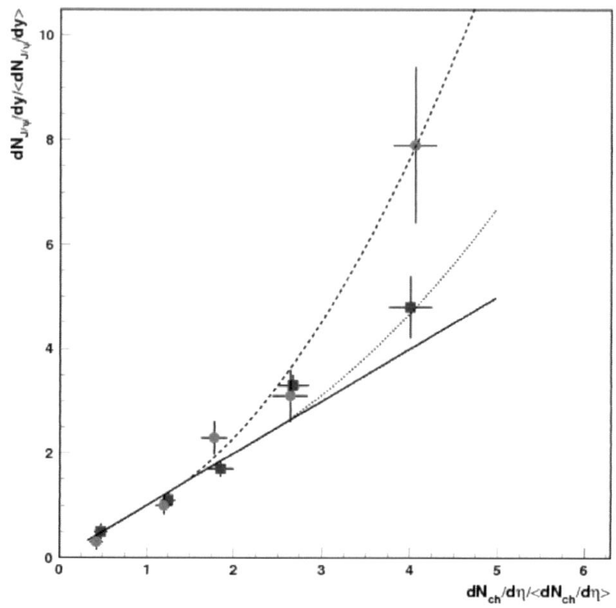

Figure 2. J/ψ yield in pp collisions at $\sqrt{s} = 7\,\text{TeV}$ as function of charged particle multiplicity at mid-rapidity, both values being normalized to the corresponding mean values in minimum bias pp collisions. Blue squares refer to J/ψ at forward rapidity, red circles to J/ψ at mid-rapidity. Left: ALICE measurement (top), compared to prediction by PYTHIA6 (bottom) [12]. Right: Prediction by percolation model [13]. The dotted line refers to J/ψ at forward rapidity, the dashed line to J/ψ at mid-rapidity, while the solid line shows the linear behavior.

2. Results

2.1. Inclusive cross section as function of transverse momentum

Figure 1 shows the inclusive J/ψ cross section in pp collisions measured by ALICE as a function of the transverse momentum. On the left, results at $\sqrt{s} = 7\,\text{TeV}$ at forward and mid-rapidity are shown, together with results from other LHC experiments. At forward rapidity the different LHC experiments are in good agreement with each other, at mid-rapidity the ALICE results complement the other experiments, since ALICE is the only experiment that can measure the cross section down to zero transverse momentum.

In Fig. 1 right, results at forward rapidity at $\sqrt{s} = 13\,\text{TeV}$ are shown, together with model predictions. In the transverse momentum range above $5\,\text{GeV}/c$ the prompt contribution to the J/ψ cross section is calculated in the NRQCD framework [7], below $7\,\text{GeV}/c$ the NRQCD model is combined with a description of the incoming protons in the Color Glass Condensate (CGC) model [8]. The non-prompt contribution is calculated in the whole p_{T} range from the FONLL model [9]. The different contributions are then summed to describe the inclusive production. The combination of these models provides a very good description of the measured cross section over the full transverse momentum range. In the overlapping region from $5\,\text{GeV}/c$ to $7\,\text{GeV}/c$ the predictions from the NRQCD model and from the combined approach including CGC are in agreement with each other.

IOP Conf. Series: Journal of Physics: Conf. Series **832** (2017) 012029 doi:10.1088/1742-6596/832/1/012029

2.2. Self-normalized yields as function of self-normalized charged particle density

Figure 2 shows the relative J/ψ yields at forward and at mid-rapidity as a function of the relative charged particle density around mid-rapidity, in pp collisions at $\sqrt{s} = 7\,\text{TeV}$. On the top left plot the ALICE measurement is shown. In both rapidities an approximate linear increase can be seen, with the hint of a stronger than linear increase at mid-rapidity. On the left bottom plot, a prediction by PYTHIA 6.4.25 is shown, using the Perugia 2011 tune, taking into account only J/ψ directly produced in hard scatterings. The model predicts a decrease of the relative J/ψ yield with relative multiplicity, contrary to the observed increase. This difference can be interpreted as an indication for the relevance of other mechanisms for heavy-quark production, such as multi-parton interactions [14].

On the right side of figure 2, a comparison of the data with a percolation model approach is shown. This model assumes a screening of particle multiplicities in regimes of high string densities. Together with the assumption of heavy quark production in initial collisions, a linear increase of relative production with multiplicity is expected at low relative multiplicities, changing into a quadratic behavior at high relative multiplicities, especially at mid-rapidity where higher densities are expected. The model is in good agreement with the data.

3. Conclusions

ALICE measured the p_{T} differential inclusive J/ψ cross section at forward and mid-rapidity in pp collisions at various LHC energies down to zero transverse momentum. At mid-rapidity ALICE's capabilities at low transverse momentum are unique among the LHC experiments. The p_{T} dependent cross section at forward rapidity at $\sqrt{s} = 13\,\text{TeV}$ can be well described with a combination of predictions from the NRQCD model, a CGC description of the incoming protons at low p_{T} and FONLL predictions for the non-prompt contribution to the cross section.

J/ψ production in pp collisions at $\sqrt{s} = 7\,\text{TeV}$ rises linearly with the mid-rapidity charged particle density, both at forward and mid-rapidity, with a hint of a stronger than linear increase at mid-rapidity. This trend hints to an important role of other mechanisms in charmonium production beside hard initial scattering processes, such as multi-parton interactions. The results are also in agreement with the percolation scenario.

References

[1] Matsui T and Satz H 1986 *Phys. Lett.* **B178** 416
[2] Halzen F 1977 *Phys. Lett.* **B69** 105 – 108 ISSN 0370-2693
[3] Einhorn M B and Ellis S D 1975 *Phys. Rev.* **D12**(7) 2007–2014
[4] Bodwin G T, Braaten E and Lepage G P 1995 *Phys. Rev.* **D51** 1125–1171 [Erratum: Phys. Rev. D55,5853(1997)] (*Preprint* hep-ph/9407339)
[5] Aamodt K *et al.* 2011 *Phys. Lett.* **B704** 442 – 455 ISSN 0370-2693
[6] Abelev B B *et al.* (ALICE) 2014 *Eur. Phys. J.* **C74** 2974 (*Preprint* 1403.3648)
[7] Ma Y Q, Wang K and Chao K T 2011 *Phys. Rev. Lett.* **106** 042002 (*Preprint* 1009.3655)
[8] Ma Y Q and Venugopalan R 2014 *Phys. Rev. Lett.* **113** 192301 (*Preprint* 1408.4075)
[9] Cacciari M, Frixione S, Houdeau N, Mangano M L, Nason P and Ridolfi G 2012 *JHEP* **10** 137 (*Preprint* 1205.6344)
[10] Porteboeuf S and Granier de Cassagnac R 2011 *Nucl. Phys. Proc. Suppl.* **214** 181–184 (*Preprint* 1012.0719)
[11] Aamodt K *et al.* (ALICE) 2008 *JINST* **3** S08002
[12] Abelev B *et al.* (ALICE) 2012 *Phys. Lett.* **B712** 165–175 (*Preprint* 1202.2816)
[13] Ferreiro E G and Pajares C 2012 *Phys. Rev.* **C86**(3) 034903
[14] Bartalini P and Fano L (eds) 2010 *Proceedings, 1st International Workshop on Multiple Partonic Interactions at the LHC (MPI08)* (*Preprint* 1003.4220)

IOP Conf. Series: Journal of Physics: Conf. Series **832** (2017) 012030 doi:10.1088/1742-6596/832/1/012030

Performance studies of D-meson tagged jets in pp collisions at $\sqrt{s} = 7$ TeV with ALICE

Salvatore Aiola, for the ALICE Collaboration
Physics Department, Yale University, New Haven, CT 06511
E-mail: salvatore.aiola@yale.edu

Abstract. We present the current status of the measurement of jets that contain a D meson (D-tagged jets) with the ALICE detector. D^0-meson candidates, identified via their hadronic decay into a $K\pi$ pair, were combined with the other charged tracks reconstructed with the central tracking system, using the anti-k_T jet-finding algorithm. The yield of D-tagged jets was extracted through an invariant mass analysis of the D-meson candidates. A Monte Carlo simulation was used to determine the detector performance and validate the signal extraction techniques.

1. Introduction

At hadron colliders, charm quarks are produced as a result of a hard scattering of partons. Like lighter quarks or gluons, charm quarks fragment into collimated sprays of hadrons called *jets*. The charm content of the jet is conserved throughout the fragmentation process, which is dominated by Quantum Chromo-Dynamics (QCD). In the final state, the charm content can be identified by looking for the presence of charmed hadrons among the jet constituents.

The measurement of the charm jet production cross section in pp collisions is a sensitive test of perturbative QCD (pQCD) calculations [1]. Heavy quarks are also an ideal probe of the Quark-Gluon Plasma (QGP) [2, 3] that is created in ultra-relativistic heavy-ion collisions. Hard scattered partons, including heavy quarks, interact with the QGP, which increases their virtuality and interferes with the parton shower (*jet quenching*) [4–6]. At low parton energy, comparable to the charm mass, the charm quark is expected to interact less strongly with the QGP [7], when compared to light quarks and gluons.

The production cross section of charmed hadrons has been measured with good accuracy at the LHC [8–11]. Measuring the kinematic observables of jets with charm content implies integrating out some of the hadronization degrees of freedom. Since hadronization is a highly non-perturbative process, known only with large uncertainties [12], using observables less dependent on this process may improve comparisons with pQCD calculations. Furthermore, the measurement of the fragmentation function (FF) of charmed hadrons can provide important insights into the charm production mechanism [13–15]. ATLAS has measured the FF of $D^{*\pm}$ mesons, observing a large discrepancy with Monte Carlo event generators [16]. ALICE has the potential to extend this measurement to low z ($0.1 < z < 0.3$), where z is the fraction of jet momentum carried by the D meson.

IOP Conf. Series: Journal of Physics: Conf. Series **832** (2017) 012030 doi:10.1088/1742-6596/832/1/012030

2. The ALICE Experiment

ALICE is the experiment dedicated to the study of heavy-ion collisions at the LHC. The central barrel detectors ($|\eta| \lesssim 1$) are located inside a large solenoid magnet, providing a field $B = 0.5$ T. The *Inner Tracking System* (ITS) is a six-layer silicon detector that allows a precise determination of the primary vertex and of displaced secondary vertices of weak decays. The main tracking detector is the *Time Projection Chamber* (TPC), which, combined with the ITS, allows reconstruction of tracks from low ($p_{\mathrm{T}} \approx 0.15$ GeV/c) to high transverse momentum ($p_{\mathrm{T}} \approx 100$ GeV/c) with good momentum resolution ($1-3\%$) and tracking efficiency ($60-80\%$). Several detectors contribute to the Particle Identification (PID) capabilities of ALICE. In this analysis, we use the $\mathrm{d}E/\mathrm{d}x$ measured by the TPC and the velocity β measured by the *Time Of Flight* detector, consisting of an array of Multigap Resistive Plate Chambers. A full description of the ALICE detector and of its performance during LHC Run-1 is available at Ref. [17].

3. Analysis procedures

The analysis relies on the well-established D-meson reconstruction techniques [8, 18], as well as jet reconstruction methods [6, 19, 20], both developed by the ALICE Collaboration during Run-1.

For this study, only charged tracks were used to reconstructed the jets (*charged jets*). Track quality cuts were applied to ensure good momentum resolution. The tightest cuts, which ensure the best possible momentum resolution at the expenses of tracking efficiency, were applied to the tracks used to identify the decay products of the D mesons. In particular, at least one space point was required in one of the two layers of the ITS closest to the beam pipe. For jet reconstruction, this last requirement was lifted to achieve a more uniform azimuthal efficiency.

D^0 mesons ($m = 1.865$ GeV/c^2, $c\tau = 123\,\mu$m) and their charge conjugates were used to tag jets with charm content. They were reconstructed via their hadronic decay: $\mathrm{D}^0 \rightarrow \pi^+\mathrm{K}^-$(BR $= 3.88\%$) [21]. The topological cuts select unlike-sign (US) pairs that form a secondary vertex displaced from the reconstructed primary vertex. PID on the D^0 candidate daughters was used to reject pairs not compatible with the πK hypothesis. The four-momenta of each D^0 candidate and all reconstructed tracks (excluding the two daughters of the D^0) were used as input to the anti-k_{T} jet-finding algorithm [22] with $R = 0.4$. In view of replicating this analysis in background-rich environments, such as heavy-ion or high pile-up pp collisions, the anti-k_{T} algorithm was chosen for its good soft-resilient properties [23]. Only jets containing the D^0 candidates were retained.

In order to extract the signal out of the combinatorial background, three different invariant mass analysis techniques were developed and their performance compared.

Invariant mass fit The invariant mass distribution of the D^0 candidates was constructed for various intervals of their associated-jet transverse momentum ($p_{\mathrm{T,ch\,jet}}$); each candidate was given a weight corresponding to the inverse of its reconstruction efficiency $\epsilon(p_{\mathrm{T,D}})$. The invariant mass distribution was fit using the sum of an exponential function (background) and a Gaussian (signal). The yield $N^{\mathrm{D^0\text{-}jet}}(p_{\mathrm{T,ch\,jet}})$ was extracted from the fit parameters.

Side-band subtraction For each $p_{\mathrm{T,D}}$ interval, the yield $N^{\mathrm{D^0\text{-}jet}}(p_{\mathrm{T,ch\,jet}}, p_{\mathrm{T,D}})$ was extracted by subtracting the $p_{\mathrm{T,ch\,jet}}$ distribution in the side bands of the invariant mass distribution ($4\sigma_{\mathrm{fit}} < |m - m_{\mathrm{fit}}| < 8\sigma_{\mathrm{fit}}$) from the $p_{\mathrm{T,ch\,jet}}$ distribution in the peak area ($|m - m_{\mathrm{fit}}| < 2\sigma_{\mathrm{fit}}$); the peak position m_{fit}, width σ_{fit} and side-band normalization factor were extracted by fitting the invariant mass distribution with an exponential + Gaussian function; the $p_{\mathrm{T,D}}$ bins were weighted by the inverse of the D^0-meson reconstruction efficiency $\epsilon(p_{\mathrm{T,D}})$ and summed over:

$$N^{\mathrm{D^0\text{-}jet}}(p_{\mathrm{T,ch\,jet}}) = \sum_{p_{\mathrm{T,D}}} \frac{1}{\epsilon(p_{\mathrm{T,D}})} \left[N_{\mathrm{peak}}^{\mathrm{D^0\text{-}jet}}(p_{\mathrm{T,ch\,jet}}, p_{\mathrm{T,D}}) - \frac{B_{\mathrm{peak}}^{\mathrm{fit}}}{B_{\mathrm{SB}}} N_{\mathrm{SB}}^{\mathrm{D^0\text{-}jet}}(p_{\mathrm{T,ch\,jet}}, p_{\mathrm{T,D}}) \right],$$

where $B_{\mathrm{peak}}^{\mathrm{fit}}$ and B_{SB} are respectively the total background in the peak area estimated by fitting the invariant mass distribution and the total background in the side bands.

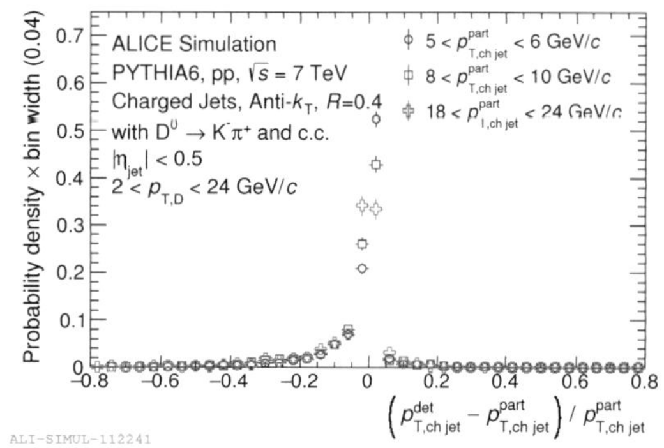

Figure 1. Probability density distribution of the jet momentum shift in $p_{\rm T,ch\,jet}$ intervals.

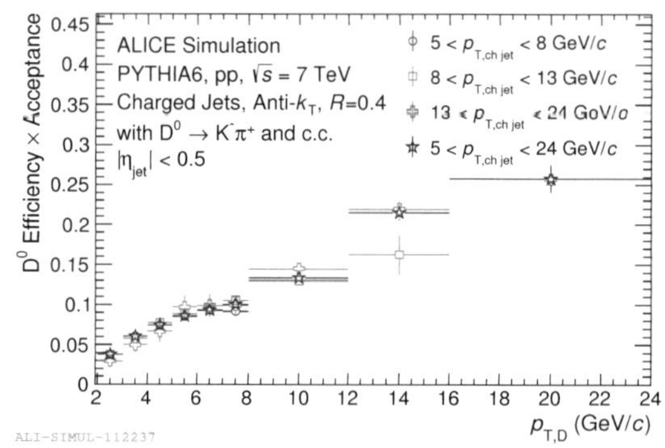

Figure 2. Efficiency × Acceptance of D^0 mesons vs. $p_{\rm T,D}$ in $p_{\rm T,ch\,jet}$ intervals.

Like-sign subtraction This method is analogous to the side-band method. In this case, the background $p_{\rm T,ch\,jet}$ distribution was provided by the like-sign (LS) πK pairs and subtracted from the corresponding unlike-sign (US) pairs:

$$N^{D^0\text{-jet}}(p_{\rm T,ch\,jet}) = \sum_{p_{\rm T,D}} \frac{1}{\epsilon(p_{\rm T,D})} \left[N^{D^0\text{-jet}}_{\rm US,peak}(p_{\rm T,ch\,jet}, p_{\rm T,D}) - \frac{B_{\rm US,SB}}{B_{\rm LS,SB}} N^{D^0\text{-jet}}_{\rm LS,peak}(p_{\rm T,ch\,jet}, p_{\rm T,D}) \right],$$

where $B_{\rm US,SB}$ and $B_{\rm LS,SB}$ are respectively the integral of the side bands in the US and in the LS invariant mass distributions.

Monte Carlo (MC) simulations were used to determine the detector performance and validate the analysis techniques. PYTHIA6 (6.4.25) [24] (tune Perugia-2011) was used to provide pp collision events at $\sqrt{s} = 7$ TeV. Particles produced by PYTHIA were propagated through the detector using the GEANT3 transport code [25].

4. Results

In order to determine the detector performance, reconstructed detector-level D-tagged jets were matched with their corresponding counterparts at generator-level. The variable $\delta_{p_{\rm T,ch\,jet}} = \left(p^{\rm det}_{\rm T,ch\,jet} - p^{\rm part}_{\rm T,ch\,jet}\right)/p^{\rm part}_{\rm T,ch\,jet}$, was computed for each matched pair, where $p^{\rm det}_{\rm T,ch\,jet}$ and $p^{\rm part}_{\rm T,ch\,jet}$ are the transverse momenta of the D-tagged jet at detector-level and at generator-level, respectively. Its probability density distribution is shown in Fig. 1 for three $p^{\rm part}_{\rm T,ch\,jet}$ intervals. No significant dependence on $p^{\rm part}_{\rm T,ch\,jet}$ was observed. The shape of the distribution features a sharp peak at zero and is skewed towards negative values, due to tracking inefficiency (higher probability of reconstructing smaller jet momenta than the generated ones). The jet momentum resolution (standard deviation of $\delta_{p_{\rm T,ch\,jet}}$) for D^0-tagged jets is approximately 11%, slightly smaller with respect to its value for inclusive jets [20]. The mean jet momentum shift is approximately -3%. The reconstruction efficiency is calculated as the ratio of the yield of reconstructed D-tagged jets over all generated D-tagged jets, as a function of generator-level observables. It is shown in Fig. 2 as a function of $p_{\rm T,D}$ for different $p_{\rm T,ch\,jet}$ intervals; it shows a strong dependence on $p_{\rm T,D}$, mainly due to varying topological cuts. No significant dependence on $p_{\rm T,ch\,jet}$ is observed in the interval $5 < p_{\rm T,ch\,jet} < 24$ GeV/c. Figure 3 shows the D-tagged jet yields as a function of $p^{\rm det}_{\rm T,ch\,jet}$ obtained using the invariant mass fit, side-band and like-sign methods, compared to the true D-tagged jet spectrum. All signal extraction methods perform well and do not show significant biases, beyond the statistical uncertainties assigned to each.

IOP Conf. Series: Journal of Physics: Conf. Series **832** (2017) 012030 doi:10.1088/1742-6596/832/1/012030

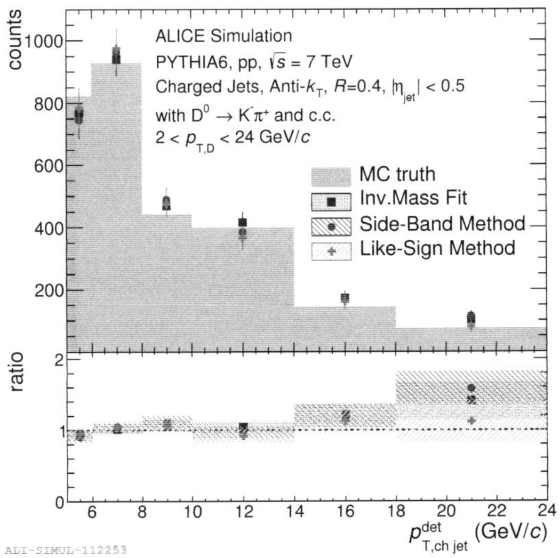

Figure 3. D^0-jet signal yield extracted using the invariant mass fit (blue squares), side-band method (red circles) and like-sign method (green triangles) compared with the true D-tagged jet spectrum (gray filled area). The bottom panel shows the ratio to the MC truth. The yields are corrected for the reconstruction efficiency but not for distortions due to detector jet-momentum resolution (10 – 15% on the yields).

5. Conclusions

ALICE has the potential for measuring jets with charm content tagged using reconstructed D mesons. The ALICE detector performance for D^0-tagged jets was assessed using MC simulations for pp collisions at $\sqrt{s} = 7$ TeV. The D^0-jet momentum resolution was determined to be about 11%; the reconstruction efficiency, independent of $p_{T,\mathrm{ch\,jet}}$, ranges from 5% to 25% as a function of $p_{T,D}$. Three methods were implemented to extract the signal: invariant mass fit, side-band and like-sign subtraction. The three methods were validated using a MC simulation and do not show biases larger than the statistical uncertainties.

References

[1] Cacciari M *et al.* 2012 *JHEP* **2012** 1–24 (arXiv:1205.6344)
[2] Adams J *et al.* (STAR Collaboration) 2005 *Nuclear Physics* A **757** 102 – 183 (arXiv:nucl-ex/0501009)
[3] Adcox K *et al.* (PHENIX Collaboration) 2005 *Nuclear Physics* A **757** 184 – 283 (arXiv:nucl-ex/0410003)
[4] Adare A *et al.* (PHENIX Collaboration) 2008 *Phys. Rev. Lett.* **101** 232301 (arXiv:0801.4020)
[5] Chatrchyan S *et al.* (CMS Collaboration) 2012 *Phys. Lett.* B **712** 176–197 (arXiv:1202.5022)
[6] Adam J *et al.* (ALICE Collaboration) 2015 *Phys. Lett.* B **746** 1 – 14 (arXiv:1502.01689)
[7] Dokshitzer Y and Kharzeev D 2001 *Phys. Lett.* B **519** 199 – 206
[8] Abelev B *et al.* (ALICE Collaboration) 2012 *JHEP* **2012** 1–30 (arXiv:1111.1553)
[9] Aaij R *et al.* (LHCb Collaboration) 2013 *Nuclear Physics* B **871** 1 – 20 (arXiv:1302.2864)
[10] Aad G *et al.* (ATLAS Collaboration) 2016 *Nuclear Physics* B **907** 717 – 763 (arXiv:1512.02913)
[11] Adam J *et al.* (ALICE Collaboration) 2016 *JHEP* **2016** 1–43 (arXiv:arXiv:1509.06888)
[12] d'Enterria D, Eskola K J, Helenius I and Paukkunen H 2014 *Nuclear Physics* B **883** 615
[13] Abe F *et al.* (CDF Collaboration) 1990 *Phys. Rev. Lett.* **64**(4) 348–352
[14] Albajar C *et al.* (UA1 Collaboration) 1990 *Phys. Lett.* B **244** 566
[15] Abelev B I *et al.* (STAR Collaboration) 2009 *Phys. Rev.* D **79**(11) 112006 (arXiv:0901.0740)
[16] Aad G *et al.* (ATLAS Collaboration) 2012 *Phys. Rev.* D **85**(5) 052005 (arXiv:1112.4432)
[17] Abelev B *et al.* (ALICE Collaboration) 2014 *Int'l Journ. of Modern Phys.* A **29** 1430044 (arXiv:1402.4476)
[18] Adam J *et al.* (ALICE Collaboration) 2016 (arXiv:1605.07569)
[19] Abelev B *et al.* (ALICE Collaboration) 2013 *Phys. Lett.* B **722** 262 (arXiv:1301.3475)
[20] Abelev B *et al.* (ALICE Collaboration) 2015 *Phys. Rev.* D **91**(11) 112012 (arXiv:1411.4969)
[21] Patrignani C *et al.* (Particle Data Group Collaboration) 2016 *Chin. Phys.* C **40** 100001
[22] Cacciari M, Salam G and Soyez G 2008 *JHEP* **04** 063
[23] Cacciari M, Rojo J, Salam G and Soyez G 2011 *The European Physical Journal* C **71** 1–21 (arXiv:1010.1759)
[24] Sjöstrand T, Mrenna S and Skands P 2006 *JHEP* **5** 1 (arXiv:hep-ph/0603175)
[25] Brun R *et al.* 1978 Tech. rep. CERN URL `https://cds.cern.ch/record/118715`

IOP Conf. Series: Journal of Physics: Conf. Series **832** (2017) 012031 doi:10.1088/1742-6596/832/1/012031

Charmonium production in pp, pPb and PbPb collisions with CMS

Andre Govinda Ståhl on behalf of the CMS collaboration
Laboratoire Leprince-Ringuet, École polytechnique, F91128 Palaiseau Cedex, France
E-mail: andre.stahl@cern.ch

Abstract.
The LHC Run 1 results of the analysis of charmonium production in pp, pPb and PbPb collisions with the CMS experiment are reported. The coherent J/ψ photoproduction cross section is measured as a function of rapidity in ultra-peripheral PbPb collisions at 2.76 TeV. The forward-backward ratio of prompt J/ψ yields in pPb collisions at 5.02 TeV is presented as a function of the event activity and p_T. The nuclear modification factor of prompt J/ψ in PbPb collisions at 2.76 TeV is shown as a function of rapidity, centrality and p_T. Finally, the ratio of $\psi(2S)$ to J/ψ yields in PbPb collisions with respect to pp collisions at 2.76 TeV is analysed in different rapidity and centrality bins.

1. Introduction

Charmonium states constitute an important probe of the Quark-Gluon Plasma (QGP), since they are produced in the early stage of heavy ion collisions and their yields are sensitive to the evolution of the medium. It has been predicted that the charmonium production can be suppress in QGP due to the Debye screening of the strong interaction between quarks [1]. Since the binding energy of charmonia decreases for higher excited states, it is expected that charmonium states dissociate sequentially in the QGP [1]. Moreover, a large amount of charm quarks are produced in PbPb collisions at LHC energies, and it becomes possible that uncorrelated charm and anticharm quarks recombine to form charmonium mesons during hadronization, enhancing their measured yields compared to those in proton-proton (pp) collisions [3]. Apart from the presence of QGP, the production of charmonium states can be altered by nuclear matter effects such as parton energy loss due to multiple scattering in the nucleus [2], or nuclear modification of the parton distribution functions (PDF) [4]. These nuclear matter effects can be studied more precisely in proton-lead (pPb) collisions and ultra-peripheral PbPb collisions (UPC), where the formation of QGP is not expected.

2. Analysis procedure and results

2.1. J/ψ in Ultra Peripheral PbPb collisions

The gluon nuclear PDF can be probed, in the range of Bjorken-x values from 10^{-5} to 10^{-2}, by photon-nucleus interactions in ultra-peripheral PbPb collisions at 2.76 TeV [6]. The photon-induced reactions can be classified as coherent if the photon interacts with the whole nucleus, or incoherent if the photon interacts with a single nucleon. Coherent and incoherent photoproduction can be accompanied by the emission of forward neutrons produced by the strong electromagnetic fields present in UPC events and the break up of the nucleus [6]. The UPC events were selected by requiring low activity in the hadron forward calorimeter, the presence of at least one neutron in the zero degree calorimeters, and two muon tracks within the invariant mass region $2.6 < m_{\mu\mu} < 3.5$ GeV/c^2. In order to extract the coherent signal contribution from background ($\gamma\gamma \to \mu^+\mu^-$ interactions and incoherent J/ψ production), the dimuon invariant mass and p_T spectra were simultaneously fitted using an unbinned maximum likelihood fit.

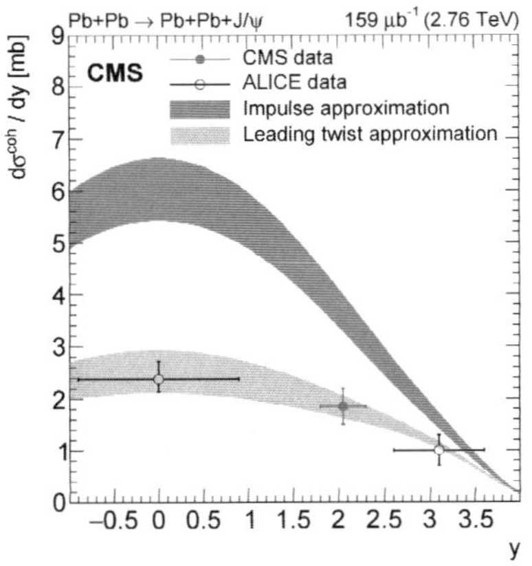

Figure 1: Differential cross section as a function of rapidity for coherent J/ψ photoproduction measured by CMS [6] (red) and ALICE [7,8] (black), in ultra-peripheral PbPb collisions at 2.76 TeV [6]. The horizontal bars represent the size of the measured rapidity bins, and the vertical error bars display the statistical and systematic uncertainties. The impulse approximation (blue) and the leading twist approximation predictions (yellow) are also included.

The CMS cross section measurement for coherent J/ψ photoproduction using 159 μb^{-1} PbPb data [6], is compared in Fig. 1 to the ALICE measurements [7,8], and to theoretical predictions. The impulse approximation model neglects all nuclear effects by estimating the coherent photoproduction in $\gamma+Pb$ collisions using data from $\gamma+p$ collisions [9], while the leading twist approximation model describes an effective nuclear gluon shadowing by implementing a gluon recombination mechanism at the partonic level using as input a diffractive proton PDF [10]. Experimental results from CMS and ALICE show a continuous decrease with rapidity, favouring the leading twist approximation.

2.2. J/ψ in pPb collisions

The prompt J/ψ production was measured in pPb collisions at 5.02 TeV [12]. Since B-mesons traverse a measurable pathlength before decaying, the prompt and non-prompt J/ψ can be separated by the pseudo-proper decay length, $l_{J/\psi} = L_{xy} m_{J/}/p_T$, where L_{xy} is the transverse distance between the primary vertex and the $\mu^+\mu^-$ vertex computed in the laboratory frame. The prompt and non-prompt J/ψ were extracted by performing an extended unbinned maximum likelihood fit simultaneously to the invariant-mass and $l_{J/\psi}$ distributions of dimuons [12]. The forward-to-backward ratio (R_{FB}) of prompt J/ψ cross sections, where the forward (backward) region is defined by the p-going (Pb-going) direction, is used to analyse the p_T and rapidity dependence of nuclear effects.

In Figure 2, the p_T dependence of the prompt $J\psi$ R_{FB} is compared, in three different rapidity regions, to the next-to-leading order (NLO) EPS09 shadowing calculation based on the Color Evaporation Model (CEM) [13]. The results, using 34.6 nb^{-1} pPb data, shows R_{FB} values approaching unity at high p_T, but lower than the model calculations at low p_T suggesting the presence of other nuclear effects apart from just nPDF modifications. Moreover, the transverse energy deposited in the forward hadronic calorimeter at $4 < |\eta| < 5.2$ is used to determine the dependence of the R_{FB} with the event activity. Fig. 3 shows a decreasing pattern of the R_{FB} with increasing E_T over all p_T and rapidity intervals.

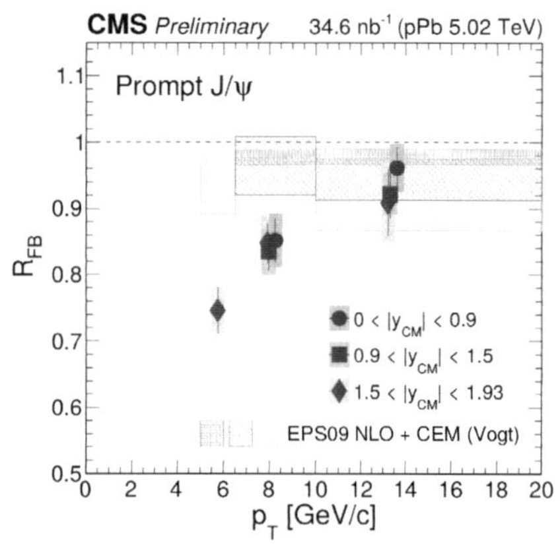

Figure 2: Prompt J/ψ R_{FB} versus p_T in three rapidity ranges [12]. The statistical (systematic) uncertainties are represented by the vertical bars (shaded boxes). The EPS09 (NLO) shadowing calculation is also shown.

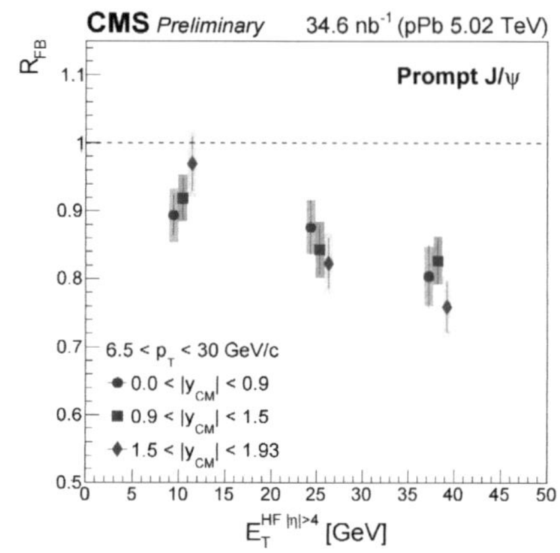

Figure 3: Prompt J/ψ R_{FB} versus E_T in three rapidity ranges [12]. The statistical (systematic) uncertainties are represented by the vertical bars (shaded boxes). Data points are slightly shifted to avoid overlaps between each other.

2.3. J/ψ and $\psi(2S)$ in PbPb collisions

The enhancement or suppression of charmonia due to QGP can be quantified with the nuclear modification factor (R_{AA}), which represents the ratio of yields in heavy ion collisions with respect to pp collisions scaled by the number of binary nucleon-nucleon interactions. The R_{AA} of prompt J/ψ was measured using pp and PbPb data at 2.76 TeV [11]. The prompt J/ψ R_{AA} distributions versus rapidity, centrality and p_T, are shown in Fig. 4. A strong suppression of prompt J/ψ yields in PbPb collisions compared to pp collisions, with no significant dependence on p_T and rapidity, is observed for more central events, suggesting the presence of hot medium effects.

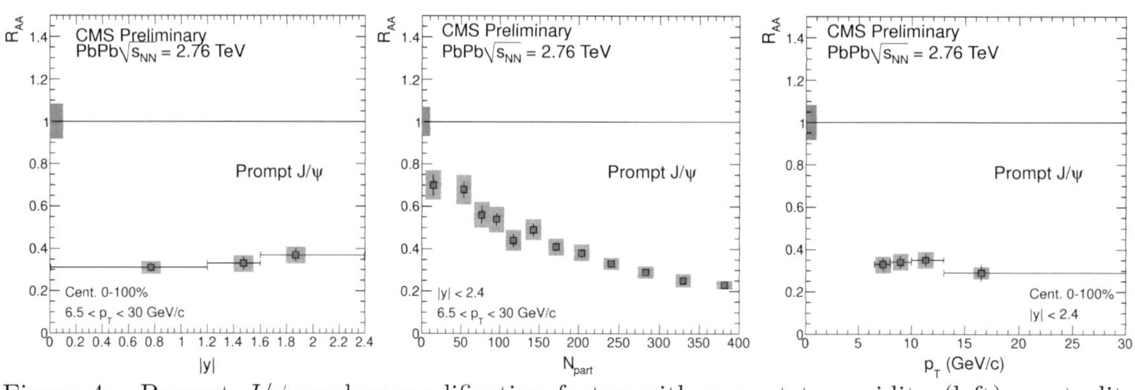

Figure 4: Prompt J/ψ nuclear modification factor with respect to rapidity (left), centrality (middle), and p_T (right) in PbPb collisions at 2.76 TeV [11]. The horizontal lines display the range of the measured bins and the gray boxes represent the magnitude of the global uncertainties.

The variation between the production of $\psi(2S)$ and J/ψ in PbPb compared to pp collisions is determined by the double ratio $(N_{\psi(2S)}/N_{J/\psi})_{PbPb}/(N_{\psi(2S)}/N_{J/\psi})_{pp}$. This double ratio was measured by CMS using PbPb and pp data at 2.76 [14]. The centrality dependence of the double ratio is presented in Fig. 5. At mid rapidity and high p_T ($|y| < 1.6$ and $6.5 < p_T < 30$ GeV/c), the $\psi(2S)$ production is more suppressed than J/ψ production in all measured centrality bins. Moreover, the opposite trend is observed at forward rapidity and intermediate p_T ($1.6 < |y| < 2.4$ and $3 < p_T < 30$ GeV/c), where the $\psi(2S)$ production is less suppressed than J/ψ in central PbPb collisions, implying the existence of other medium effects apart from sequential melting.

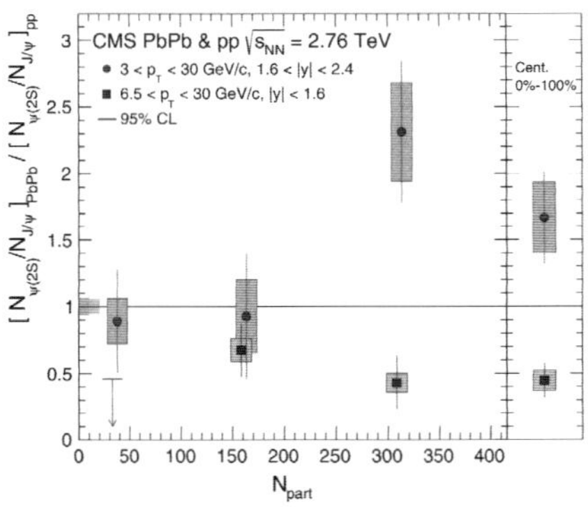

Figure 5: Double ratio of $\psi(2S)$ over J/ψ yields as a function of centrality at forward rapidity (red) and mid rapidity (blue) [14]. The pp uncertainties are shown by boxes at unity while the statistical uncertainties are displayed by bars. The centrality-integrated results are presented in the right panel.

3. Summary

The coherent J/ψ photoproduction cross section in PbPb UPC is in good agreement with the leading twist approximation, indicating that shadowing effects are needed to describe the data. The R_{FB} of J/ψ in pPb collisions is consistent with unity at high p_T and decreases at low p_T, suggesting the existence of additional nuclear effects apart from just nPDF modifications. The prompt J/ψ R_{AA} is shown to be lower than unity and has a decreasing centrality dependence, expected from QGP. The double ratio of prompt $\psi(2S)$ to J/ψ yields is observed to be above unity for central PbPb collisions in the low p_T and forward rapidity region, implying the presence of physics processes beyond the sequential melting scenario.

References

[1] Matsui T and Satz H 1986 *Phys. Lett.* **B178** 416–422
[2] Peigne S and Peshier A 2008 *Phys. Rev.* **D77** 114017 (*Preprint* 0802.4364)
[3] Zhao X and Rapp R 2011 *Nucl. Phys.* **A859** 114–125 (*Preprint* 1102.2194)
[4] Andronic A *et al.* 2016 *Eur. Phys. J.* **C76** 107 (*Preprint* 1506.03981)
[5] Chatrchyan S *et al.* (CMS) 2008 *JINST* **3** S08004
[6] Khachatryan V *et al.* (CMS) 2016 *Submitted to: Phys. Lett. B* (*Preprint* 1605.06966)
[7] Abelev B, Adam J and more 2013 *Physics Letters B* **718** 1273 – 1283 ISSN 0370-2693 URL http://www.sciencedirect.com/science/article/pii/S0370269312012257
[8] Abbas E, Abelev B and etal 2013 *The European Physical Journal C* **73** 2617 ISSN 1434-6052 URL http://dx.doi.org/10.1140/epjc/s10052-013-2617-1
[9] Guzey V, Strikman M and Zhalov M 2014 *Eur. Phys. J.* **C74** 2942 (*Preprint* 1312.6486)
[10] Guzey V, Kryshen E, Strikman M and Zhalov M 2013 *Phys. Lett.* **B726** 290–295 (*Preprint* 1305.1724)
[11] 2012 J/psi results from CMS in PbPb collisions, with 150mub-1 data Tech. Rep. CMS-PAS-HIN-12-014 CERN Geneva URL https://cds.cern.ch/record/1472735
[12] 2015 Jpsi production in pPb collisions Tech. Rep. CMS-PAS-HIN-14-009 CERN Geneva URL http://cds.cern.ch/record/2044699
[13] Vogt R 2015 *Phys. Rev. C* **92**(3) 034909 URL http://link.aps.org/doi/10.1103/PhysRevC.92.034909
[14] Khachatryan V *et al.* (CMS) 2014 *Phys. Rev. Lett.* **113** 262301 (*Preprint* 1410.1804)

IOP Conf. Series: Journal of Physics: Conf. Series **832** (2017) 012032 doi:10.1088/1742-6596/832/1/012032

Measurements of Open Heavy Flavor Production in Semi-leptonic Decay Channels at the STAR experiment

Shenghui Zhang

University of Science and Technology of China (USTC) & University of Illinois at Chicago (UIC)

E-mail: zhang08@mail.ustc.edu.cn

Abstract. In these proceedings, we will present the new results of the Non-Photonic Electron (NPE) production from semi-leptonic decays of open heavy flavor hadrons at the STAR experiment. Firstly we will report the updated results on NPE production in p+p collisions at $\sqrt{s}$ = 200 GeV with much improved precision and wider kinematic coverage than previous measurements. Secondly we will report measurements of the nuclear modification factor, R_{AA}, for NPE production in Au+Au collisions at $\sqrt{s_{NN}}$ = 200 GeV with this new p+p reference. The NPE R_{AA} shows large suppression at high transverse momenta (p_T) in central collisions, which reduces gradually towards more peripheral collisions, and an enhancement at low p_T with large systematic uncertainties from the p+p reference. We will compare NPE R_{AA} to the R_{AA} of charm decayed electrons, D^0 mesons and light hadrons in Au+Au collisions as well as to NPE and D^0 mesons in central U+U collisions and show that they are all consistent within the uncertainties. Finally we will report the results on measurements of NPE from open bottom hadron decays and discuss the prospect of measuring NPE from open bottom and charm hadron decays separately utilizing the Heavy Flavor Tracker.

1. Introduction

Heavy quarks are predominantly produced in hard scatterings at early stages of high-energy heavy-ion collisions due to their large masses. Studies of interactions between heavy quarks and the Quark-Gluon Plasma (QGP) in various collision goemetries, especially for charm and bottom quarks separately, can provide new insights to the properties of QGP. The heavy quark production in p+p collisions provides a baseline to similar measurements in heavy-ion collisions and is expected to be well described by perturbative Quantum Chromodynamics (pQCD) calculations. Thus measurements of heavy quark production in both p+p and Au+Au collisions are crucial. Heavy quark production can be studied via measuring the electrons from semi-leptonic decays of heavy flavor hadrons, also known as Non-Photonic Electrons (NPE). In these proceedings, we focus on the measurements of NPE in p+p and Au+Au collisions at $\sqrt{s_{NN}}$ = 200 GeV at the STAR experiment.

2. Analysis

2.1. *Track selection and electron identification*

STAR uses three main detectors, the Time Projection Chamber (TPC), the Time Of Flight (TOF) detector and the Barrel Electromagnetic Calorimeter (BEMC), to reconstruct charged tracks and perform Paticle IDentification (PID). The TPC, covering full azimuth within pseudo-rapidity range of $|\eta| < 1$, provides tracking, momentum determination and PID via measuring ionization energy loss. The BEMC, covering $|\eta| < 1$ and full azimuth, can trigger on and identify high-p_T eletrons via the ratio of energy to momentum measurements. The TOF covers $|\eta| < 1$ within full azimuth and provides PID capability through measured time-of-flight. Electrons are identified using the TPC combined with the TOF at low p_T and the BEMC at high p_T.

2.2. *NPE cross-section extraction*

The NPE raw yield is extracted using the equation $N_{NPE} = N_{inc} * purity - N_{phe}/\epsilon_{phe}$, where N_{inc} is the number of inclusive electron candidates, $purity$ the fraction of real electrons in the inclusive electron sample, N_{phe} the number of photonic electron candidates obtained via selecting the invariant mass of e^+e^- pairs, and ϵ_{phe} the photonic electron reconstruction efficiency. There are primarily two types of photonic electron background: gamma conversion, π^0 and η Dalitz decays. The final yields of NPE need to be corrected for the detector acceptance and efficiency.

3. NPE measurement in p+p and Au+Au collisions at $\sqrt{s_{NN}}$ = 200 GeV

3.1. *NPE p_T spectrum*

The measured NPE cross-section for p+p collisions at $\sqrt{s}$ = 200 GeV from 2012 data (red points) is shown in the upper panel of Fig. 1, along with the previous STAR result from 2005 and 2008 data (black points) [1] and the FONLL (Fixed-Order Next-to-Leading Logarithm) pQCD calculation (blue curves) [2]. The lower panel shows the ratio of data to FONLL calculation. The new measurement extends to both the lower and higher p_T regions compared to the previous STAR measurement with greatly reduced uncertainty and is consistent with the FONLL prediction. Due to the large uncertainties in the gluon density function and the dramatic increase of strong coupling constant towards low p_T, pQCD calculations have little predictive power for the charm cross-section at low p_T [3].

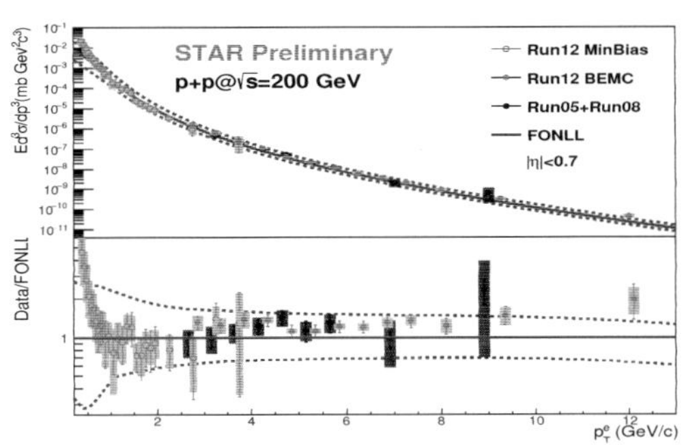

Figure 1: *Top* : The NPE cross-section in p+p collisions. *Bottom* : Ratio of data to the pQCD calculation.

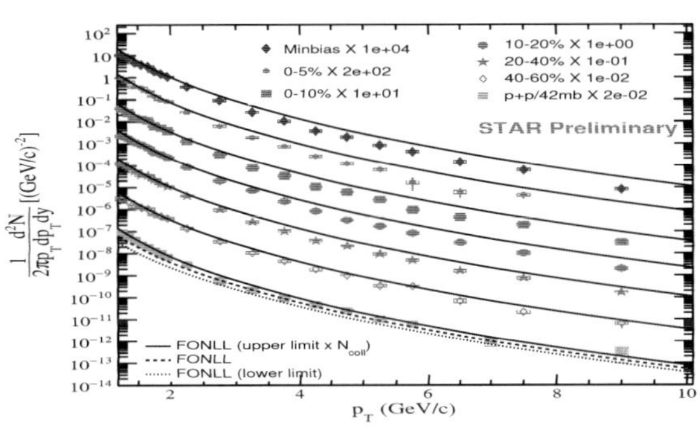

Figure 2: The NPE invariant yields in different centralites for Au+Au collisions.

The measured NPE invariant yields for Au+Au collisions at $\sqrt{s_{NN}}$ = 200 GeV from 2010 data (red points) are shown in Fig. 2 for different centralities, along with a comparison to the FONLL pQCD calculation scaled by the number of nucleon-nucleon collisions (N_{coll}) (black curves). In central collisions, there are significant differences between Au+Au measurements and the scaled FONLL calculation, indicating existence of hot medium effects. From central to peripheral collisions, the difference is getting smaller, which is consistent with the expectation of less QGP effects in peripheral collisions.

3.2. NPE R_{AA}

The R_{AA} is obtained by taking the ratio of the NPE yield in Au+Au collisions to that in p+p collisions normalized by N_{coll}. In Fig. 3, there are four panels showing R_{AA} in different centralites. We observe an enhancement at low p_T across all centralities, with large systematic uncertainties from p+p reference, and strong suppression at high p_T in central collisions, which reduces gradually towards peripheral collisions. Different model calculations are compared with our data in the 0-10% centrality. The DGLV model with only radiative energy loss [4] underestimates the suppression of NPE. With the addition of the collisional energy loss, the model calculation agrees with our data. The other models, Collisional dissociation [6], Min He et al. [7, 8], and Gossiaux et al. [5, 9, 10] have some difficulties to describe the data. In low p_T, NPE R_{AA} is consistent with R_{AA} for charm decayed electrons ($c \to e$) (blue line) extracted via decaying D^0 mesons [11] according to the measured p_T spectrum which is shown in upper panel of Fig. 5 (black points) into electrons through PYTHIA. As we can see from upper panels of Fig. 5, bottom decayed electrons ($b \to e$) have larger contribution to NPE when $p_T > 3$ GeV/c. So NPE R_{AA} is not consistent with R_{AA} for $c \to e$ in high p_T.

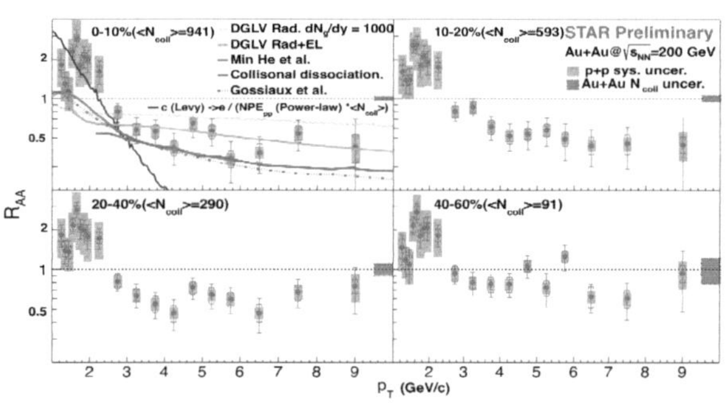

Figure 3: NPE R_{AA} as a function of p_T in different centralities of Au+Au collisions.

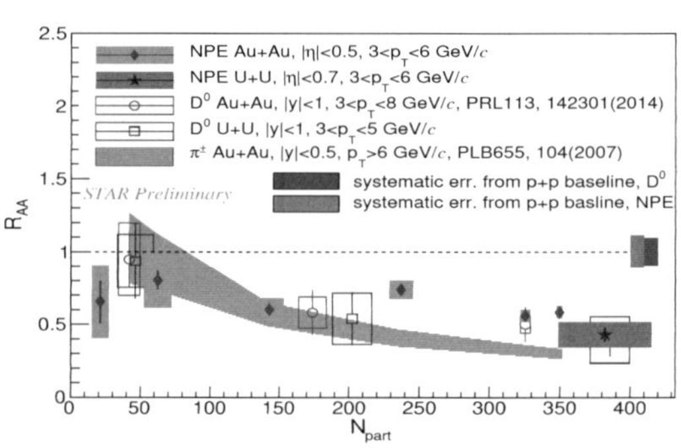

Figure 4: R_{AA} as a function of the number of participating nucleons (N_{part}).

Figure 4 shows R_{AA} as a function of the number of participating nucleons (N_{part}). We compare NPE R_{AA} with D^0 and light hadron R_{AA} [12] in different heavy-ion collision systems and find that the suppression of NPE at high p_T in central Au+Au collisions is similar to those of D^0 mesons and light hadrons in Au+Au collisions as well as NPE and D^0 mesons in central U+U collisions.

3.3. Bottom decayed electron R_{AA}

The invariant yields and R_{AA} of $b \to e$ are shown in upper and lower panels of Fig. 5. The equation $N_{b \to e} = N_{NPE} - N_{c \to e}$ is used to extract the invariant yields of $b \to e$ (blue lines), where $N_{c \to e}$ is the number of $c \to e$ (red lines), and N_{NPE} the number of measured NPE (red points). Using the $b \to e$ FONLL calculation scaled by N_{coll} as the reference, we can obtain the R_{AA} of $b \to e$. The difference from Levy and Power-law function fits to the D^0 p_T spectrum is taken as the uncertainty. In peripheral collisions, $b \to e$ R_{AA} is consistent with no suppression. In 0-80% and 0-10% central collisions, $b \to e$ R_{AA} shows an indication of suppression and is similar to D^0 R_{AA} within large uncertainties.

The Heavy Flavor Tracker (HFT) has been installed into STAR and taken data since 2014. It provides excellent track impact parameter resolution [13], enabling separation of NPE from D and B mesons via their different decay lengths.

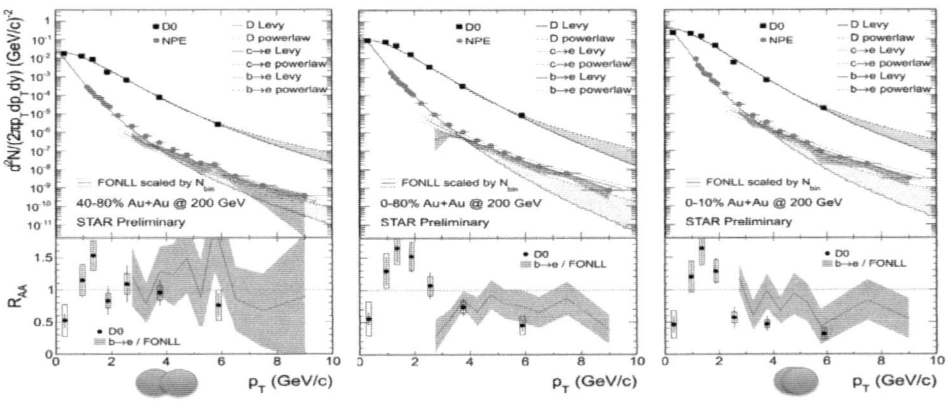

Figure 5: The invariant yields and R_{AA} of bottom decayed electrons.

4. Summary and Outlook

In these proceedings we present the NPE production cross-section in p+p collisions at $\sqrt{s}$ = 200 GeV, which is measured over a broad p_T range 0.3-12 GeV/c with significantly improved precision than previous measurements. The new result confirms the pQCD calculation and provides further constraints on theoretical calculations. Using the new p+p reference, the NPE R_{AA} in Au+Au collisions at $\sqrt{s_{NN}}$ = 200 GeV is presented. We observe large suppression at high p_T in central collisions, which is consistent with substantial energy loss of heavy quarks in dense matter, and an enhancement at low p_T, which is consistent with charm decayed electron R_{AA}, suggesting recombination of charm quarks with light quarks in the medium of strong radial flow. Finally, the perspective for the separation of charm and bottom contributions to NPE in Au+Au collisions with HFT data is given.

5. Acknowledgments

We express gratitude to Heavy Ion Nuclear Physics groups at USTC and UIC for their support. This work was supported in part by the Major State Basic Research Development Program in China with grant no. 2014CB845402 and the National Natural Science Foundation of China with grant no. 11375184.

References

[1] A. et al. (STAR Collaboration), Phys. Rev. D 83 (2011) 052006.
[2] R. E. Nelson, R. Vogt, A. D. Frawley, Phys. Rev. C 87 (2013) 014908.
[3] R. Vogt, Eur. Phys. J. Special Topics 155, 213 (2008).
[4] M. Djordjevic, M. Gyulassy, R. Vogt, S. Wicks, Physics Letters B 632 (1) (2006) 81 86.
[5] P. B. Gossiaux, J. Aichelin, T. Gousset, V. Guiho, Journal of Physics G: Nuclear and Particle Physics 37 (9) (2010) 094019.
[6] R. Sharma, I. Vitev, B. -W. Zhang, Phys. Rev. C 80 (2009) 054902.
[7] M. He, R. J. Fries, R. Rapp, Phys. Rev. C 86 (2012) 014903.
[8] S. Cao, G. -Y. Qin, S. A. Bass, Phys. Rev. C 88 (2013) 044907.
[9] P. B. Gossiaux, J. Aichelin, Phys. Rev. C 78 (2008) 014904.
[10] J. Aichelin, P. B. Gossiaux, T. Gousset, ActaPhys. Polon. B 43 (2012) 655 662.
[11] L. Adamczyk, Phys. Rev. Lett. 113 (2014) 142301.
[12] Physics Letters B 655 (34) (2007) 104 113.
[13] The star heavy flavor tracker, technical design report (2011).

IOP Conf. Series: Journal of Physics: Conf. Series **832** (2017) 012033 doi:10.1088/1742-6596/832/1/012033

Open heavy-flavour and electroweak boson measurements via the (di-)muonic decay channel with ALICE at the LHC

Kgotlaesele Johnson Senosi
for the ALICE collaboration
University of Cape Town and iThemba Laboratory of Accelerator Based Sciences

Abstract. Heavy flavours (charm and beauty) and electroweak bosons (W and Z) are produced in initial hard partonic scatterings. The former interact strongly with the medium formed in ultra-relativistic heavy-ion collisions throughout its evolution, thus making them well suited to investigate its properties. Furthermore, heavy-flavour measurements in proton-nucleus collisions can be used to investigate initial-state effects whereas in proton-proton (pp) collisions they are considered an important test for perturbative Quantum ChromoDynamics (pQCD) predictions. In addition, open heavy-flavour measurements in pp collisions are used as a reference for proton-lead (p–Pb) and lead-lead (Pb–Pb) collisions. On the other hand, electroweak bosons and their leptonic decay products only interact weakly with the QCD matter and thus are suitable probes to test the validity of binary-collision scaling of hard processes. Moreover, their measurements in p–Pb collisions could help to constrain nuclear parton distribution functions. The ALICE muon spectrometer allows the measurement of open heavy flavour, W- and Z-boson production at forward rapidity ($-4.0 < \eta < -2.5$) exploiting their (di)muonic decay channel. In this talk the results obtained with the LHC Run I data in pp, p–Pb and Pb–Pb collisions will be discussed and compared with theoretical predictions.

1. Introduction

Due to their large masses, open heavy flavours (charm and beauty) and electroweak bosons (W and Z) are produced in initial hard partonic processes. The former traverse and interact strongly with the medium formed in heavy-ion collisions and their energy is reduced. The modification of the transverse momentum (p_{T}) distribution is quantified via nuclear modification factor (R_{AA} or R_{pPb}) as a function of p_{T} [1]. The R_{AA} value is equal to unity if no nuclear modification is present while R_{AA} smaller than unity can be due to partonic energy loss such as radiative energy loss (medium induced gluons radiation) [2, 3, 4] and elastic scattering (collisional energy loss) [2, 5]. Initial-state effects [6, 7] or cold nuclear matter (CNM) effects, studied in proton-lead (p–Pb) collisions, complicate the interpretation of any deviation from unity of the R_{AA} in terms of energy loss effects, therefore, for a thorough treatment, heavy-ion collisions need to be studied in conjunction with proton-proton and p–Pb collisions. In proton-proton collisions, open heavy flavour production is used to test theoretical models based on perturbative Quantum ChromoDynamics (pQCD) and they are used as reference measurement for heavy-ion collisions. Since open heavy flavours experience the whole evolution of the medium, the question is whether they participate in the collective expansion of the medium or do they have a non-zero azimuthal anisotropy (v_2).

The isospin dependent production of W- and Z-bosons makes them good probes to constrain parton distribution functions (PDFs) at high momentum transfer (Q) [12]. In high-energy heavy-ion collisions, these electroweak bosons decay before the formation of the QGP. Their weak coupling nature in the leptonic decay channel makes them good probes to study CNM effects. Therefore, they give access to the initial-state properties in nuclear collisions and their yields provide a benchmark to validate the binary-scaling of hard processes [12]. This scaling was verified by the CMS [8, 9] and ATLAS [10, 11] collaborations for photons, W and Z bosons at mid-rapidity.

The production of open heavy flavours, W- and Z-bosons in proton-lead collisions (p–Pb) provides an excellent tool to study cold nuclear matter (CNM) effects [6, 7, 12]. This control experiment (p–Pb) is used to understand whether the effects seen in heavy-ion collisions are a combination of initial and final state effects.

ALICE (A Large Ion Collider Experiment) collected data in pp, Pb-Pb and p-Pb collisions at center-of-mass energies of 7 and 2.76, 5.02 and 2.76 TeV, respectively, at the CERN Large Hadron Collider (LHC). Details of the ALICE detector can be found in ref. [13, 14]. The analysis procedure of open heavy flavour measurement at forward rapidity can be found in [15] and [16] for pp collisions at 2.76 TeV and 7 TeV, in [17] for Pb-Pb collisions at 2.76 TeV and in [18] for p–Pb collisions at 5.02 TeV. The analysis procedure of electroweak bosons is detailed in [19].

2. Results: open heavy flavour

Shown in Figure 1 are plots of the R_{AA} (left) and azimuthal anisotropy (right) of open heavy flavour decay muons as function of p_T in the most central and semi-central collisions, compared to transport models: MC@sHQ+EPOS, BAMPS and TAMU which include radiative and collisional energy loss (for details see Ref. [17] and references therein). It is seen in Figure 1 that open heavy flavour production is effectively modified/reduced in the medium (left plot) and also a non-zero v_2 suggests that they participate in the collective expansion of the medium (right plot).

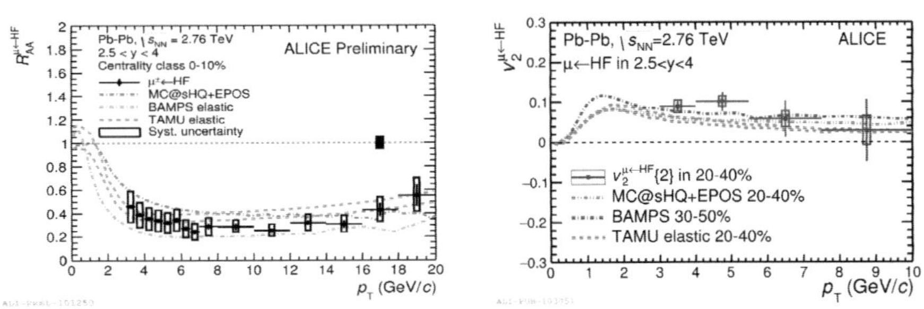

Figure 1. The R_{AA} (left) and azimuthal anisotropy (right) of open heavy flavour muons as functions of p_T in the most central and semi-central Pb-Pb collisions at $\sqrt{s_{NN}} = 2.76$ TeV compared to theoretical models [17].

The R_{pPb} of open heavy flavour decay muons in p–Pb collisions is shown in Figure 2 at backward (left) and forward (right) rapidity. The R_{pPb} is consistent with unity in both rapidity intervals indicating that the effects seen in Pb–Pb collisions are dominated by final state or medium effects.

3. Results: electroweak bosons

In Figure 3 (left), the measured cross sections of W bosons ($\sigma_{\mu^{\pm}\leftarrow W^{\pm}}$) at forward and backward rapidity in p–Pb collisions are compared with pQCD predictions with CT10 PDFs [20],

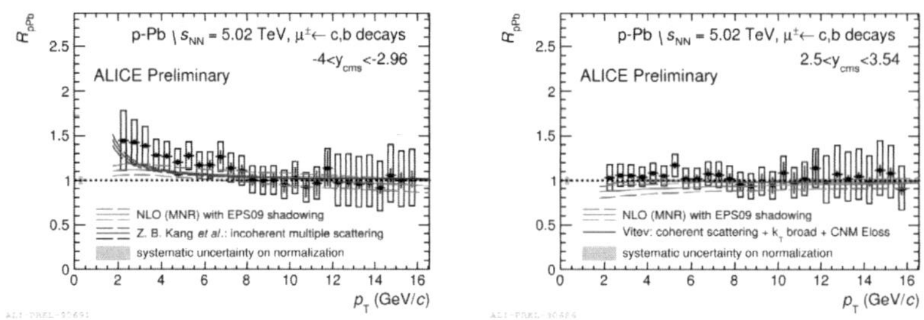

Figure 2. The R_{pPb} of open heavy flavour decay muons at backward (left) and forward (right) rapidity in p-Pb collisions. The measurements are compared with theoretical calculations [18].

with and without EPS09 parametrization [21] for nuclear shadowing. Theoretical calculations agree with the measured cross sections within uncertainties. In Figure 3 (right),

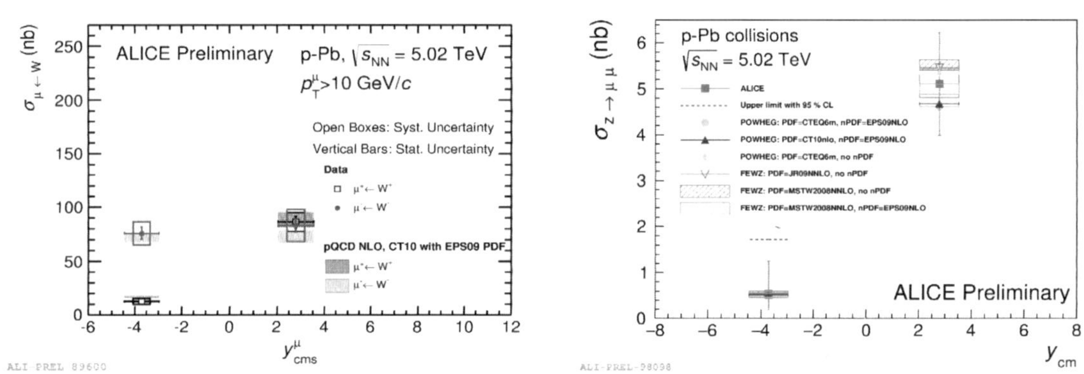

Figure 3. The measured production cross section of muons from W-boson (left) and Z-boson (right) compared with theoretical predictions with and without nuclear PDFs.

the Z-boson cross sections at forward and backward rapidity are compared with next-to-next-to leading order (NNLO) Fully Exclusive W and Z (FEWZ) [22] predictions with CT10, CTEQ6m [23], JR09NNLO [24] and MSTW2008NNLO [25] with and without nuclear PDFs. Theoretical calculations agree with the measured cross sections within uncertainties.

In ALICE the binary scaling of hard processes is tested by evaluating the yield of muons from W-boson decays per event in selected event activity (or centrality) intervals, renormalized by $\langle N_{\mathrm{coll}}\rangle$ [26]. The values of $\langle N_{\mathrm{coll}}\rangle$ are obtained using methods described in [26]. Figure 4 shows the results for yield normalized to $\langle N_{\mathrm{coll}}\rangle$ for both backward (right) and forward (left) rapidity in p-Pb collisisions. The $\langle N_{\mathrm{coll}}\rangle$-normalized yield is independent of event activity and compatible among estimators within uncertainties which is consistent with the results by CMS and ATLAS, indicating binary scaling of hard processes in heavy ion collisions.

4. Conclusions

The $R_{\mathrm{AA/pPb}}$ of open heavy flavour decay muons has been measured in Pb-Pb and p–Pb collisions, respectively, and the results show that the effects seen in Pb-Pb are due to final state effects. The azimuthal anisotropy of open heavy flavour decay muons shows that these

IOP Conf. Series: Journal of Physics: Conf. Series **832** (2017) 012033 doi:10.1088/1742-6596/832/1/012033

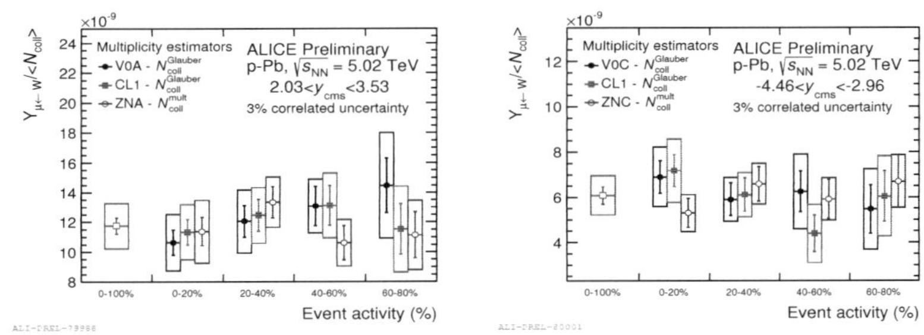

Figure 4. Yield of muons from W-boson decays at forward (left) and backward (right) rapidity as a function of event activity in p-Pb collisions. The green points represent event activity integrated points while black, red and blue represent cases where event activity is estimated with information from V0A, SPD and ZDC, respectively [26].

heavy particles participate in collective expansion of the medium. Theoretical calculations describe the measurements within uncertainties. The production cross sections of W and Z bosons in p–Pb collisions have been measured in two rapidity intervals and compared with theoretical calculations with and without nuclear PDFs parametrized by EPS09. These theoretical calculations agree with the measurements within uncertainties. The measured yield of muons from W-boson decays normalized to $\langle N_{\text{coll}} \rangle$ as function of event-activity shows that the W-boson production scales with $\langle N_{\text{coll}} \rangle$.

References

[1] Miller M *et al.* (2007) *Annu. Rev. Nucl. Part. Sci.* 57, 205
[2] Djordjevic M *et al.* (2007) *Nucl. Phys.* **A783** 493
[3] Dokshitzer Y *et al.* (2001) *Phys.Lett.* **B519** 199-206
[4] Wang E *et al.* (2004) *Phys. Rev. Lett.* **93** 072301
[5] Mustafa M (2005) *Phys. Rev.* **C72** 014905
[6] Kharzeev D *et al.* (2003) *Phys.Lett.* **B561** 93-101
[7] Armesto N (2006) *J.Phys.* **G32** R367-R394
[8] Chatrchyan S *et al.* (CMS) 2012 *Phys. Lett.* **B715** 66–87 (*Preprint* 1205.6334)
[9] Chatrchyan S *et al.* (CMS) 2015 *JHEP* **03** 022 (*Preprint* 1410.4825)
[10] Aad G *et al.* (ATLAS) 2015 *Eur. Phys. J.* **C75** 23 (*Preprint* 1408.4674)
[11] Aad G *et al.* (ATLAS) 2013 *Phys. Lett.* **110**022301 (*Preprint* 1210.6486)
[12] Paukkunen H *et al.* (2011) *JHEP* **03** 071 (*Preprint* 1010.5392)
[13] Aamodt K *et al.* (ALICE) 2008 *JINST* **3** S08002
[14] Abelev B *et al.* (ALICE) 2014 *Int.J.Mod.Phys.* **A29** 1430044 (*Preprint* 1402.4476)
[15] Abelev B (ALICE) 2012 *Phys. Rev. Lett.* **B109** 112301
[16] Abelev B (ALICE) 2012 *Phys.Lett.* **B708** 265-275
[17] Abelev B (ALICE) 2016 *Phys. Lett.* **B753** 41 - 56, and references therein
[18] Li S (ALICE) 2014 *Nucl. Phys.* **A931** 546-551
[19] Senosi K (ALICE) 2016 *arXiv*:1604.03713 (*Preprint* 1604.03713)
[20] Lai H *et al.* (2010) *Phys. Rev.* **D82** 074024 (*Preprint* 1007.2241)
[21] Eskola K, *et al.* (2009) *JHEP* **04** 065 (*Preprint* 0902.4154)
[22] Gavin R *et al.* (2011) *Comput. Phys. Commun.* **182** 2388–2403 (*Preprint* 1011.3540)
[23] Nadolsky P *et al.* (2008) *Phys. Rev. D* **78**(1) 013004
[24] Jimenez-Delgado P *et al.* (2009) *Phys. Rev.* **D79** 074023 (*Preprint* 0810.4274)
[25] Martin A *et al.* (2009) *Eur. Phys. J.* **C63** 189–285 (*Preprint* 0901.0002)
[26] Adam J *et al.* (ALICE) 2015 *Phys. Rev.* **C91** 064905 (*Preprint* 1412.6828)

IOP Conf. Series: Journal of Physics: Conf. Series **832** (2017) 012034 doi:10.1088/1742-6596/832/1/012034

Quark mass effects in quark number susceptibilities

Thorben Graf[1], **Peter Petreczky**[2]

[1] SUBATECH, UMR 6457, Universite de Nantes, Ecole des Mines de Nantes, IN2P3/CNRS. 4 rue Alfred Kastler, 44307 Nantes cedex 3, Franc

[2] Physics Department, Brookhaven National Laboratory, Upton, NY 11973, USA

E-mail: thorben_graf@gmx.de

Abstract. The quark degrees of freedom of the QGP with special focus on mass effects are investigated. A next-to-leading-order perturbation theory approach with quark mass dependence is applied and compared to lattice QCD results.

1. Introduction

The experiments at the Relativistic Heavy Ion Collider (RHIC) at Brookhaven National Laboratory as well as at the Large Hadron Collider (LHC) in Geneva showed clear evidences for the creation of a novel state of matter that is usually refered to as Quark-Gluon-Plasma (QGP) (see e.g. Ref. [1] for a recent review). According to asymptotic freedom, at sufficiently high T the quarks and gluons in this medium are weakly coupled and enable the description of the underlying physics by perturbation theory. On the other hand Lattice Monte Carlo simulations are the method of choice to fully capture the thermodynamic effects of strongly interacting matter at least for nonvanishing chemical potential. In this work these two approaches are used to investigate important thermodynamical quantities such as quark number fluctuations that also can be accessed by experiments. The focus lies on the incorporation of quark masses in the calculations.

2. Calculation

For the perturbative calculations we resort to a next-to-leading order setup including mass dependence that was intensively described in Ref. [2]. The lattice data to which the perturbative results are compared to can be found in Refs. [3–5].

A few words about the chosen accuracy in perturbation theory are in order. The convergence of the weak coupling expansion is very bad. This behaviour is already known for decades. The free energy of hot QCD can in fact be separated into contributions from momentum scales T, gT and g^2T [6], that are called hard, soft and supersoft modes. This is why the pressure of hot QCD is usually written as $p_{\text{QCD}} \equiv p_{\text{hard}} + p_{\text{soft}}$. It can be shown that p_{hard} is the contribution from the scale $\sim T$ and has a well behaved perturbative expansion. This is not the case for p_{soft} where the contributions from the soft and supersoft scales gT, g^2T are included. However, in this work the impact of the quark masses is studied. The convergence of the expansion in terms of the mass dependece is different, which was nicely illustrated in Ref. [7]. This is why the accuracy of the perturbative order is restricted to g^2 here. Accidentaly, the g^2 perturbative result is close to the full non-perturbative result and therefore it serves as a reasonable starting point to study quark mass effects.

3. Results

In all numerical perturbative calculations the light quark masses are set to $m_u = 2.3$ MeV and $m_d = 4.8$ MeV. The setup of Ref. [2] includes a running prescription of the coupling constant. Such a behaviour usually appears at NNLO so that higher order effects are partially taken into account though the calculation of the thermodynamic potential is at NLO. As in Ref. [8] we choose $\alpha_s(\Lambda) = \frac{4\pi}{\beta_0 L}\left[1 - 2\frac{\beta_1}{\beta_0^2}\frac{\ln L}{L}\right]$, with Λ being the renormalization scale and $L = 2\ln(\Lambda/\Lambda_{\overline{\text{MS}}})$, $\beta_0 = 11 - 2N_f/3$ and $\beta_1 = 51 - 19N_f/3$. For the heavier quarks we take into account the running $m_s(\Lambda) = \hat{m}_s\left(\frac{\alpha_s}{\pi}\right)^{4/9}\left[1 + 0.895062\frac{\alpha_s}{\pi}\right]$ and $m_c(\Lambda) = \hat{m}_c\left(\frac{\alpha_s}{\pi}\right)^{12/25}\left[1 + 1.01413\frac{\alpha_s}{\pi}\right]$. The scale $\Lambda_{\overline{\text{MS}}}$ is fixed by requiring $\alpha_s \simeq 0.336$ $\Lambda = 1.5$ GeV [9]; one obtains $\Lambda_{\overline{\text{MS}}} \simeq 343$ MeV for 2+1 flavors and $\Lambda_{\overline{\text{MS}}} \simeq 285$ MeV for 2+1+1 flavors. Requiring $m_s = 93.6$ at $\Lambda = 2$ GeV [10] for the strange quark mass, leads to $\hat{m}_s = 252$ MeV. For calculations that include m_c, we proceeded in the following way. For the calculations we use either 2+1 flavor running or 2+1+1 flavor running depending on the temperature. At $\Lambda = m_c^{\text{pole}} \approx 1.7$ GeV we change between the different number of flavors. Therefore the 2+1-running α_s is again fixed by $\alpha_s \simeq 0.336$ $\Lambda = 1.5$ GeV. The $\Lambda_{\overline{\text{MS}}}$ for the 2+1+1-running is determined by matching them at the scale of the c-quark pole mass, which means, that $\alpha_s^{2+1}(\Lambda = m_c^{\text{pole}}) = \alpha_s^{2+1+1}(\Lambda = m_c^{\text{pole}})$. It yields $\Lambda_{\overline{\text{MS}}}^{2+1} \simeq 343$ MeV and $\Lambda_{\overline{\text{MS}}}^{2+1+1} \simeq 285$ MeV. Of course for m_c the analogous condition must be fulfilled, $m_c^{2+1}(\Lambda = m_c^{\text{pole}}) = m_c^{2+1+1}(\Lambda = m_c^{\text{pole}})$, which means for $\hat{m}_c^{2+1} = 3231$ MeV and $\hat{m}_c^{2+1+1} = 3507$ MeV. With these conventions, the only freedom left is the choice of Λ which is chosen to be $\Lambda = 2\pi T$. For this choice of the renormalization scale the matching point of 2+1 and 2+1+1 flavor running corresponds to roughly $T = 271$ MeV. The band uncertainties in our plots shown below result from the variation of Λ by a factor of two.
We first calculated the pressure and investigated the impact of 2+1+1 compared to 2+1 flavors. The lattice data lay completely in the uncertainty band of the perturbative calculation. However, at low temperatures this uncertainty was quite large, so that one can conclude that for temperatures $T \gtrsim 300$ MeV both methods are in good agreement. The contribution of the charm quarks to the pressure revealed the Boltzmann-suppression for low temperatures, as expected. In this approximation the distribution functions are simply replaced by the Boltzmann distribution, and at leading order the partical charm pressure is $p_c(T, \mu_c) = f(T)\cdot\cosh(\mu_c/T)$, where μ_c is the quark chemical potential of the charm-quark.
In Ref. [2] it was pointed out that the susceptibilities are the thermodynamic quantities that are the most sensitive ones to the quark mass effects. This is why we are especially interested in investigations of these quantities. We compute quark number susceptibilities, which are defined as derivatives with respect to the corresponding quark chemical potentials $\vec{\mu} \equiv (\mu_u, \mu_d, \ldots, \mu_{N_f})$ as

$$\chi_{ijk\ldots}(T) \equiv \left.\frac{\partial^{i+j+k+\ldots}p(T,\vec{\mu})}{\partial\mu_u^i\partial\mu_d^j\partial\mu_s^k\ldots}\right|_{\vec{\mu}=0}. \tag{1}$$

From the perturbative point of view the quark mass depedence of the quark number susceptibilities has not been investigated. We start with the ratio of the strange- over up-quark susceptibility of second order that is depicted in Fig. 1 (a) and compared with lattice data of Ref. [4]. The trend is obviously described by both of the approaches in the same way even at temperatures down to $T = 350$ MeV. Only at even lower temperatures there is a slight deviation that is catched by the uncertainties though. Note that previous perturbative calculations would only result in a horizontal line at 1 because there is no difference between χ_u^2 and χ_s^2. The same ratio but this time at fourth order is shown in Fig. 1 (b). Again the trend of the lattice data [5] is in good agreement with the perturbative results. For low temperatures, $T < 400$ MeV, there

IOP Conf. Series: Journal of Physics: Conf. Series **832** (2017) 012034 doi:10.1088/1742-6596/832/1/012034

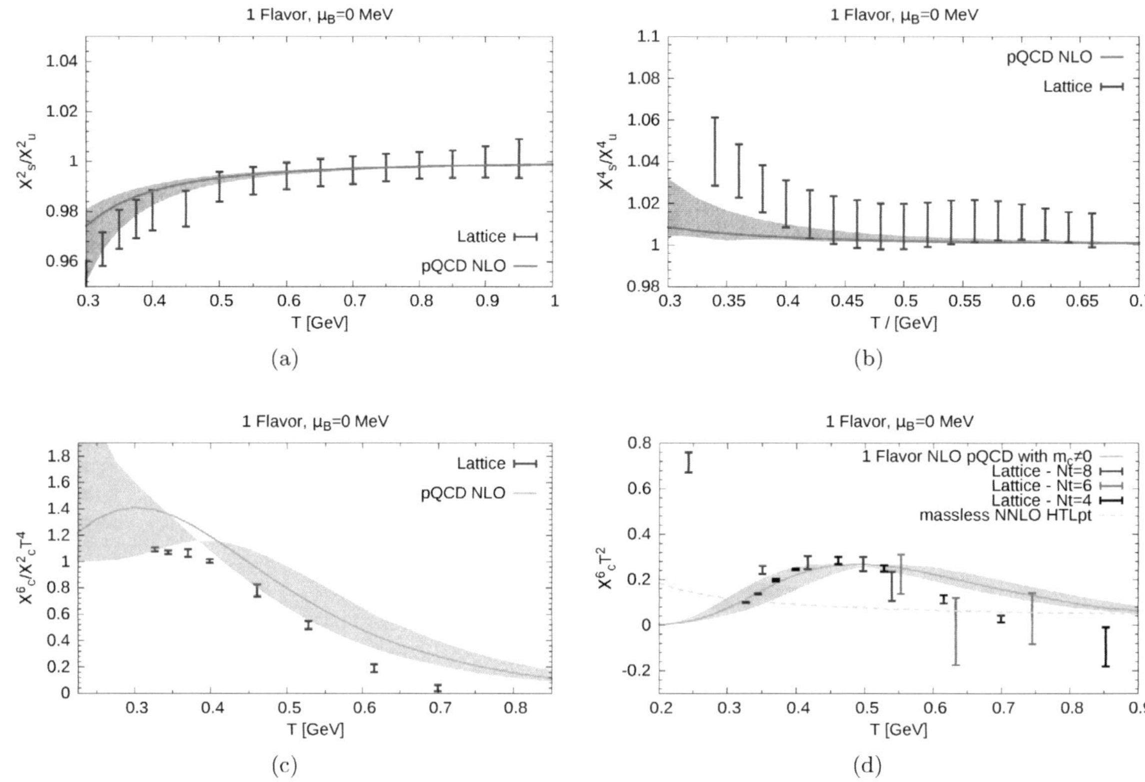

Figure 1. Comparison of the ratios χ_s^2/χ_u^2 (a), χ_s^4/χ_u^4 (b) and $\chi_c^6/\chi_c^2 \cdot T^4$ (c) at $\mu_B = 0$ with lattice QCD results [3–5]. The comparison of $\chi_c^6 \cdot T^2$ at $\mu_B = 0$ with lattice QCD results [3] and HTLpt results [11] is illustrated in (d).

is an increasing deviation. The tiny maximun provided by the lattice data for temperatures of roughly $T = 580$ MeV cannot be reproduced by the perturbative setup.

In Fig. 1 (c) the ratio of the sixth- and second order charm-quark susceptibilities is studied. Having a look at the expression for p_c we expect the ratio to be close to 1 at least at leading order. Obviously the perturbative result in Fig. 1 (c) is slightly above 1. This behaviour can be explained if we have a more detailed look at the NLO expressions. At NLO products of distribution functions appear. If the Boltzmann-approximation is applied a term proportional to $\cosh(2\beta\mu_c)$ and a term independent of μ_c are arising. For the normalized ratio $\chi_c^6/\chi_c^2 \cdot T^4$ this means it is proportional to $\frac{\cosh(\beta\mu_c)+2^6\cosh(2\beta\mu_c)}{\cosh(\beta\mu_c)+2^2\cosh(2\beta\mu_c)}$ which is obviously bigger than 1. In general the comparison of this ratio with lattice data is successful because the trend is described analogously even when the perturbative results are a little bit above the lattice data.

The sixth order charm-quark susceptibility is interesting by itself because lattice simulations showed that there is a maximum around $T = 450$ MeV that cannot be reproduced by even resummed three-loop perturbative calculations without mass dependence [11]. This is why it was not clear so far if the bump is caused by nonvanishing bare quark masses or by higher order effects beyond three-loop. Our results (solid line) are shown in Fig. 1 (d) and compared to lattice data at different temporal spacings and NNLO HTLpt (dashed line). Obviously our

IOP Conf. Series: Journal of Physics: Conf. Series **832** (2017) 012034 doi:10.1088/1742-6596/832/1/012034

approach is able to recover the maximum that is also described by the lattice calculations. We did calculations at LO where the bump also appeared, but it lay outside of the lattice error bars. For the NLO result the bump is within the error bars of lattice data at $N_t = 8$. It is also clearly visible that the Boltzmann-approximation holds for our calculation at low temperatures because of the large charm-quark mass of roughly $m_c = 1.2$ GeV at a temperature of $T = 200$ MeV.

4. Summary

The effects of heavy quark flavors on the pressure and susceptibilities have been investigated. A perturbative and a lattice Monte Carlo approach were compared in order to see if mass effects can be reproduced in the same way. The considered ratios of the susceptibilities show a good agreement between the two methods. The investigation included calculations of the quark number susceptibility up to the sixth order which confirmed the behaviour already seen in previous lattice calculations. An obvious following project would be the inclusion of higher orders in the perturbative setup.

Acknowledgments

The authors want to thank Moritz Greif, Swagato Mukherjee, Robert D. Pisarski and Sayantan Sharma for fruitful discussions. T.G. was supported by the Helmholtz International Center for FAIR, the Helmholtz Graduate School HGS-HIRe, the Service pour la science et la technologie près l'Ambassade de France en Allemagne and Campus France. T.G. is grateful for the kind hospitality of the Nuclear Theory Group at Brookhaven National Laboratory, where part of this work was carried out. P.P. was supported by U.S. Department of Energy under contract No. DE-SC0012704

References

[1] Bala R, Bautista I, Bielcikova J and Ortiz A 2016 *Int. J. Mod. Phys.* **E25** 1642006 (*Preprint* 1605.03939)
[2] Graf T, Schaffner-Bielich J and Fraga E S 2016 *Eur. Phys. J.* **A52** 208 (*Preprint* 1507.08941)
[3] Petreczky P, Hedge P and Velytsky A 2009 *PoS (Beijing)* LAT2009 159 (*Preprint* 0911.0196)
[4] Bazavov A, Ding H T, Hedge P, Karsch F, Miao C, Mukherjee S, Petreczky P, Schmidt C and Velytsky A 2013 *Phys. Rev.* **D88** 094021 (*Preprint* 1309.2317)
[5] Ding H T, Mukherjee S, Ohno H, Petreczky P and Schadler H P 2015 *Phys. Rev.* **D92** 074043 (*Preprint* 1507.06637)
[6] Braaten E and Nieto A 1996 *Phys. Rev.* **D53** 3421-3437 (*Preprint* hep-ph/9510408)
[7] Laine M and Schroder Y 2006 *Phys. Rev.* **D73** 085009 (*Preprint* hep-ph/0603048)
[8] Fraga E S and Romatschke P 2005 *Phys. Rev.* **D71** 105014 (*Preprint* hep-ph/0412298)
[9] Bazavov A, Brambilla N, Garcia i Tormo X, Petreczky P, Soto J and Vairo A 2014 *Phys. Rev.* **D90** 074038 (*Preprint* 1407.8437)
[10] Chakraborty B, Davies C T H, Galloway B, Knecht P, Koponen J, Donald G, Dowdall R, Lepage G and McNeile C 2015 *Phys. Rev.* **D91** 054508 (*Preprint* 1408.4169)
[11] Haque N, Bandyopadhyay A, Andersen J O, Mustafa M G, Strickland M and Su N 2014 *JHEP* **05** 027 (*Preprint* 1402.6907)

IOP Conf. Series: Journal of Physics: Conf. Series **832** (2017) 012035 doi:10.1088/1742-6596/832/1/012035

Overview of electromagnetic probe production in ultra-relativistic heavy ion collisions

Jean-François Paquet

Department of Physics and Astronomy, Stony Brook University, Stony Brook, New York 11794, USA

E-mail: jean-francois.paquet@stonybrook.edu

Abstract. An introductory overview of electromagnetic probe production in ultra-relativistic heavy ion collisions is provided. Experimental evidence supporting the production of thermal photons and dileptons in heavy ion collisions at the Relativistic Heavy Ion Collider (RHIC) and the Large Hadron Collider (LHC) are reviewed. Thermal electromagnetic probe production from hydrodynamical models of collisions is discussed.

1. Introduction

The production of a deconfined nuclear plasma in heavy ion collisions currently under investigation at RHIC and the LHC provides a window into the many-body properties of Quantum Chromodynamics. The exceedingly short lifetime of this deconfined plasma prevents the use of external probes to study its properties, which must rather be inferred from the imprint left by the plasma on particle flying out of the collision region.

The standard picture of the evolution of a heavy ion collision begins with a relatively rapid approach to local equilibrium ($\tau \sim 0.1 - 1$ fm) for the mid-rapidity region of the collision. This locally quasi-equilibrated plasma undergoes a rapid hydrodynamical expansion until its local density becomes too small to maintain local equilibrium ($\tau \sim 10$ fm). This last stage is thought to happen below confinement, allowing the late evolution of the medium to be described in terms of hadronic degrees of freedom. While multiple uncertainties remain in this picture of the evolution of the deconfined plasma, in particular for the earliest stages of the collision, the overall success of hydrodynamic models of heavy ion collisions in describing a wide array of soft hadron measurements suggests that the main features of this picture are credible[1] (see e.g. Refs. [1, 2] for a recent overview of hydrodynamical modelling of heavy ion collisions).

The deconfined plasma created at RHIC and the LHC does not generally exceed ~ 10 fm in size in the direction transverse to the collision axis. Such a transverse size is small compared to the mean-free-path of energetic photons and dileptons[2]. This leads to the conclusion that the

[1] This work focuses on heavy ion collisions at the LHC as well as the top RHIC energy (Au-Au at $\sqrt{s_{NN}} =$ 200 GeV), where evidence for the above picture of the evolution of collisions are most convincing at the moment.

[2] The equilibration time for photons and dileptons is of order [3, 4] $\tau_{eq} \sim \frac{f_B^{(0)}(k)/(2\pi)^3}{d^3\Gamma/dk^3}$ where $f_B^{(0)}(k)$ is the Bose-Einstein distribution (which, up to a degeneracy factor, is the equilibrium distribution of photons and dileptons), and $d^3\Gamma/dk^3$ is the thermal photon/dilepton emission rate. For soft photons of $1-3$ GeV of energies and typical $150-400$ MeV plasma temperatures, $\tau_{eq} \sim 100-500$ fm, about an order of magnitude larger than the lifetime of the plasma.

IOP Conf. Series: Journal of Physics: Conf. Series **832** (2017) 012035 doi:10.1088/1742-6596/832/1/012035

deconfined plasma, while electrically charged, is optically thin for such probes. Electromagnetic probes produced at any point during the lifetime of the plasma can thus escape with minimal interaction.

Photons and dileptons produced by the expanding deconfined medium are referred to as "thermal", since the plasma they are emitted from is close to local thermal equilibrium. The rate of electromagnetic probe production and the momentum distribution of these probes varies according to the local properties of the plasma — the plasma's temperature for example, or the extent of its local deviation from equilibrium. A number of theoretical studies have shown the sensitivity of thermal electromagnetic probes to the properties of the deconfined nuclear matter produced in heavy ion collisions (see Refs. [5, 6] for a recent overview). The enhanced dependence of electromagnetic observables, compared to hadronic observables, to the earlier stage of the plasma evolution is of particular interest (see e.g. Ref. [7]). In consequence, experimental measurements of thermal electromagnetic probes can provide valuable information about the properties of hot deconfined nuclear matter.

Yet this information is not always easily accessible, since thermal electromagnetic probes cannot be singled out experimentally from other sources of electromagnetic radiation, and these other sources often outshine the thermal signal. To better understand the different sources of photons and dileptons present in heavy ion collisions, and how they compete with thermal probes in measurements, a brief overview of experimental electromagnetic observables is first provided. The production of thermal photons and dileptons from a hydrodynamical model is discussed next.

2. Experimental observables and sources of electromagnetic probes

Photon and dilepton production in relativistic *proton-proton* (p-p) collisions is in general well understood theoretically [8, 9]. One expects every source of electromagnetic radiation present in p-p collisions to also be produced in heavy ion collisions, in addition to sources such as thermal photons and dileptons that result from the formation of a deconfined plasma. The term "cold electromagnetic sources" can be used to describe emissions already present in p-p collisions, while "hot sources" can be used for sources produced in connection with the hot plasma. This classification is not always clear-cut but is still useful. In the simplest cases, specific cold electromagnetic sources in heavy ion collisions are related to the same sources in proton-proton collisions by binary nucleon collision scaling: that is, for certain sources of radiation, a heavy ion collision can be seen as an incoherent superposition of proton-proton collisions[3]. Relativistic proton-*nucleus* collisions are used frequently as an intermediate step to calibrate cold sources of photons and dileptons from p-p collisions to heavy ion ones (e.g. Ref. [10]).

Dileptons One of the large "cold" sources of dileptons is hadronic decays (e.g. $\pi^0 \to \gamma e^+ e^-$, $\eta \to \gamma e^+ e^-$, $\omega \to e^+ e^-$) , with different hadrons contributing in different windows of dilepton invariant mass $M_{l^+l^-}$. The contribution to the dilepton invariant mass spectra of decays channels with three-body final states (e.g. $\pi^0 \to \gamma e^+ e^-$) dies off as $M_{l^+l^-}$ exceeds the mass difference between the decaying hadron (e.g. π^0) and the daughter particle accompanying the lepton pair (e.g. γ). Hadronic decays with only e^+e^- in the final state (e.g. $\omega \to e^+ e^-$) rather produce a peak in the spectra at an invariant mass close to the decaying hadron's mass. These different hadronic decay channels, combined with branching fractions that vary considerably from channel to channel, produce a complicated but fairly well-understood background which partially hide the thermal dilepton signal. Dileptons from charm quark decays are also a significant background over a wide invariant mass range ($\sim 0-4$ GeV). Dileptons from the hard initial nucleon-nucleon

[3] The small difference between protons and neutrons is typically referred to as "isospin effect", and can usually be corrected for in theoretical calculations.

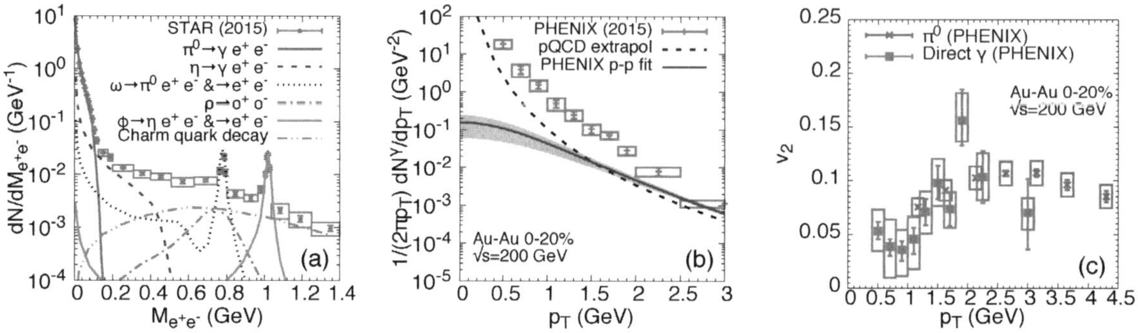

Figure 1: (a) Hadronic and charm quark dilepton decays compared to the dilepton invariant mass spectra from the STAR collaboration, for minimum bias Au+Au collisions at $\sqrt{s_{NN}} =$ 200 GeV [11] (b) Direct photon spectra in 0-20% centrality Au+Au collisions at $\sqrt{s_{NN}} =$ 200 GeV [12], compared to the prompt photon background as predicted by a fit [12] from the PHENIX collaboration to direct photon measurements in p-p collisions and to a perturbative QCD calculation [13] (c) Direct photon v_2 [14] in the same collision system as (b), compared to the v_2 of π^0 [15].

scattering — Drell-Yan dileptons — are typically a small contribution for low values of $M_{l^+l^-}$. The hadronic and charm dilepton backgrounds are shown in Figure 1(a) for Au+Au collisions at $\sqrt{s_{NN}} = 200$ GeV with the STAR detector's acceptance cuts [11]. These backgrounds are compared with the total (inclusive) dilepton measurement from STAR [11], and show an excess of dileptons in the $0.3-0.6$ GeV and $0.8-1.0$ GeV invariant mass regions. This excess is not seen in proton-proton collisions [16, 17], and is interpreted as a signal from the deconfined plasma produced in heavy ion collisions. Since dilepton decays from certain hadrons (in particular the ρ meson) and from charm quarks are affected by the presence of the hot plasma in heavy ion collisions, the difference between dilepton measurements in Au+Au collisions and the "cold background" shown on Fig. 1(a) is understood to be a combination of both a thermal dilepton signal and a modification of cold dilepton sources (see e.g. Ref. [18] for a recent overview).

The dilepton spectra can also be investigated as a function of the transverse momentum p_T of the lepton pair. Since different sources of dileptons dominate the signal in different ranges of transverse momentum p_T and invariant mass $M_{l^+l^-}$, it is possible to construct a variety of observables more and less sensitive to thermal dileptons and the different backgrounds. One can look at the invariant mass spectra with various cuts in p_T, or at the p_T spectra with various cuts in $M_{l^+l^-}$ (see e.g. Ref. [11]). Beyond the spectra, one can look at momentum anisotropies of dileptons, although this is a difficult measurement and only limited data are available at the moment [19].

Photons For photons, hadronic decays are the largest cold source of radiation at low transverse momentum ($p_T^\gamma \sim 0 - 4$ GeV), with $\pi^0 \to \gamma\gamma$ dominating the signal, followed by $\eta \to \gamma\gamma$. The main difference with dileptons is that the simpler kinematic of photon decays make it possible to subtract statistically the contribution of hadronic decays from photon measurements. The resulting observables, called "direct photons", provides a better access to the thermal photon signal.

The single significant "cold" source of photons present in direct photon measurements are prompt photons . These photons are produced in hard initial nucleon scattering, at the very initial impact of the colliding ions, and are known to scale with the number of binary collisions (discussed above). Prompt photons can be produced either when a hard parton from each nucleon produces a final state photon directly (e.g. $q + \bar{q} \to g + \gamma$), or when a hard final state parton is produced (e.g. $q + q \to q + q$) and later radiates a photon (so-called *fragmentation photons*). These processes have been computed perturbatively at next-to-leading order in the strong coupling constant α_s [20, 21]. In proton-proton collisions, these perturbative QCD calculations of prompt photons have been found to be in very good agreement with direct photon measurements at both RHIC and the LHC [22, 23, 24]. This dominance of prompt photons in *proton-proton* collisions implies that their contribution in *heavy ion* collisions can be obtained using either direct photons measurements from p-p collisions or perturbative QCD calculations of prompt photons. Both methods give essentially the same result at high p_T ($p_T^\gamma \gtrsim 5$ GeV), and prompt photons describe very well the direct photon signal at large p_T^γ in heavy ion collisions [23, 25, 26].

At low p_T^γ, the contribution of prompt photons to heavy ion collision is a more difficult question. Direct photon measurements in p-p collisions are not yet very precise at low p_T^γ ($p_T^\gamma \lesssim 1.5$ GeV). On the other hand, it is unclear how reliable are perturbative QCD calculations for small photon energies. Moreover, prompt photon production is expected to be different in heavy ion collisions than it is in p-p collisions, because final state parton interactions with the deconfined plasma modifies fragmentation photon production [27, 28, 29, 30, 31]. This complication does not appear at high p_T^γ, where the fragmentation photon contribution is small.

At the moment, the contribution of prompt photons to heavy ion collisions is typically estimated with binary-nucleon-collision-scaled direct photon measurements from proton-proton collisions, or with perturbative QCD calculations without parton energy loss. Such estimates of prompt photons are shown in Figure 1(b), and are compared with direct photon measurements in Au+Au collisions at RHIC. One can see that while the two estimates of prompt photons in heavy ion collisions differ[4] below $p_T^\gamma \sim 1.5$ GeV, they are both considerably lower than the direct photon measurements. This means that direct photon measurements in heavy ion collisions are showing a considerable excess above prompt photons, an excess generally attributed to thermal photons.

Measurements are also available for the momentum anisotropy v_n of direct photons, obtained as for the spectra by a statistical subtraction of decay photons. The measured direct photon v_n is found to be of the same order as the pion v_n, as shown on Fig. 1(c). This is a very large v_n, considering that prompt photons are expected to have a momentum anisotropy around an order of magnitude lower than pions [29]. This is again seen as a strong signal of thermal photon production. A closer look at the momentum anisotropy development of thermal photons is provided in the next section.

One should note that other sources of electromagnetic probes in heavy ion collisions have been suggested. A recent overview can be found in Rcfs. [5, 6]. The importance of such sources is still under investigation. The rest of this work focuses on thermal photons.

3. Thermal emission

Thermal electromagnetic probes are those emitted by the deconfined plasma during its phase of hydrodynamical expansion, while the plasma is close to local thermal equilibrium.

[4] The large difference between the two prompt photon estimates at low p_T^γ is straightforward to understand. The lower estimate is from a fit [12] by the PHENIX collaboration to direct photons measurements in p-p collisions. This fit still is poorly constrained by measurements below $p_T^\gamma \sim 1.5$ GeV. Moreover, the fit and its uncertainty band rely on the assumption that prompt photon production saturates for p_T^γ lower than $1 - 2$ GeV. On the other hand, the prompt photon estimate based on perturbative QCD assumes that the power-law dependence of the spectra observed at high p_T^γ continues at low p_T^γ [13]. This last scenario may considerably overestimates low p_T^γ prompt photon production. New low p_T^γ measurements of direct photons in p-p collisions, together with calculations of prompt photons that include the effect of the hot deconfined plasma on fragmentation photons, will help clarify the situation.

IOP Conf. Series: Journal of Physics: Conf. Series **832** (2017) 012035 doi:10.1088/1742-6596/832/1/012035

Hydrodynamical models of heavy ion collisions provide a description of the deconfined nuclear plasma, from hyperbolic time $\tau \sim 0.1 - 1$ fm to ~ 10 fm, in terms of macroscopic variables: the local temperature $T(X)$ of the plasma, the flow velocity profile $u^\mu(X)$ and the extent of the plasma's deviation from equilibrium, as quantified by the shear stress tensor $\pi^{\mu\nu}(X)$ and bulk pressure $\Pi(X)$. Given local values for T, u^μ, $\pi^{\mu\nu}$ and Π, the production of thermal photons and dileptons in a small spacetime volume $\Delta^4 X$ is equal to the thermal electromagnetic emission rate $d^4\Gamma_{\gamma/l^+l^-}(K^\mu, u^\mu, T, \pi^{\mu\nu}, \Pi)/d^4k$ multiplied by $\Delta^4 X$. The total thermal electromagnetic probe signal is obtained by integrating the emission rate over the spacetime profile provided by hydrodynamics:

$$\frac{d^4N_{\gamma/l^+l^-}}{d^4k} = \int d^4X \frac{d^4\Gamma_{\gamma/l^+l^-}}{d^4k}(K^\mu, u^\mu(X), T(X), \pi^{\mu\nu}(X), \Pi(X)) \quad (1)$$

where the integral over X is usually limited to the spacetime volume that has not reached kinetic freeze-out (which usually means the region with $T > T_{FO}$ when the kinetic freeze-out criteria for the hydrodynamical model is the temperature T_{FO}). The thermal or quasi-thermal photon/dilepton emission rate $d^4\Gamma_{\gamma/l^+l^-}/d^4k$ is constrained theoretically from different models of hadronic and partonic degrees of freedom. Recent summaries of the thermal emission rates, including discussions of the effects of shear and bulk viscosities, can be found in e.g. Refs. [5, 6, 7, 13]. Uncertainties remain in $d^4\Gamma_{\gamma/l^+l^-}/d^4k$, especially in and slightly above the cross-over temperature region ($T \sim 150 - 300$ MeV), and a better understanding of the thermal electromagnetic rates is still an active field of research.

Photons and dileptons inherit a small momentum anisotropy from the emission rate $d^4\Gamma_{\gamma/l^+l^-}/d^4k$, which is not isotropic when non-equilibrium corrections are taken into account. However, as for soft hadrons, the momentum anisotropy of thermal electromagnetic probes originates overwhelmingly from a different mechanism: the asymmetry of the plasma's flow velocity profile $u^\mu(X)$. Initially ($\tau \sim 0.1 - 1$ fm), flow velocities are thought to be small and similar in size along the short and long axis of the almond-shape energy deposition in heavy ion collisions. However, after ~ 10 fm of hydrodynamic evolution, the flow is much stronger along the short axis of the energy deposition. This flow anisotropy gives thermal photons and dileptons emitted at late times a v_n similar to hadrons, while electromagnetic probes emitted at early times have much smaller momentum anisotropies. To obtain the large v_2 of direct photons seen in Fig. 1(c) from a hydrodynamical calculation of thermal photons, one needs rapid development and saturation of the flow velocity anisotropy in heavy ion collisions, as well as a significant production of thermal photons at late times. Since prompt photons are also present and have a small momentum anisotropy, thermal photons must also outshine them in order for a large v_2 to be seen in the sum of their contributions.

At the moment, the combination of a prompt photon background with hydrodynamical calculations of thermal photons generally underestimate measurements of the direct photon spectra and v_2 at RHIC and the LHC [5, 6]. On the other hand, these calculations describe many *qualitative* features observed in measurements, such as an exponential p_T^γ-dependence of the direct photon spectra at low p_T^γ, and a large thermal photon v_2 that agrees well with measurements before prompt photons are added. In this sense, targeted improvements to current hydrodynamic calculations, such as a better description of the early stage of the plasma, possible changes in the thermal emission rates in the cross-over region and a revisiting of low p_T^γ prompt photon production could prove sufficient to resolve the tension observed with measurements. Nevertheless, the presence of additional sources of photons still cannot be ruled out and investigations in that direction are on-going.

While thermal photons have been studied over the past decade with increasingly sophisticated hydrodynamic models, hydrodynamical studies of thermal dileptons have been fewer, possibly due to the limited availability of measurements. A recent hydrodynamic calculation [32] found general good agreement with the low-$M_{e^+e^-}$ invariant mass spectra measured by the STAR collaboration. This result is in line with earlier calculations made with a simpler description of the plasma evolution [18].

It is not clear why hydrodynamic calculations of dileptons appear to agree with measurements, while tension is observed with direct photons. An important step will be to evaluate both thermal photons and dileptons from the same modern hydrodynamic simulation, with similarly constrained thermal emission rates, to provide for a more consistent comparison with measurements. It must also be emphasised that the dilepton invariant mass spectra is integrated in $p_T^{e^+e^-}$, which greatly reduces its sensitivity to the flow velocity profile of the plasma. In this sense, the photon and dilepton measurements currently available are sensitive to different features of the deconfined plasma.

As more systematic comparisons of hydrodynamical models with electromagnetic and hadronic observables are made, it should become clearer if agreement with measurements can be improved by modifying the spacetime description of the plasma evolution, or by including new sources of radiation, or even simply by a better understanding of the thermal emission rates.

Acknowledgments

I thank Chun Shen and Gojko Vujanovic, along with the organizers and participants of the Hot Quarks 2016 workshop, for interesting discussions and feedback. This work was supported by the U.S. D.O.E. Office of Science, under Award No. DE-FG02-88ER40388.

References

[1] Denicol G S 2016 *Nucl. Phys.* **A956** 67–74
[2] Derradi de Souza R, Koide T and Kodama T 2016 *Prog. Part. Nucl. Phys.* **86** 35–85
[3] Kapusta J I, Lichard P and Seibert D 1991 *Phys. Rev.* **D44** 2774–2788 [Erratum: Phys. Rev.D47,4171(1993)]
[4] Wong C Y 1995 *Introduction to high-energy heavy ion collisions*
[5] Shen C 2016 *Nuclear and Particle Physics Proceedings* **276278** 72–77
[6] Shen C 2016 *Nucl. Phys.* **A956** 184–191
[7] Vujanovic G, Paquet J F, Denicol G S, Luzum M, Jeon S and Gale C 2016 *Phys. Rev.* **C94** 014904
[8] Aurenche P, Fontannaz M, Guillet J P, Pilon E and Werlen M 2006 *Phys. Rev.* **D73** 094007
[9] McGaughey P L, Moss J M and Peng J C 1999 *Ann. Rev. Nucl. Part. Sci.* **49** 217–253
[10] Adare A *et al.* (PHENIX) 2013 *Phys. Rev.* **C87** 054907
[11] Adamczyk L *et al.* (STAR) 2015 *Phys. Rev.* **C92** 024912
[12] Adare A *et al.* (PHENIX) 2015 *Phys. Rev.* **C91** 064904
[13] Paquet J F, Shen C, Denicol G S, Luzum M, Schenke B, Jeon S and Gale C 2016 *Phys. Rev.* **C93** 044906
[14] Adare A *et al.* (PHENIX) 2015 (*Preprint* 1509.07758)
[15] Adare A *et al.* (PHENIX) 2012 *Phys. Rev. Lett.* **109** 122302
[16] Adare A *et al.* (PHENIX) 2009 *Phys. Lett.* **B670** 313–320
[17] Adamczyk L *et al.* (STAR) 2012 *Phys. Rev.* **C86** 024906
[18] Rapp R 2013 *Adv. High Energy Phys.* **2013** 148253
[19] Adamczyk L *et al.* (STAR) 2014 *Phys. Rev.* **C90** 064904
[20] Aurenche P, Baier R, Fontannaz M and Schiff D 1988 *Nucl. Phys.* **B297** 661–696
[21] Aversa F, Chiappetta P, Greco M and Guillet J P 1989 *Nucl. Phys.* **B327** 105
[22] Adare A *et al.* (PHENIX) 2012 *Phys. Rev.* **D86** 072008
[23] Chatrchyan S *et al.* (CMS) 2012 *Phys. Lett.* **B710** 256–277
[24] Aad G *et al.* (ATLAS) 2014 *Phys. Rev.* **D89** 052004
[25] Adam J *et al.* (ALICE) 2016 *Phys. Lett.* **B754** 235–248
[26] Adamczyk L *et al.* (STAR) 2016 (*Preprint* 1607.01447)
[27] Fries R J, Muller B and Srivastava D K 2003 *Phys. Rev. Lett.* **90** 132301
[28] Turbide S, Gale C, Jeon S and Moore G D 2005 *Phys. Rev.* **C72** 014906
[29] Turbide S, Gale C, Frodermann E and Heinz U 2008 *Phys. Rev.* **C77** 024909
[30] Arleo F, Eskola K J, Paukkunen H and Salgado C A 2011 *JHEP* **04** 055
[31] Zakharov B G 2004 *JETP Lett.* **80** 1–6 [Pisma Zh. Eksp. Teor. Fiz.80,3(2004)]
[32] Vujanovic G, Young C, Schenke B, Rapp R, Jeon S and Gale C 2014 *Phys. Rev.* **C89** 034904

IOP Conf. Series: Journal of Physics: Conf. Series **832** (2017) 012036 doi:10.1088/1742-6596/832/1/012036

Low-Mass Dielectron Production in p-Pb and Pb–Pb Collisions with ALICE

Carsten Klein (for the ALICE Collaboration)
Institut für Kernphysik, Goethe-Universität Frankfurt am Main, Germany
E-mail: cklein@ikf.uni-frankfurt.de

Abstract. The production of low-mass dielectrons is measured with ALICE at the LHC in pp, p-Pb and Pb–Pb collisions. The dielectron yield in p-Pb collisions in comparison to the expected hadronic cocktail shows an overall reasonable agreement. The production of direct virtual photons, visible as an excess over the hadronic cocktail, in Pb–Pb collisions is measured in two transverse momentum ranges. A feasibility study for the future LHC Run 3 indicates that it should be possible to extract an effective temperature with a combined statistical and systematic uncertainty of around 30%.

1. Introduction

Electron-positron pairs (dielectrons) are an ideal probe to investigate every stage of a heavy-ion collision. Since they interact only electromagnetically they pass through the produced quark-gluon plasma (QGP) with negligible final state interaction. Dielectrons provide information on the properties of light vector mesons in the vacuum and in the medium. In addition they are produced by direct virtual photons decays which internally convert to electron-positron pairs. Proton-proton collision measurements serve as a baseline for understanding hot medium effects in lead-lead collisions in comparison to proton-lead collisions which help to disentangle cold nuclear matter effects from those of the QGP.

Dielectron measurements in ALICE are performed with the central barrel detectors around midrapidity. In this work, a comparison of a hadronic cocktail with data in p-Pb collisions and a virtual direct photon measurement in Pb–Pb collisions are presented. Prospects of a future measurement in Pb–Pb after the ALICE upgrade for LHC Run 3 are also discussed.

2. Dielectron Production in p-Pb collisions at $\sqrt{s_{\mathrm{NN}}} = 5.02$ TeV

The analysis presented is based on 10^8 minimum bias p-Pb events. Selection criteria based on the specific energy loss in the time projection chamber and the silicon inner tracking system, together with a measurement of the time of flight give a clean electron sample while maintaining a high detection efficiency [6]. Fiducial selections of $p_{\mathrm{T}} > 200$ MeV/c and $\eta < 0.8$ are applied. Dielectron spectra of pairs with unlike sign $N^{\mathrm{same}}_{\mathrm{ULS}}$ and like sign $N^{\mathrm{same}}_{\mathrm{LS}} = 2 \cdot \sqrt{N_{++} \cdot N_{--}}$ are created. The unlike sign spectrum consists of the actual signal and combinatorial background. The like sign distribution serves as an approximation of the combinatorial background. The acceptance difference of unlike sign and like sign pairs is corrected with $R = N^{\mathrm{mix}}_{\mathrm{ULS}}/N^{\mathrm{mix}}_{\mathrm{LS}}$ calculated with a mixed event technique. The signal yield is obtained with $N^{\mathrm{same}}_{\mathrm{ULS}} - N^{\mathrm{same}}_{\mathrm{LS}} \cdot R$.

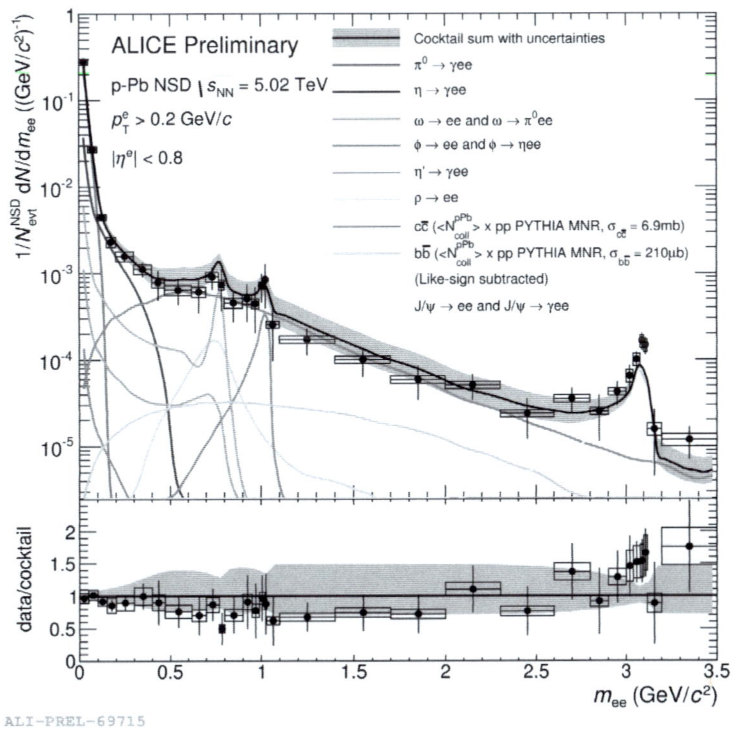

Figure 1. Invariant mass distribution for p-Pb collisions at $\sqrt{s_{\rm NN}} = 5.02$ TeV compared to a hadronic cocktail.

The dielectron invariant mass spectrum obtained in p-Pb collisions at $\sqrt{s_{\rm NN}} = 5.02$ TeV is compared to a hadronic cocktail. The transverse momentum spectrum of charged $\pi^{+/-}$ serves as an estimate for the π^0 spectrum. Spectra of other light mesons are obtained assuming $m_{\rm T}$-scaling. The contributions from semileptonic decays of open heavy-flavour mesons and from J/ψ are calculated using PYTHIA [1] tuned to ALICE measurements in pp and p-Pb collisions.

Figure 1 shows the resulting invariant mass spectrum. The grey band corresponds to the uncertainty of the cocktail calculation. Data and cocktail are in reasonable agreement over the whole mass range. The pair transverse momentum spectrum in the invariant mass region of $0.14 < m_{\rm ee} < 0.75$ GeV/c^2 is shown in Figure 2. No clear enhancement over the cocktail is observed, unlike measurements in Au+Au collisions reported at RHIC [2][3]. Figure 3 shows the intermediate mass region where cocktail and data are also in reasonable agreement.

3. Virtual direct photon production in Pb–Pb collisions at $\sqrt{s_{\rm NN}} = 2.76$ TeV

The analysis presented here is based on $17 \cdot 10^6$ central Pb–Pb events at $\sqrt{s_{\rm NN}} = 2.76$ TeV recorded in 2011. A fiducial selection of $p_{\rm T} > 400$ MeV/c and $\eta < 0.8$ is applied. Electron identification is similar to the p-Pb analysis with a higher focus on purity.

Figure 4 shows the measured dielectron invariant mass distribution in the transverse momentum interval $2 < p_{\rm T}^{\rm ee} < 4$ GeV/c together with a hadronic cocktail, including semileptonic decays from open heavy flavour mesons, and the contribution of direct virtual photons. A fit to the data is performed with the function $f = r \cdot f_{\rm dir} + (1-r) \cdot f_{\rm LF} + f_{\rm HF}$ in the mass range $0.1 < m_{\rm ee} < 0.3$ GeV/c^2 where $f_{\rm dir}$, $f_{\rm LF}$ and $f_{\rm HF}$ describe the contributions from virtual

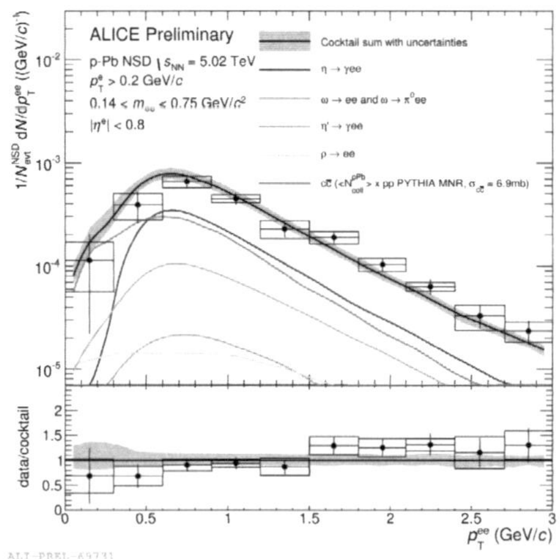

Figure 2. Transverse momentum spectrum for $0.14 < m_{\rm ee} < 0.75$ GeV$/c^2$ for p-Pb collisions at $\sqrt{s_{\rm NN}} = 5.02$ TeV.

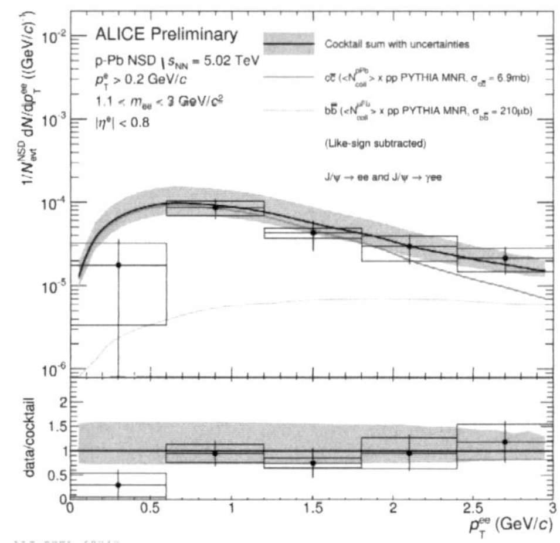

Figure 3. Transverse momentum spectrum for $1.1 < m_{\rm ee} < 3.0$ GeV$/c^2$ of p-Pb collisions at $\sqrt{s_{\rm NN}} = 5.02$ TeV.

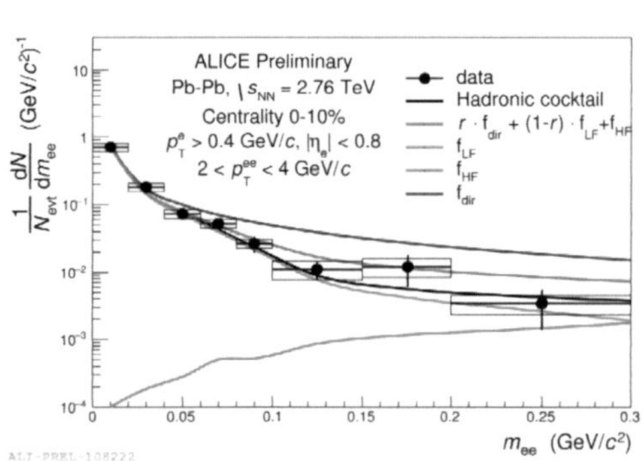

Figure 4. Measured invariant mass yield with scaled hadronic cocktail and direct photon contribution of Pb–Pb collisions.

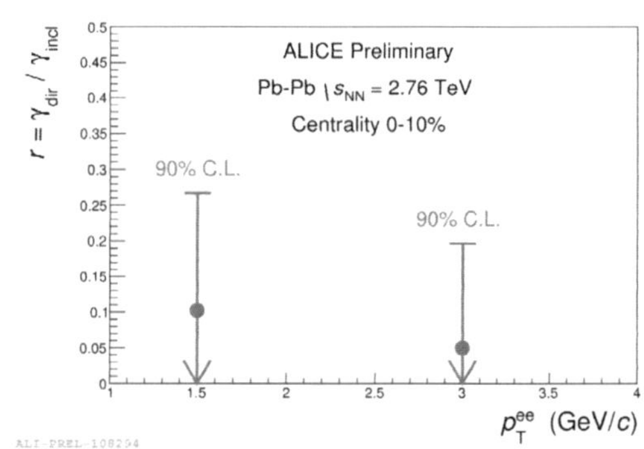

Figure 5. Upper limit of direct photon over inclusive photon production in central Pb–Pb collisions at $\sqrt{s_{\rm NN}} = 2.76$ TeV.

direct photons, light-flavour and heavy-flavour hadrons, respectively. The fit parameter r can be interpreted as the ratio of the direct photon over the inclusive photon yield. To extract upper limits for r, 10^5 invariant mass distributions were simulated with the data points being shifted and sampled according to their systematic and statistical uncertainty and the respective fit parameter r for each simulation is extracted. The distribution of the r values is fitted with a gaussian. The r value at 90% of its integral corresponds to the extracted upper limit. The fit procedure was also performed in a second transverse momentum range from $1 < p_{\rm T}^{\rm ee} < 2$ GeV$/c$. The result as a function of transverse momentum is shown in Figure 5.

4. Upgrade study for future dielectron measurement in Pb–Pb collisions

One of the major physics cases for ALICE in LHC Run 3 will be a precise measurement of the dielectron invariant mass spectrum in Pb–Pb collisions at $\sqrt{s_{\rm NN}} = 5.5$ TeV [4]. A simulation was performed to assess the possible precision of a measurement of the early effective temperature

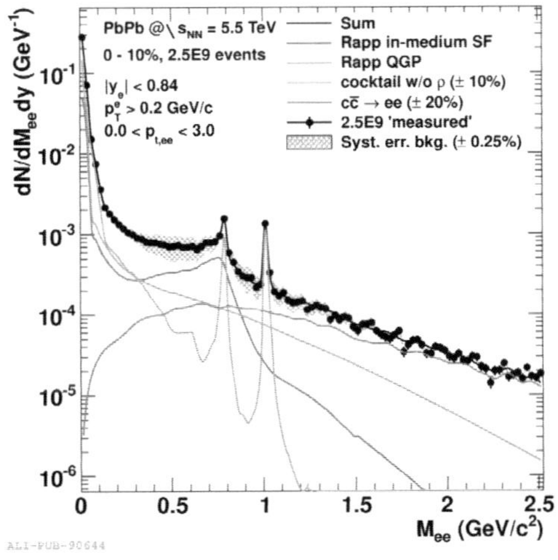

Figure 6. Simulated invariant mass distribution for central Pb-Pb collisions with $\sqrt{s_{\mathrm{NN}}} = 5.5$ TeV with an upgraded ALICE setup.

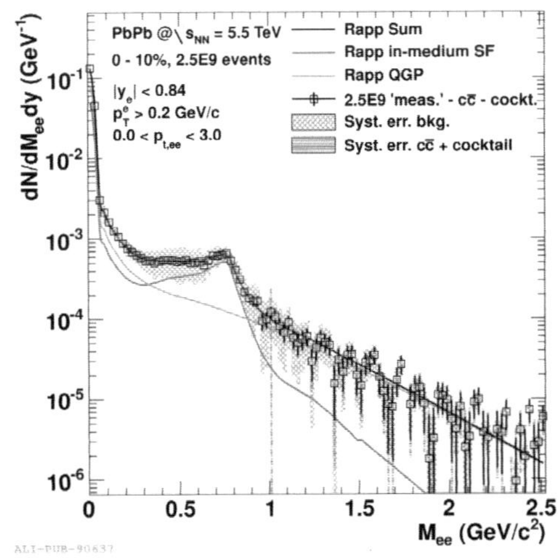

Figure 7. Simulated dielectron excess yield after subtraction of hadronic cocktail and contribution from open charm decay.

of the QGP. This temperature is described by the inverse slope parameter of an exponential fit to the invariant mass distribution in the intermediate mass region $1.1 < m_{\mathrm{ee}} < 2.0$ GeV/c^2. To simulate the spectrum the current ALICE setup was used as a baseline and then modified in accordance with plans of the upgrade and to match a reduced magnetic field in the ALICE solenoid. Based on the simulated signal and background distributions a statistical significance is calculated. This significance is used for a Poisson sampling of the data points around a model input including contributions from an expected hadronic cocktail, from open charm meson decays [1] and QGP-related contributions like a medium-modified ρ and thermal radiation [5]. Figure 6 shows the simulated invariant mass distribution for the 0-10% most central Pb–Pb collisions with $2.5 \cdot 10^9$ events. Figure 7 shows the excess spectrum after subtracting the hadronic cocktail and the charm contribution. The aforementioned fit to this excess yield indicates that the upgraded ALICE setup will allow for a measurement of the QGP temperature with 10% statistical and 20% systematic uncertainty. The systematic uncertainty contains the uncertainty of the background subtraction and the uncertainty of the hadronic cocktail subtraction including the subtraction of the contribution from semileptonic open charm decays.

5. References

[1] T. Sjstrand, S. Mrenna and P. Skands, JHEP05 (2006) 026
[2] E. Atomssa et al. (PHENIX Collaboration) Phys.Rev. C93 (2016) no.1, 014904
[3] L. Adamczyk et al.(STAR Collaboration) Phys. Rev. C 92, 024912 (2015)
[4] B. Abelev et al. (ALICE Collaboration), J. Phys. G 41 (2014) 087002.
[5] R. Rapp and J. Wambach, Eur. Phys. J. A 6 (1999) 415 and private communication
[6] K. Aamodt et al. (ALICE collaboration), JINST 3 (2008) S08002

IOP Conf. Series: Journal of Physics: Conf. Series **832** (2017) 012037 doi:10.1088/1742-6596/832/1/012037

Non-equilibrium dilepton production in hadronic transport approaches

Jan Staudenmaier[1,2], **Janus Weil**[1], **Hannah Petersen**[1,2,3]

1 Frankfurt Institute for Advanced Studies (FIAS), Ruth-Moufang-Straße 1, 60438 Frankfurt am Main

2 Institut für Theoretische Physik, Johann Wolfgang Goethe-Universität, Max-von-Laue-Str. 1, 60438 Frankfurt am Main, Germany

3 GSI Helmholtzzentrum für Schwerionenforschung, Planckstr. 1, 64291 Darmstadt, Germany

E-mail: staudenmaier@fias.uni-frankfurt.de

Abstract. In this work the non-equilibrium dilepton production from a hadronic transport approach (SMASH) is presented. The dilepton emission from the hadronic stage is of interest for current HADES results measured at GSI in the beam energy range from 1.25 - 3.5 GeV. Also at high collision energies (RHIC/LHC) the later dilute stages of the reaction are dominated by hadronic dynamics. The newly developed hadronic transport approach called SMASH (=Simulating Many Accelerated Strongly-interacting Hadrons) is introduced first. After explaining the basic interaction mechanisms, a comparison of elementary cross sections for pion production to experimental data is shown. The dilepton production within SMASH is explained in detail. The main contribution to the dilepton spectra in the low energy regime of GSI/FAIR/RHIC-BES originates from resonance decays. Results of the dilepton production with SMASH such as invariant mass spectra are shown.

1. Introduction

Lepton pairs are clean probes for strongly-interacting matter, since they only interact electromagnetically. They are directly emitted from the hot and dense medium created in a heavy-ion collision, whereas hadronic probes are rescattered or absorbed. It is therefore possible to extract medium properties and medium modifications of resonances from dileptons over the whole lifetime of such collisions. Dileptons are measured for low beam energies by the HADES collaboration at GSI [1] and for high beam energies by STAR and PHENIX at RHIC [2, 3] or ALICE at LHC [4].

At the moment it is not possible to describe heavy-ion collisions from first principles, since descriptions of highly-dynamical many-body systems directly based on QCD are not feasible. Therefore, often a combination of effective approaches is used, where the early stage is described by models based on fluctuating color fields or strings, the hot and dense stage is governed by relativistic dissipative hydrodynamics with an equation of state from QCD lattice calculations and the late, dilute stage is described by hadronic transport approaches. Hadronic transport approaches do not rely on the assumption of equilibrium and are able to describe low beam energy collisions. For this work a new hadronic transport approach, SMASH ("Simulating Many Accelerated Strongly-Interacting Hadrons"), is employed that constitutes an effective numerical solution of the relativistic Boltzmann equation.

IOP Conf. Series: Journal of Physics: Conf. Series **832** (2017) 012037 doi:10.1088/1742-6596/832/1/012037

2. Model Description

SMASH [5, 6] has been developed with the intention to provide a standard reference for the description of a hadronic system with vacuum properties. A geometric collision criterion is employed, where two particles interact, if the distance of closest approach is smaller than a so called *interaction distance* that depends on the total cross section. This ansatz is known from other transport models [7]. Low-energy nucleus-nucleus collisions, as well as infinite matter calculations can be performed with the approach. It is also possible to set it up as an afterburner for hydrodynamic simulation.

All well-established hadrons from the PDG [8] up to mass of 2 GeV are included in the model. The $\pi, \eta, \rho, \omega, \phi, f_2, \sigma, K$ mesons and the $N, \Delta, \Lambda, \Sigma, \Xi, \Omega$ baryonic states are incorporated.

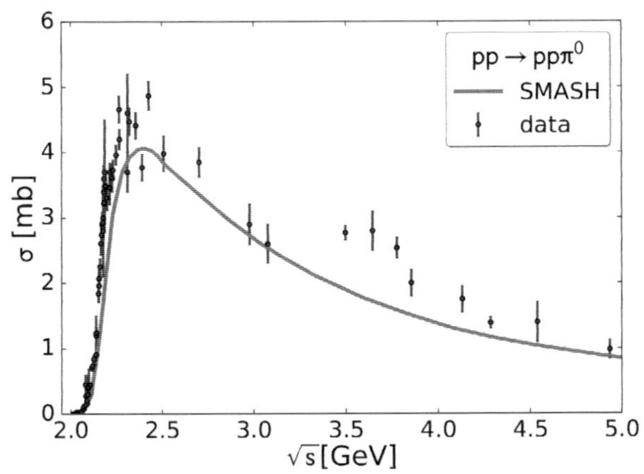

Figure 1. Cross section of the process $pp \rightarrow pp\pi^0$ for different $\sqrt{s}$ energies. Compared to data taken from [9].

In the few GeV energy regime the excitation and decay of resonances dominate the hadronic cross section. Therefore, all interactions fall in one of 4 categories: elastic scatterings, inelastic scatterings, decays and resonance absorption. In order to conserve detailed balance only $1 \leftrightarrow 2$ or $2 \leftrightarrow 2$ processes are allowed.

Figure 1 shows an example for hadronic cross sections. The cross section for the production of one π^0 in a proton-proton collision is shown for different $\sqrt{s}$. The dominant contribution to this cross section is the excitation of a nucleon state to a Δ resonance. Only in the higher energy tail additional contributions from other Δ^* and N^* states contribute. Overall the π^0 production agrees well with the experimental data. However, a slight systematic undershoot can be observed, which might be due to the negligence of non-resonant contributions to pion production.

In Figure 2 a result for a heavy-ion collision at low energies is displayed. The figure shows the m_T spectrum for π^+ in different rapidity windows for a CC collision at a kinetic energy of 1 AGeV. Although at high m_T the spectrum is slightly below the data , there is overall a good agreement with the experimental data.

3. Dilepton Production in SMASH

Pairs of electrons or muons that are produced together are called dileptons. In SMASH all dileptons are produced through decays. Although di-muons can be produced, in this work the focus is only on di-electrons. Those originate either from direct decays of the vector mesons ρ, ω and ϕ or from Dalitz decays from the pseudoscalar mesons π^0 and η or the ω and Δ resonances.

In experiments as well as in theory a challenge for dileptons as an observable is that their decays are very rare. Because these are electromagnetic decays, typical branching ratios are of

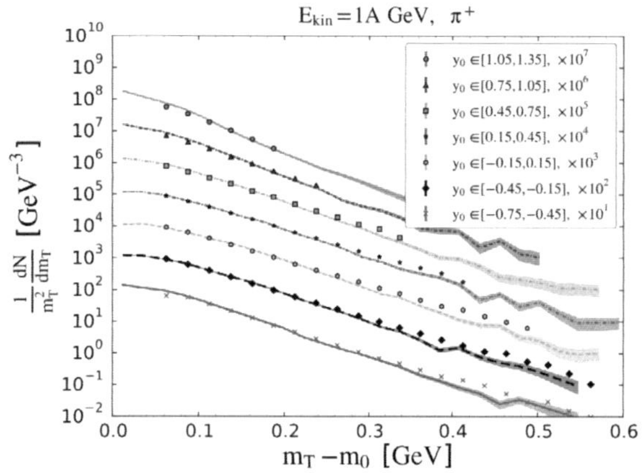

Figure 2. π^+ m_T-spectra in different rapidity windows for Carbon-Carbon collisions at $E_{\text{kin}} = 1A\,\text{GeV}$ compared with HADES data [10].

the order of 10^{-5}. Therefore in transport models like SMASH the so called *Time-Integration-Method* or also called *Shining-Method* [11, 12, 13] is used. The idea is to integrate the decay probability for the dilepton decay over the whole lifetime of the decaying resonance. In a timestep-based transport approach this can be achieved by radiating (*shining*) dileptons during every time step and weighting those pairs according to their decay probability. In this way much more lepton pairs are produced and statistics are improved.

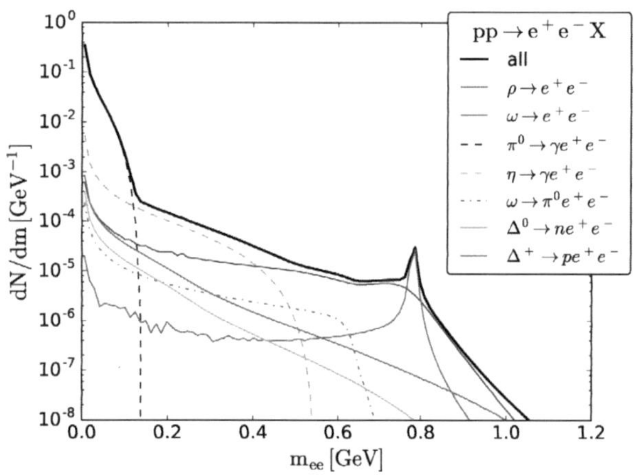

Figure 3. Invariant mass spectrum of di-electrons produced by pp collisions at $E_{\text{kin}} = 3.5\,\text{GeV}$.

The invariant mass spectrum of the dileptons consists of all the different decay channel contributions. Figure 3 shows the invariant mass spectrum of di-electrons produced by a proton-proton collision with a kinetic energy of 3.5 GeV. The spectrum is missing a contribution from the ϕ decay, since SMASH currently lacks a production mechanism for ϕ mesons in pp collisions. The dominant contribution for low invariant masses is the pion Dalitz decay, whereas in the intermediate mass region the direct decays of vector mesons are dominant. Those contributions also reveal features of the spectral function of the vector mesons. Both, ω and ρ contributions peak at the resonance pole masses and the ω peak is sharper, because the resonance has a smaller width. Overall the spectrum looks qualitatively as expected.

Since the rapidity spectrum in Figure 4 does not have a selection in invariant mass, it is clearly visible that the π decay is the biggest source of dileptons. Figure 4 shows the rapidity spectrum of di-electrons of proton-proton collision at $E_{\text{kin}} = 3.5\,\text{GeV}$. The whole spectrum is shifted to a finite rapidity, since a fixed-target setup was used. The vertical dashed line is at the mid-rapidity value of the pp system before the collision. Every spectrum peaks at this value,

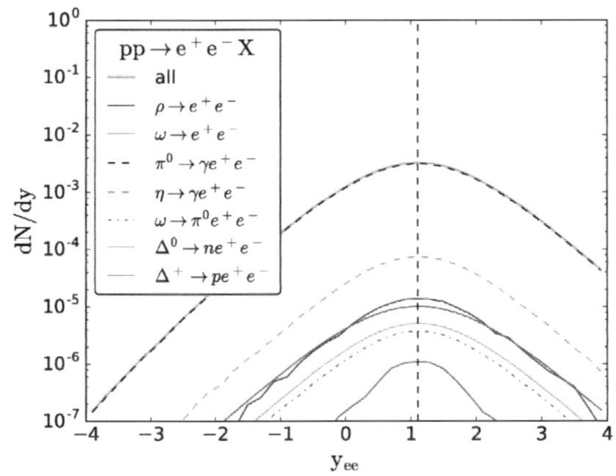

Figure 4. Rapidity spectrum of di-electrons produced by pp collisions at $E_{\text{kin}} = 3.5\,\text{GeV}$. Vertical dashed line is at mid-rapidity of the pp system in a fixed target setup.

which can be regarded as a check that the kinematic evolution works in a reasonable way.

4. Summary and Outlook

One can conclude that SMASH reproduces the hadronic cross sections well and agrees with experimental data on pion production. An extensive description and further comparison with data can be found in [5]. The dilepton production within the approach results in spectra matching the expectations qualitatively. The next step is to validate the dilepton yields by comparing with results for low-energy proton-proton and nucleus-nucleus collision measured by HADES.

Another interesting possibility left for the future is to employ the non-equilibrium dilepton production from a hadronic transport approach combined with a hydrodynamic approach (hybrid model) to explore the behaviour of electromagnetic radiation in hydrodynamics relative to transport approaches.

Acknowledgments

J.S. and H.P. acknowledge funding of a Helmholtz Young Investigator Group VH-NG-822 from the Helmholtz Association and GSI. This work was supported by the Helmholtz International Center for the Facility for Antiproton and Ion Research (HIC for FAIR) within the framework of the Landes-Offensive zur Entwicklung Wissenschaftlich- Oekonomischer Exzellenz (LOEWE) program launched by the State of Hesse. Computational resources have been provided by the Center for Scientific Computing (CSC) at the Goethe- University of Frankfurt.

References

[1] Agakishiev G *et al.* (HADES) 2012 *Eur. Phys. J.* **A48** 64 (*Preprint* 1112.3607)
[2] Adamczyk L *et al.* (STAR) 2015 *Phys. Rev.* **C92** 024912 (*Preprint* 1504.01317)
[3] Adare A *et al.* (PHENIX) 2016 *Phys. Rev.* **C93** 014904 (*Preprint* 1509.04667)
[4] Koehler M K (ALICE) 2014 *Nucl. Phys.* **A931** 665–669 (*Preprint* 1407.7809)
[5] Weil J *et al.* 2016 (*Preprint* 1606.06642)
[6] Weil J, Staudenmaier J and Petersen H 2016 *J. Phys. Conf. Ser.* **742** 012034 (*Preprint* 1604.07028)
[7] Bass S A *et al.* 1998 *Prog. Part. Nucl. Phys.* **41** 255–369
[8] Olive K A *et al.* (Particle Data Group) 2014 *Chin. Phys.* **C38** 090001
[9] Baldini A 1988 *Landolt-Boernstein. New Series* **1/12B**
[10] Agakishiev G *et al.* (HADES) 2009 *Eur. Phys. J. A* **40** 45–59 (*Preprint* 0902.4377)
[11] Heinz U W and Lee K S 1992 *Nucl. Phys.* **A544** 503–508
[12] Schmidt K, Santini E, Vogel S, Sturm C, Bleicher M and Stocker H 2009 *Phys. Rev.* **C79** 064908 (*Preprint* 0811.4073)
[13] Weil J, van Hees H and Mosel U 2012 *Eur. Phys. J.* **A48** 111

IOP Conf. Series: Journal of Physics: Conf. Series **832** (2017) 012038 doi:10.1088/1742-6596/832/1/012038

Impact of early stage non-equilibrium dynamics on photon production in relativistic heavy ion collisions

L Oliva[1,2], **M Ruggieri**[3], **S Plumari**[1,2], **F Scardina**[1,2] **and V Greco**[1,2]

[1] Department of Physics and Astronomy, University of Catania, Via S. Sofia 64, 95123 Catania, Italy
[2] INFN-Laboratori Nazionali del Sud, Via S. Sofia 62, 95123 Catania, Italy
[3] College of Physics, University of Chinese Academy of Sciences, Yuquanlu 19A, Beijing 100049, China

E-mail: `lucia.oliva@lns.infn.it`

Abstract. In this study we discuss our results on the spectrum of photons emitted from the quark-gluon plasma produced in heavy ion collisions at RHIC energies. Simulating the space-time evolution of the fireball by solving the relativistic Boltzmann transport equation and including two-particle scattering processes with photon emission allows us to make a first step in the description of thermal photons from the QGP as well as of those produced in the pre-equilibrium stage. Indeed, we consider not only a standard Glauber initial condition but also a model in which quarks and gluons are produced in the very early stage through the Schwinger mechanism by the decay of an initial color-electric field. In the latter approach relativistic kinetic equations are coupled in a self-consistent way to field equations. We aim at spotting the impact of early stage non-equilibrium dynamics on the photon production.

1. Introduction

Radiation of photons is considered an important and efficient probe to scrutinise the properties of the quark-gluon plasma. While hadrons are emitted from the freeze-out surface after suffering intense scatterings, photons are radiated during the whole spacetime history of the expanding fireball and, due to their electromagnetic nature, leave the system almost undisturbed, reaching the detector with the unaltered imprint of the circumstances of their production, which takes place from the initial stage up to the final one.

In heavy ion collisions, besides photons coming from hadron decay, there are photons emerging directly from collision processes, called *direct photons*, whose main categories are: *prompt* photons, which originate from initial hard scatterings; *pre-equilibrium* photons, produced before medium thermalization; *thermal* photons from qgp as well as from hadron gas. While thermal and prompt photons are quite well described, a good knowledge of photon emission in the initial out-of-equilibrium stage of heavy ion collision is still missing. Then, it is of particular interest to compute photon production in the initial stage of the system produced by the collision.

In the standard picture that describes the high energy nuclear collisions, after the two incoming nuclei had passed one through each other, a particular configuration of longitudinal color-electric and color-magnetic fields, named *Glasma*, is produced.

We model early time dynamics of relativistic heavy ion collisions assuming a longitudinal color-electric field which decays to gluon and quark-antiquark pairs through the Schwinger

IOP Conf. Series: Journal of Physics: Conf. Series **832** (2017) 012038 doi:10.1088/1742-6596/832/1/012038

effect; we couple in a self-consistent way the evolution equation of the initial field to the relativistic transport equation, which describes the evolution of the many-particle system and how it achieves an isotropic and thermalized state by means of collisions. In this talk we present our recent results about photon emission from the quark-gluon plasma, investigating the effect of pre-equilibrium dynamics on photon spectrum.

2. Modelling pre-equilibrium dynamics of relativistic heavy ion collisions

In this section we briefly explain how we simulate in our model the early stages of ultra-relativistic heavy ion collision; see Ref. [1] for further details.

2.1. Abelian flux tube model and Schwinger mechanism

We implement the Abelian Flux Tube (AFT) model [1, 2] with an initial condition constituted by a purely longitudinal color-electric field, neglecting for simplicity color-magnetic fields which should be considered in a more appropriate description of the Glasma state. We also assume the initial field is boost-invariant along the longitudinal direction and we ignore initial transverse fields that are produced by quantum fluctuations. The initial field then decays into particle quanta by the Schwinger mechanism, which predicts pair formation in a strong electric field [3].

In a static box the decay probability of a color-electric field E in massless particles is

$$\frac{dN_{jc}}{d\Gamma} \equiv p_0 \frac{dN_{jc}}{d^4x d^2p_T dp_z} = \frac{\mathcal{E}_{jc}}{4\pi^3} \left| \ln\left(1 \pm e^{-\pi p_T^2/\mathcal{E}_{jc}}\right)\right| \delta(p_z) p_0, \tag{1}$$

where the plus (minus) sign corresponds to the creation of a boson (fermion-antifermion) pair, the momentum components p_T and p_z refer to each of the two particles created from the vacuum and $p_0 = \sqrt{p_T^2 + p_z^2}$ is the single particle kinetic energy. $\mathcal{E}$ is the effective force acting on the tunnelling pair and depending on color and flavor; it is given by $\mathcal{E}_{jc} = (g|Q_{jc}E| - \sigma_j)\,\theta\,(g|Q_{jc}E| - \sigma_j)$, where σ_j corresponds to the string tension depending on the flavor considered. The Q_{jc} are color-flavor charges which in the case of quarks are $Q_{j1} = 1/2$, $Q_{j2} = -1/2$, $Q_{j3} = 0$, for $j = 1, N_f$; for antiquarks, corresponding to negative values of j, the color-flavor charges are just minus the corresponding charges for quarks; finally for gluons (labelled with $j = 0$) the charges are $Q_{01} = 1$, $Q_{02} = 1/2$, $Q_{03} = -1/2$, $Q_{04} = -Q_{01}$, $Q_{05} = -Q_{02}$, $Q_{06} = -Q_{03}$. We have only six gluons corresponding to the non-diagonal color generators, because we assume that the initial color field is polarized along the third direction of adjoint color space; the two gluon fields corresponding to the diagonal color generators do not couple with the background field, hence they cannot be produced by the Schwinger mechanism. Eq. 1 is then easily generalized to the case of an expanding box, see [1, 2].

Relativistic transport theory and field equations

The dynamics of the many-particle system produced by the decay of the color-electric field is described by the relativistic Boltzmann transport equation:

$$\left(p^\mu \partial_\mu + gQ_{jc} F^{\mu\nu} p_\nu \partial_\mu^p\right) f_{jc}(x,p) = \frac{dN_{jc}}{d\Gamma} + C_{jc}[f], \tag{2}$$

where $f_{jc}(x,p)$ is the distribution function for flavor j and color c and $F^{\mu\nu}$ is the color-electromagnetic tensor. On the right hand side $dN/d\Gamma$ is the invariant source term, which accounts for creation of pairs given by Eq. 1, and $C[f]$ represents the collision integral, which describes how f changes due to collision processes and is responsible for a finite value of the viscosity. We solve numerically the Boltzmann equation on a tridimensional lattice and using the test particle method to sample $f(x,p)$; the collision integral is computed by means of a stochastic algorithm [4, 5, 6, 7, 8, 9, 10]. With this theoretical approach one can follow the entire dynamical evolution of the system produced in relativistic heavy ion collisions. Altough usual inputs of a transport code are cross sections of a fixed set of microscopic processes, in our model we start from a fixed viscosity over entropy density ratio $\eta/s \neq 0$ and compute cross sections according to the Chapman-Enskog equation [8].

Assuming an Abelian dynamics, the evolution equations for the color-electric field are the Maxwell equations:

$$\nabla \cdot \mathbf{E} = \rho, \qquad \frac{\partial \mathbf{E}}{\partial t} = -\mathbf{J}, \tag{3}$$

where t is the time in the laboratory frame. The initial field is only longitudinal, whereas transverse components of $\mathbf{E}$ will be generated by transverse currents according to Eq. 3. The total current is $\mathbf{J} = \mathbf{J}_D + \mathbf{J}_M$, where the *displacement current* $\mathbf{J}_D$ is due to the dipoles created by the Schwinger effect and the *matter current* $\mathbf{J}_M$ depends on the motion of the charges. Since color charges and currents depend on parton distribution functions as well as the kinetic equations, we solve field and particles equations self-consistently, taking into account the back-reaction of particles on the field. We notice that, even if the equation of motion of the classical field is abelian, there is a sign of the non-abelian nature of QCD because the background field interacts with gluon quanta via $\mathbf{J}_D$.

3. Effect of pre-equilibrium on photon spectrum

In order to simulate photon production we add to the collision integral of Eq. 2 the processes with a photon in the final state. In this talk we focus on the $2 \to 2$ scatterings, i.e. the QCD Compton scattering $q/\bar{q} + g \to q/\bar{q} + \gamma$ and the quark-antiquark annihilation $q + \bar{q} \to g + \gamma$, leaving the inclusion of $2 \to 3$ processes and of the LPM effect [11] to future works.

We are interested in grasping the effect of pre-equilibrium stage on direct photon spectrum. For this purpose we compare simulations starting at $t_0 = 0.01\,\mathrm{fm}/c$ with the model described in the previous section to simulations with an equilibrium initial condition. The latter is based on the Glauber model, with an x-space distribution given by the a standard mixture $0.85 N_{part} + 0.15 N_{coll}$ and a p-space thermalized spectrum in the transverse plane at a time $t_0 = 0.6\,\mathrm{fm}/c$ with a maximum initial temperature $T_0 = 340\,\mathrm{MeV}$. Following Ref. [9] we call this case *Th-Glauber*, while we refer to simulations with the Abelian flux tube model as *AFT-Schwinger*. In this study we focus on RHIC collision at an impact parameter $b = 7.5\,\mathrm{fm}$.

In the AFT-Schwinger simulation we assume the initial chromo-electric field is boost invariant in the longitudinal direction and smooth in transverse plane, with the specific configuration given by a Glauber-type distribution:

$$E_z^0(x, y) = E_{max}^0 \left(c_c\, \rho_{coll}(x, y) + c_p\, \rho_{part}(x, y) \right), \tag{4}$$

with $c_p = 1 - c_c$. The two free parameters E_{max}^0 and c_c are fixed in order to match at $t_0 = 0.6\,\mathrm{fm}/c$ particle multiplicity and eccentricity with those of the Th-Glauber case at initial time. For RHIC collision at $b = 7.5\,\mathrm{fm}$ we obtain an initial longitudinal field with $E_{max}^0 = 3.3\,\mathrm{GeV}^2$ and $c_c = 0.6$.

In Fig. 1 we plot our result for photon spectrum (main panel) and photon number (inset panel) obtained in the Th-Glauber and in the AFT-Schwinger cases. We show also the curve obtained with the AFT-Schwinger model in which photon production is artificially started at 0.6 fm/c; comparing the latter with the Th-Glauber case, we notice that both photon spectrum and photon number are similar in the two simulations, meaning that the system evolves in the same way after 0.6 fm/c with Glauber and non-equilibrium initial conditions. Thus, we can consider the differences between AFT-Schwinger and Th-Glauber curves as due to pre-equilibrium photons. We see that pre-equilibrium photons enhance the total photon number of about 30%, giving a significant contribution at higher transverse momentum, in a p_T region where thermal emission (both from QGP and from hadron gas) becomes less important.

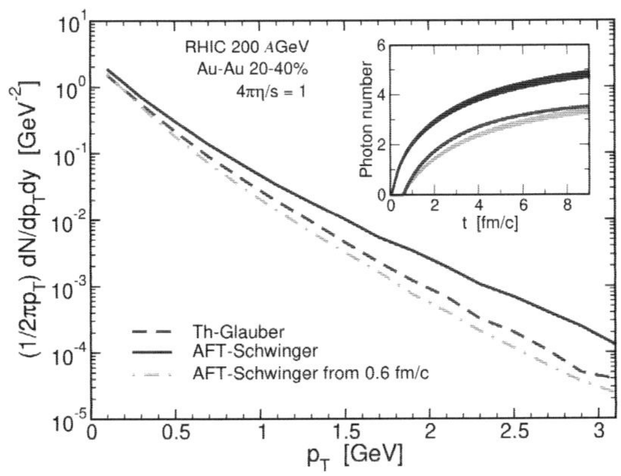

Figure 1. Photon spectrum at midrapidity $|y| < 0.5$ (main panel) and total photon number (inset panel) obtained by AFT-Schwinger (solid maroon line) and Th-Glauber (dashed blue line) simulations. The dot-dashed orange curve corresponds to an AFT-Schwinger simulation in which photon production is enabled from $t = 0.6\,\mathrm{fm/c}$.

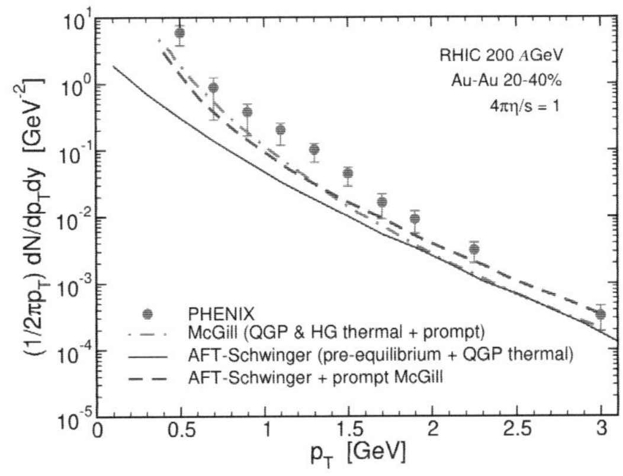

Figure 2. Comparison of the final photon spectrum obtained by Paquet et al. from the McGill University [13] (dot-dashed red line) with our results. In the dashed indico curve McGill's prompt photons are added to our AFT-Schwinger result (thin solid maroon line). Experimental data are from PHENIX Collaboration [12].

We stress that we do not compute photon production by integrating photon rate over a spacetime volume, as it is done in hydrodynamics calculations. Indeed, we use scattering matrix elements in the collision integral; this allows to follow photon production since the very early stage as soon as particles pop up from the vacuum.

In Fig. 2 we compare the direct photon spectrum obtained by our AFT-Schwinger simulation with experimental data [12] and with the result of Ref. [13]. The latter includes not only QGP thermal photons but also thermal hadronic emission as well as prompt photons, while does not take into account photon production before QGP equilibration. In the figure we plot also our result in which prompt photons from Ref. [13] are added. We notice that, although we need to include in our simulations higher order photon emission processes as well as hadronic thermal production which would enhance the direct photon yield mainly at $p_T < 1.5\,\mathrm{GeV}$, on the other hand our model of the early stages of relativistic heavy ion collision allows to describe pre-equilibrium photons which may reduce the tension between theory and experimental data.

References

[1] Ruggieri M, Puglisi A, Oliva L, Plumari S, Scardina F and Greco V 2015 *Phys. Rev.* C **92** 064904
[2] Ryblewski R and Florkowski W 2013 *Phys. Rev.* D **88** 034028
[3] Schwinger J S 1951 *Phys. Rev.* **82** 664-79
[4] Xu Z, Greiner C and Stocker H 2008 *Phys. Rev. Lett.* **101** 082302
[5] Xu Z and Greiner C 2009 *Phys. Rev.* C **79** 014904
[6] Bratkovskaya E L, Cassing W, Konchakovski V P and Linnyk O 2011 *Nucl. Phys.* A **856** 162-82
[7] Ferini G, Colonna M, Di Toro M and Greco V 2009 *Phys. Lett.* B **670** 325-9
[8] Plumari S, Puglisi A, Scardina F and Greco V 2012 *Phys. Rev.* C **86** 054902
[9] Ruggieri M, Scardina F, Plumari S and Greco V 2013 *Phys. Lett.* B **727** 177-81
[10] Ruggieri M, Scardina F, Plumari S and Greco V 2014 *Phys. Rev.* C **89** 054914
[11] Arnold P, Moore G D and Yaffe L G 2001 *JHEP* **0112** 009 *and references therein*
[12] Adare A *et al* 2015 *Phys. Rev.* C **91** 064904
[13] Paquet J F, Shen C, Denicol G S, Luzum M, Schenke B, Jeon S and Gale C 2016 *Phys. Rev.* C **93** 044906

IOP Conf. Series: Journal of Physics: Conf. Series **832** (2017) 012039 doi:10.1088/1742-6596/832/1/012039

Exploring the Nuclear Phase Diagram with Beam Energy Scans

Stephen Horvat

E-mail: stephen.horvat@yale.edu

Abstract. The nuclear phase diagram is mapped using beam energy scans of relativistic heavy-ion collisions. This mapping is possible because different collision energies develop along different trajectories through the phase diagram. High energy collisions will evolve though a crossover phase transition according to lattice QCD, but lower collision energies may traverse a first order phase transition. There are hints for this first order phase transition and its critical endpoint, but further measurements and theoretical guidance is needed. In addition to mapping the phase transition, beam energy scans allow us to see if we can turn off the signatures of deconfinement. If an observable is a real signature for the formation of the deconfined state called quark-gluon plasma, then it should turn off at sufficiently low collision energies. In this summary talk I will show the current state of the field using beam energy scan results from RHIC and SPS, I will show where precise theoretical guidance is needed for understanding recent measurements, and I will motivate the need for more data and new measurements from FAIR, NICA, RHIC, and the SPS.

1. Introduction

When the SPS began operations it reached collision energies where many theorists expected the onset of deconfinement [1]. Early observables for deconfinement were compelling, but many in the field thought that they could also be explained by other phenomena. Taking the "horn" in K^+/π^+ produced in heavy-ion collisions as a function of $\sqrt{s_{\rm NN}}$ as an example, several models that did not include a deconfined partonic phase were able to explain the data [2]. It was not until RHIC observed jet energy-loss signatures at higher energies as well as strong azimuthal anisotropy in particle distributions that the majority of the heavy-ion community became convinced that a strongly interacting QGP phase had been produced [3, 4, 5, 6]. The SPS, RHIC, and the LHC have gone on to provide many additional measurements that constrain the properties of, and the evolution of, the produced medium.

Central collisions at the LHC and at the highest RHIC energies form regions with very high initial energy and entropy densities. These overlap regions evolve hydrodynamically to follow different trajectories through the QCD phase diagram depending on their initial conditions [7, 8], where lower energy collisions follow a trajectory with a higher baryon chemical potential [9]. While lattice simulations of the QCD equation of state predict a crossover phase transition from a hadronic gas to partonic degrees of freedom for small values of baryon chemical potential as the temperature is increased above a critical value, at higher values of the baryon chemical potential their may be a first order phase transition [10]. One of the purposes of beam energy scans (BESs) in heavy ion collisions is to identify the order of the phase transition and locate its corresponding critical point if it exists. The other purpose is to test that signatures of

IOP Conf. Series: Journal of Physics: Conf. Series **832** (2017) 012039 doi:10.1088/1742-6596/832/1/012039

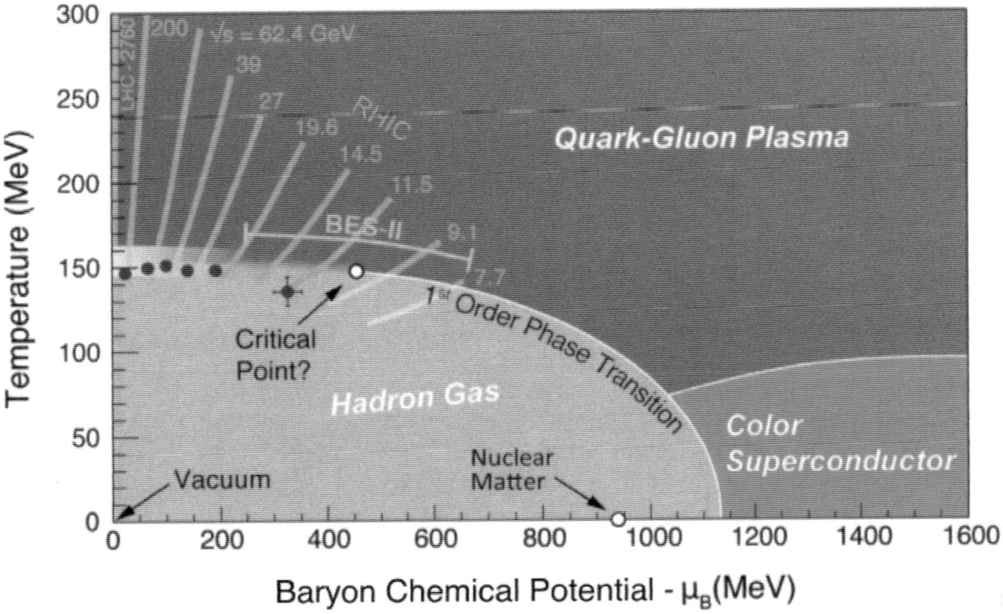

Figure 1. A cartoon of the QCD phase diagram [7].

deconfinement 'turn off' as the collision energy is decreased and to constrain the conditions required to form QGP.

2. Turning off signatures of QGP formation

There are many proposed signatures for the formation of QGP, but they generally fall into three classes: probes produced early in the collision that go on to interact with the medium, radiation from the medium, and final state particles that are formed from constituents of the medium. For the first class we consider as early probes high transverse momentum (p_T) partons that interact strongly with QGP losing energy in a process called jet-quenching.

The nuclear modification factor, R_{AA}, is the number of particles measured at a particular p_T in a heavy ion collision divided by what we would expect from independent nucleonic scatterings. To model these independent nucleonic scatterings we estimate how many of them there were by using a Glauber Monte Carlo [12] and then scale the number of particles measured at the same p_T in proton-proton collisions up by this number. If $R_{AA} = 1$ then either there are no modifications from colliding heavy nuclei relative to colliding protons, or there is a balancing of enhancement and suppression effects. However, if $R_{AA} \ll 1$ at high-p_T then this may be explained by high momentum particles losing momentum in QGP. Figure 2 shows R_{AA} for several collision energies from the SPS, RHIC, and the LHC [11]. While RHIC and the LHC measured large suppression at high-p_T, the SPS found $R_{AA} = 1$ for pions at mid-rapidity. Here we want to separate the idea of turning off QGP formation from the idea of turning off a signature for QGP formation. Depending on the measurement's sensitivity to QGP the signature may "turn off' even though QGP is still being formed. The relative time spent in a partonic phase as well as the spacial extent of the partonic phase at hadronization can be expected to be reduced as collision energies are reduced. The best signatures for QGP formation will be those that are maximally sensitive to the partonic phase and minimally effected by other phenomena. In the case of R_{AA}, measurements of strong enhancement at $\sqrt{s_{NN}} = 7.7$ GeV demonstrate that

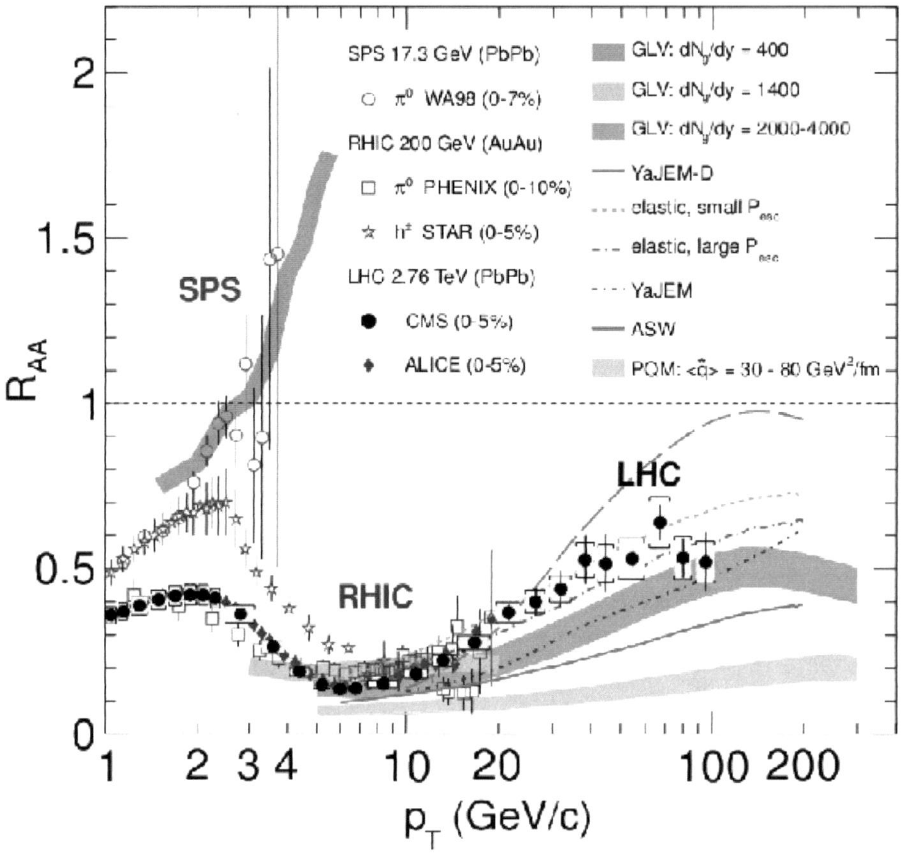

Figure 2. World data for $R_{\rm AA}$ [11].

$R_{\rm AA}$ is strongly effected by enhancement phenomena that make this observable less sensitive to QGP formation at SPS and RHIC BES energies [13]. Alternative observables for high-$p_{\rm T}$ probes suggest deconfinement may persist at least down to $\sqrt{s_{\rm NN}} = 14.5$ GeV [13, 14].

3. First Order Phase Transition

While the previous signatures give information about what phases the medium evolves through, it is also interesting to determine the nature of the phase transition. While lattice calculations tell us that the transition at small baryon chemical potential (μ_B) is a crossover, there have been suggestions that there may be a first or second order phase transition at higher μ_B [15, 16, 17, 18, 19]. It has been shown that the pressure and speed of sound in the medium would have minima near a first order phase transition and that this would lead to a characteristic shape in the directed flow measured in heavy-ion collisions [20, 21, 22, 23]. Measurements from STAR are shown in Fig. 3 where we see trends that are consistent with the predicted trends [24].

PHENIX accessed the same underlying physical processes using a different observable with Hanbury Brown and Twist (HBT) data that were collected by ALICE, PHENIX, and STAR [25]. Here the $(\mathrm{R_{out}})^2 - (\mathrm{R_{side}})^2$ in Fig. 4(a) is related to emission duration while the quantity in Fig. 4(b) is related to the speed of sound [24, 26].

IOP Conf. Series: Journal of Physics: Conf. Series **832** (2017) 012039 doi:10.1088/1742-6596/832/1/012039

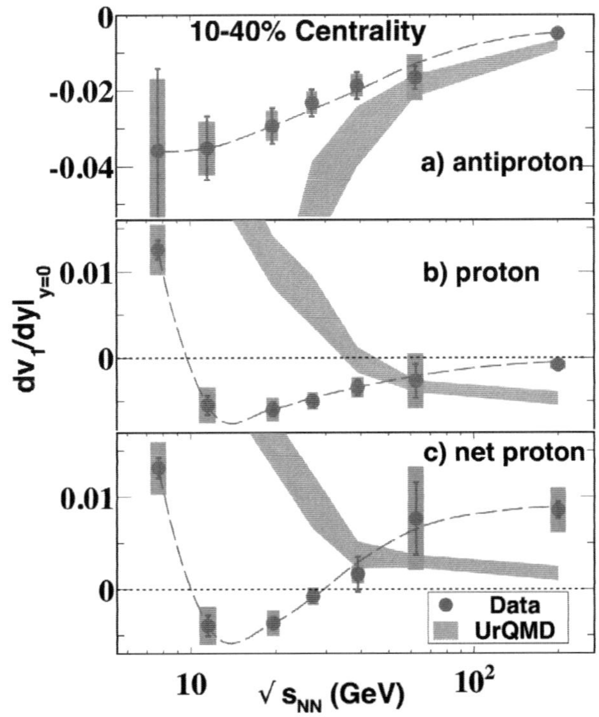

Figure 3. The slope of v_1 versus rapidity as a function of $sqrts_{\rm NN}$ for mid-central (10-40%) collisions for antiprotons (a), protons (b), and net protons (c), and compared to UrQMD [24].

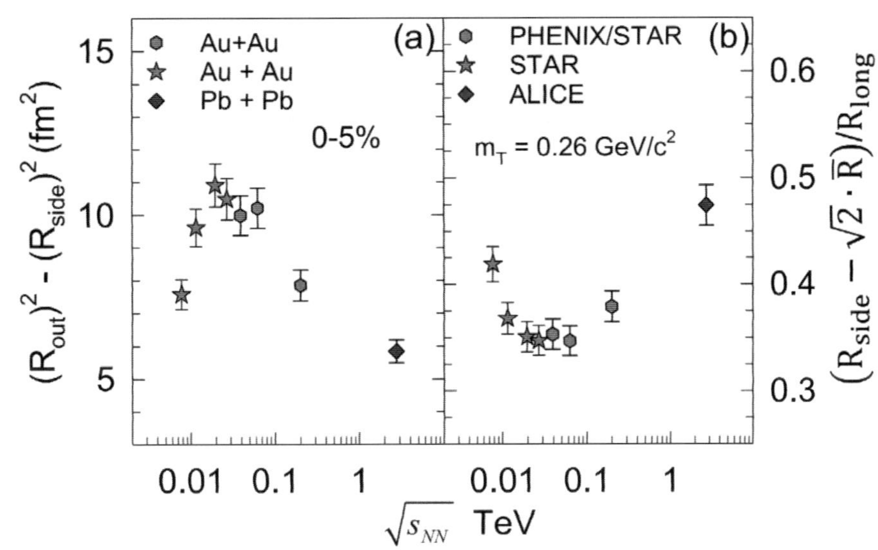

Figure 4. Two non-monotonic arangements of HBT data are shown for 0-5% centrality. The left panel is related to emmision duration and has a maximum in the range of $\sqrt{s_{\rm NN}}$ = 19.6 - 39 GeV. The right panel is related to the speed of sound and has a minimum in the range of $\sqrt{s_{\rm NN}}$ = 19.6 - 62.4 GeV [25].

IOP Conf. Series: Journal of Physics: Conf. Series **832** (2017) 012039 doi:10.1088/1742-6596/832/1/012039

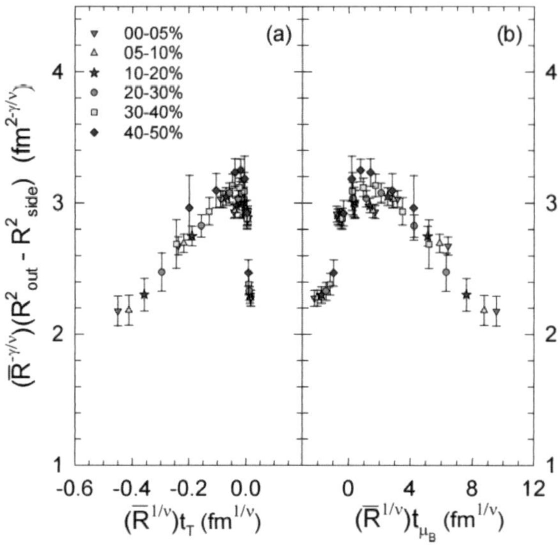

Figure 5. Finite scaling of HBT data using critical exponents collapses the data for multiple energies and centralities onto a single curve with the critical point estimated at $T^{cep} \approx 165$ MeV and $\mu_B^{cep} \approx 95$ MeV [26].

4. Critical End Point

Near a critical point there would be a divergence of susceptibilities and conserved quantities would have increased fluctuations [27]. PHENIX and STAR have measured the energy dependence of higher moments of net charge [28, 29]. While these results did not locate the critical point, more data and larger acceptance will be available for RHIC BES II.

Figure 5 shows a measurement that does suggest both the existence and the location of the critical point by doing a finite size scaling of HBT data using critical exponents [26]. This analysis found critical exponents $\nu \approx 0.66$ and $\gamma \approx 1.2$ and the location of the critical point to be $T^{cep} \approx 165$ MeV and $\mu_B^{cep} \approx 95$ MeV. The idea here is that with the proper description of the data, the data will all point toward the critical point even if the trajectories of individual collisions, such as those at the LHC, do not pass particularly close to the critical point. This is similar to the method that was used earlier to determine the location of the liquid-vapor critical point [30]. This work was carried further by showing that these same critical exponents can be used to scale the moments of net proton distributions [31].

5. Summary

With a couple of possible exceptions [26, 31] these signatures do not characterize the order of the phase transition or describe the location of a critical point but with more observables, data, and detector upgrades the SPS, RHIC, FAIR and NICA are poised to gather the data over a broad range of energies and collision systems. In order to coordinate the work of theorists who are working to describe in quantitative detail the different stages of evolution in heavy-ion collisions at low energies, and to provide direction to experimentalists, a new theory collaboration has been formed called the Beam Energy Scan Theory (BEST) collaboration. The next few years will see BES data being collected at four different experimental facilities and the coordination of BES theorists in order to map the QCD phase diagram.

References

[1] K. Kajantie. Quark Matter '84, Helsinki, 17–21 June 1984. *Nucl. Phys.*, A956:907–910, 2016.

[2] Jajati K. Nayak, S. Banik, and Jan-e Alam. The horn in the kaon to pion ratio. *Phys. Rev.*, C82:024914, 2010.

[3] I. Arsene et al. *Nucl.Phys.*, A **757**:1–27, 2005.

[4] B.B. Back et al. *Nucl.Phys.*, A **757**:28–101, 2005.

[5] John Adams et al. *Nucl.Phys.*, A **757**:102–183, 2005.

[6] K. Adcox et al. *Nucl.Phys.*, A **757**:184–283, 2005.

[7] Ani Aprahamian et al. Reaching for the horizon: The 2015 long range plan for nuclear science. 2015.

[8] Chiho Nonaka and Masayuki Asakawa. Hydrodynamical evolution near the QCD critical end point. *Phys. Rev.*, C71:044904, 2005.

[9] J. Cleymans, H. Oeschler, K. Redlich, and S. Wheaton. Status of chemical freeze-out. *J. Phys.*, G32:S165–S170, 2006.

[10] Z. Fodor and S. D. Katz. Critical point of QCD at finite T and mu, lattice results for physical quark masses. *JHEP*, 04:050, 2004.

[11] Serguei Chatrchyan et al. Study of high-pT charged particle suppression in PbPb compared to *pp* collisions at $\sqrt{s_{NN}} = 2.76$ TeV. *Eur. Phys. J.*, C72:1945, 2012.

[12] Michael L. Miller, Klaus Reygers, Stephen J. Sanders, and Peter Steinberg. *Ann.Rev.Nucl.Part.Sci.*, **57**:205–243, 2007.

[13] Stephen P. Horvat. Measurement of hadron suppression and study of its connection with vanishing v_3 at low $\sqrt{s_{\rm NN}}$ in Au+Au collisions with STAR. *Nucl. Phys.*, A956:838–841, 2016.

[14] James D. Brandenburg. Identified Light and Strange Hadron Spectra at $\sqrt{s_{NN}}$=14.5 GeV and Systematic Study of Baryon/Meson Effect at Intermediate Transverse Momentum with STAR at RHIC BES I. *Nucl. Phys.*, A956:445–448, 2016.

[15] Juergen Berges and Krishna Rajagopal. Color superconductivity and chiral symmetry restoration at nonzero baryon density and temperature. *Nucl. Phys.*, B538:215–232, 1999.

[16] Yoshitaka Hatta and Takashi Ikeda. Universality, the QCD critical / tricritical point and the quark number susceptibility. *Phys. Rev.*, D67:014028, 2003.

[17] Mikhail A. Stephanov. QCD phase diagram and the critical point. *Prog. Theor. Phys. Suppl.*, 153:139–156, 2004. [Int. J. Mod. Phys.A20,4387(2005)].

[18] Masayuki Asakawa and Chiho Nonaka. Critical end point and its consequences. *Nucl. Phys.*, A774:753–756, 2006.

[19] Philippe de Forcrand, Jens Langelage, Owe Philipsen, and Wolfgang Unger. Lattice QCD Phase Diagram In and Away from the Strong Coupling Limit. *Phys. Rev. Lett.*, 113(15):152002, 2014.

[20] Dirk H. Rischke, Yaris Pursun, Joachim A. Maruhn, Horst Stoecker, and Walter Greiner. The Phase transition to the quark - gluon plasma and its effects on hydrodynamic flow. *Heavy Ion Phys.*, 1:309–322, 1995.

[21] L. P. Csernai and D. Rohrich. Third flow component as QGP signal. *Phys. Lett.*, B458:454, 1999.

[22] J. Brachmann, S. Soff, A. Dumitru, Horst Stoecker, J. A. Maruhn, W. Greiner, L. V. Bravina, and D. H. Rischke. Antiflow of nucleons at the softest point of the EoS. *Phys. Rev.*, C61:024909, 2000.

[23] Horst Stoecker. Collective flow signals the quark gluon plasma. *Nucl. Phys.*, A750:121–147, 2005.

[24] L. Adamczyk et al. Beam-Energy Dependence of the Directed Flow of Protons, Antiprotons, and Pions in Au+Au Collisions. *Phys. Rev. Lett.*, 112(16):162301, 2014.

[25] A. Adare et al. Beam-energy and system-size dependence of the space-time extent of the pion emission source produced in heavy ion collisions. 2014.

[26] Roy A. Lacey. Indications for a Critical End Point in the Phase Diagram for Hot and Dense Nuclear Matter. *Phys. Rev. Lett.*, 114(14):142301, 2015.

[27] Misha A. Stephanov, K. Rajagopal, and Edward V. Shuryak. Event-by-event fluctuations in heavy ion collisions and the QCD critical point. *Phys. Rev.*, D60:114028, 1999.

[28] J. T. Mitchell. The RHIC Beam Energy Scan Program: Results from the PHENIX Experiment. *PoS*, CPOD2013:003, 2013.

[29] L. Adamczyk et al. Beam energy dependence of moments of the net-charge multiplicity distributions in Au+Au collisions at RHIC. *Phys. Rev. Lett.*, 113:092301, 2014.

[30] J. B. Elliott et al. The Liquid to vapor phase transition in excited nuclei. *Phys. Rev. Lett.*, 88:042701, 2002.

[31] Roy A. Lacey, Peifeng Liu, Niseem Magdy, B. Schweid, and N. N. Ajitanand. Finite-Size Scaling of Non-Gaussian Fluctuations Near the QCD Critical Point. 2016.

IOP Conf. Series: Journal of Physics: Conf. Series **832** (2017) 012040 doi:10.1088/1742-6596/832/1/012040

Phase Structure and Dynamics of QCD– A Functional Perspective

Nils Strodthoff

Nuclear Science Division, Lawrence Berkeley National Laboratory, Berkeley, CA 94720, USA

E-mail: `nstrodthoff@lbl.gov`

Abstract. The understanding of the phase structure and the fundamental properties of QCD matter from its microscopic description requires appropriate first-principle approaches. Here I review the progress towards a quantitative first-principle continuum approach within the framework of the Functional Renormalization group established by the fQCD collaboration. I focus on recent quantitative results for quenched QCD and Yang-Mills in the vacuum before addressing the calculation of dynamical quantities such as spectral functions and transport coefficients in this framework.

1. Introduction

The understanding of the phase structure of QCD from first principles has been on the forefront of research in hadron physics for several decades [1]. While Lattice QCD provides a reliable approach at vanishing and small chemical potentials, the fermion sign problem makes reliable predictions at large chemical potentials, $\mu_B \gtrsim T$, very difficult [2]. The latter are however required to make quantitative predictions about the existence and location of a possible critical point in the phase diagram, a question which has been one of the driving forces for theoretical as well as experimental efforts in the field. The second major challenge is to gain a quantitative understanding of the fundamental properties of QCD matter from a microscopic description. This includes observables connected to the hadron spectrum such as pole masses, decay constants, form factors or scattering amplitudes but also real-time observables like elementary spectral functions and transport coefficients with particular relevance for phenomenological applications. Such real-time observables are notoriously difficult to obtain in Euclidean approaches due to the necessity of performing an analytic continuation from Euclidean to Minkowski signature.

Both fundamental challenges require first-principle approaches to QCD in order to make quantitative predictions. The fQCD collaboration [3] aims to provide such a quantitative continuum approach to QCD, cf. [4] for an overview, [5, 6] for recent quantitative works in quenched QCD and Yang-Mills (YM) theory and [7] for an exploratory study of the unquenched system. This implies in particular to renounce the use of phenomenological input and to restrict the input parameters to those of QCD itself, namely the strong coupling and the quark mass at perturbative momentum scales.

IOP Conf. Series: Journal of Physics: Conf. Series **832** (2017) 012040 doi:10.1088/1742-6596/832/1/012040

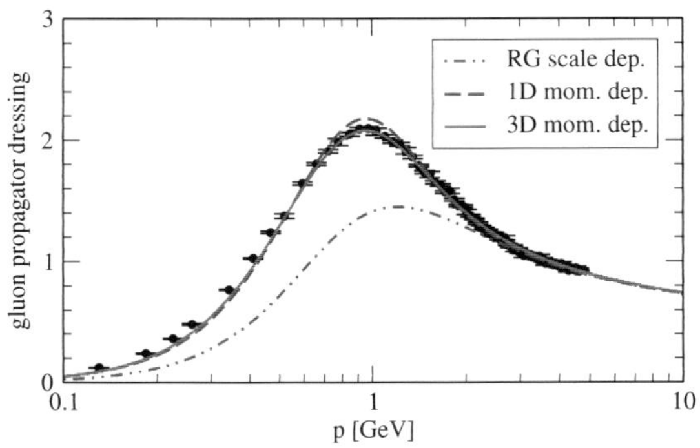

(a) Gluon propagator in Yang-Mills theory in comparison to lattice results (black dots); figure taken from [6].

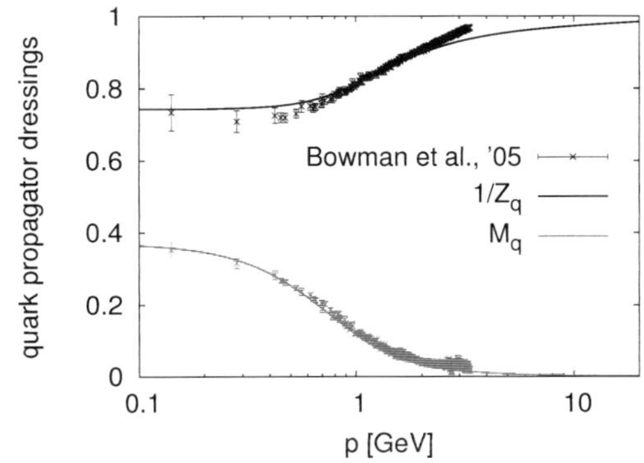

(b) Quark propagator in quenched QCD in comparison to lattice results; figure taken from [5].

Figure 1. Euclidean propagator results from first-principle continuum QCD

2. Euclidean correlation functions from continuum QCD

Functional continuum approaches exploit relations between off-shell Green's functions. In the following I will focus on the Functional Renormalization group (FRG) which represents a powerful non-perturbative tool that implements the idea of the Wilson RG of gradually integrating quantum fluctuations momentum shell by momentum shell in a formulation in momentum space, see [8, 9, 10, 11, 12] for QCD-related reviews. Technically this is achieved by adding an infrared regulator to the theory which acts like a mass term thereby suppressing fluctuations below a certain cutoff scale. Consequently all objects in the theory carry a dependence on this artificial regulator scale. The central object in this approach is the scale-dependent analogue of the effective action Γ_k whose scale dependence is governed by a one loop equation,

$$k\partial_k\Gamma_k = \frac{1}{2}\mathrm{Tr}\frac{1}{\Gamma_k^{(2)} + R_k}\, k\partial_k R_k\,, \tag{1}$$

where $\Gamma_k^{(2)}$ denotes the full scale-, momentum- and field-dependent inverse propagator and R_k is the regulator term. Flow equations for higher n-point function can then be derived by taking functional derivatives with respect to the fields in (1). These coupled systems of functional equations then have to be integrated from some large UV scale down to $k = 0$ where the regulator term is removed and the effective action is recovered. The practical application requires the use of appropriate non-perturbative expansion schemes to render the infinite tower of coupled equations tractable. One example for such a scheme is the vertex expansion corresponding to an expansion of the effective action in terms of 1-particle irreducible vertex functions.

This was put into practice in a calculation of the transverse correlation functions in Landau-gauge Yang-Mills theory [6] by self-consistently solving the system of propagators and primitively divergent vertex functions. The resulting gluon propagator is found to be in good agreement with lattice results, see Fig. 1a. This study is complemented by a study of the quenched matter system using only gluon and ghost propagators as input [5]. Again the quark propagator is found in very good agreement with the lattice, see Fig. 1b. To achieve this quantitative agreement it is required to consider a large truncation scheme including in particular a full tensor basis for the quark-gluon vertex as well as non-classical higher-order quark-gluon interaction vertices, which is only possible with the help of appropriate (computer-)algebraic tools [13, 14]. The emergence of effective mesonic degrees of freedom is captured by means of the dynamical hadronisation

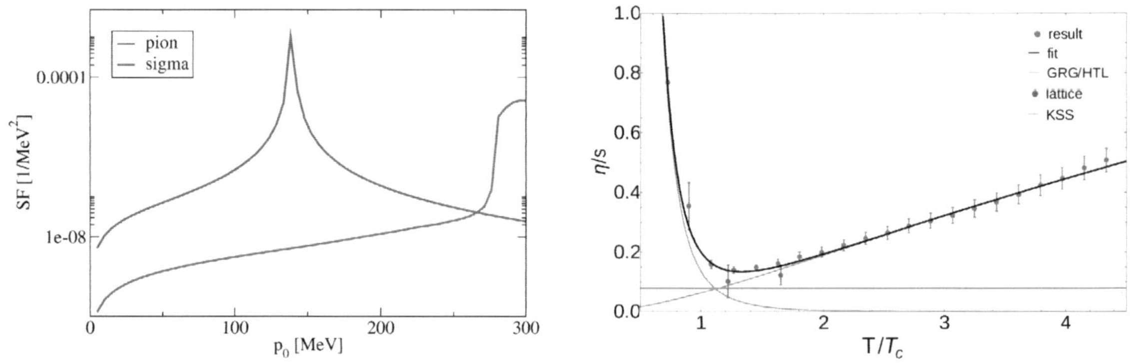

(a) Self-consistent spectral functions in the $O(N)$ model; figure adapted from [17].

(b) Shear viscosity over entropy density ratio η/s in YM theory; figure taken from [18].

Figure 2. Exemplary real-time results from functional approaches

technique, see [15] and [5] for the momentum-dependent generalization, which recasts resonant structures in four-fermi interactions in terms of effective mesonic interactions.

First quantitative results for the coupled system of QCD correlation functions in the vacuum are currently in preparation and will soon be extended to finite temperature. The further extension to moderately large chemical potentials will then allow for example studies of fluctuation observables, see [16] for a recent model study in this direction.

3. Direct calculation of real-time observables

Apart from the promising self-consistent solutions of Euclidean correlation functions, there has been tremendous progress in the direct calculation of real-time observables within the FRG. The central difficulty in calculating real-time observables from a Euclidean approach is the necessity of performing an analytic continuation in the external momentum. The standard procedure is to use Euclidean data and to infer the analytic continuation numerically e.g. by means of the Maximum Entropy Method (MEM). However, this procedure goes along with large systematic errors, which urges for a direct calculation of real-time observables.

The main idea of the approach initiated in [19] is the fact that the external momentum appears, very much like the chemical potential, as an external parameter in the functional equations. This allows an analytic continuation on the level of the equation itself and hence a direct calculation of elementary spectral functions anchored to a quantitative Euclidean baseline. This procedure has been applied successfully in scalar models [19] and Yukawa systems [20, 21, 22]. The application to numerical propagators as prerequisite for studies of spectral functions in full QCD required a significant extension of the original formalism [23]. The applicability of this procedure to include also momentum-dependent vertices and the full (complex) momentum dependence has recently been demonstrated in [17] at the example of the first self-consistent direct calculation of spectral functions in the $O(N)$ model, see Fig. 2a.

Elementary spectral functions do not only provide physical insights by itself but furthermore provide input for the calculation of transport coefficients by means of Kubo formulae, which relate the latter to correlation functions of the energy-momentum tensor. These can in turn be calculated within a finite diagrammatic expansion in terms of full propagators and vertices, which represent the only input of the calculation. This procedure has been applied to the calculation of the shear viscosity over entropy density ratio η/s in Yang-Mills theory [24] using gluonic spectral functions obtained via MEM as input. This allows to determine the temperature dependence of η/s over wide range of temperatures, see Fig. 2b. The asymptotic behavior is found to be in agreement with perturbative results whereas the low temperature regime is well-described by a power-law decay attributed to a glueball resonance gas similar to the hadron resonance gas in QCD. Based on a direct sum of these two contributions we provided a first estimate for η/s in QCD [24]. This represents the first step towards a direct determination of transport coefficients from the microscopic dynamics as urgently required input for hydrodynamical calculations.

4. Summary and Outlook

In this contribution I reviewed the recent progress on the QCD phase diagram and the direct calculation of real-time observables within the framework of the FRG. Firstly, there are by now quantitative results available for Yang-Mills theory and QCD with extensions to finite temperature and density on the way. These were obtained by self-consistently solving the systems of correlation functions renouncing further phenomenological input. The extension to finite chemical potential will allow for example the calculation of fluctuation observables such as cumulants of the net-baryon distribution or couplings [25] possibly coupled to the hydrodynamic evolution and including non-equilibrium effects [26]. Secondly, there has been tremendous progress in a direct calculation of real-time observables in this framework. In combination with the quantitative Euclidean results from above this approach has very bright prospects for a direct calculation of elementary spectral functions which can then in a second step be used for the calculation of transport coefficients in full QCD.

Acknowledgments

The author thanks the fQCD collaboration and in particular N. Christiansen, A. K. Cyrol, M. Mitter and J. M. Pawlowski for joint work on topics discussed above. This work was supported by the DFG under grant no. Str1462/1-1 and by the Office of Nuclear Physics in the US Department of Energy's Office of Science under Contract No. DE-AC02-05CH11231.

References

[1] Fukushima K and Hatsuda T 2011 *Rept.Prog.Phys.* **74** 014001 (*Preprint* 1005.4814)
[2] de Forcrand P 2009 *PoS* **LAT2009** 010 (*Preprint* 1005.0539)
[3] fQCD Collaboration, J. Braun, L. Corell, A. K. Cyrol, W.-j. Fu, M. Leonhardt, M. Mitter, J. M. Pawlowski, M. Pospiech, F. Rennecke, N. Strodthoff, N. Wink
[4] Pawlowski J M 2014 *Nucl.Phys.* **A931** 113–124
[5] Mitter M, Pawlowski J M and Strodthoff N 2015 *Phys.Rev.* **D91** 054035 (*Preprint* 1411.7978)
[6] Cyrol A K, Fister L, Mitter M, Pawlowski J M and Strodthoff N 2016 *Phys. Rev.* **D94** 054005 (*Preprint* 1605.01856)
[7] Braun J, Fister L, Pawlowski J M and Rennecke F 2016 *Phys. Rev.* **D94** 034016 (*Preprint* 1412.1045)
[8] Berges J, Tetradis N and Wetterich C 2002 *Phys. Rept.* **363** 223–386 (*Preprint* hep-ph/0005122)
[9] Pawlowski J M 2007 *Annals Phys.* **322** 2831–2915 (*Preprint* hep-th/0512261)
[10] Gies H 2012 *Lect.Notes Phys.* **852** 287–348 (*Preprint* hep-ph/0611146)
[11] Schaefer B J and Wambach J 2008 *Phys.Part.Nucl.* **39** 1025–1032 (*Preprint* hep-ph/0611191)
[12] Braun J 2012 *J.Phys.* **G39** 033001 (*Preprint* 1108.4449)
[13] Huber M Q and Braun J 2012 *Comput.Phys.Commun.* **183** 1290–1320 (*Preprint* 1102.5307)
[14] Cyrol A K, Mitter M and Strodthoff N 2016 (*Preprint* 1610.09331)
[15] Gies H and Wetterich C 2002 *Phys.Rev.* **D65** 065001 (*Preprint* hep-th/0107221)
[16] Fu W j, Pawlowski J M, Rennecke F and Schaefer B J 2016 (*Preprint* 1608.04302)
[17] Strodthoff N 2016 (*Preprint* 1611.05036)
[18] Christiansen N, Haas M, Pawlowski J M and Strodthoff N 2015 *Phys. Rev. Lett.* **115** 112002 (*Preprint* 1411.7986)
[19] Kamikado K, Strodthoff N, von Smekal L and Wambach J 2014 *Eur.Phys.J.* **C74** 2806 (*Preprint* 1302.6199)
[20] Tripolt R A, Strodthoff N, von Smekal L and Wambach J 2014 *Phys.Rev.* **D89** 034010 (*Preprint* 1311.0630)
[21] Yokota T, Kunihiro T and Morita K 2016 *PTEP* **2016** 073D01 (*Preprint* 1603.02147)
[22] Jung C, Rennecke F, Tripolt R A, von Smekal L and Wambach J 2016 (*Preprint* 1610.08754)
[23] Pawlowski J M and Strodthoff N 2015 *Phys. Rev.* **D92** 094009 (*Preprint* 1508.01160)
[24] Haas M, Fister L and Pawlowski J M 2014 *Phys.Rev.* **D90** 091501 (*Preprint* 1308.4960)
[25] Bzdak A, Koch V and Strodthoff N 2016 (*Preprint* 1607.07375)
[26] Herold C, Nahrgang M, Yan Y and Kobdaj C 2016 *Phys. Rev.* **C93** 021902 (*Preprint* 1601.04839)

IOP Conf. Series: Journal of Physics: Conf. Series **832** (2017) 012041 doi:10.1088/1742-6596/832/1/012041

Susceptibilities from a black hole engineered EoS with a critical point

Israel Portillo

Department of Physics, University of Houston, Houston, TX 77204, USA

E-mail: iportillovazquez@gmail.com

Abstract. Currently at the Beam Energy Scan at RHIC experimental efforts are being made to find the QCD critical point. On the theoretical side, the behavior of higher-order susceptibilities of the net-baryon charge from Lattice QCD at $\mu_B = 0$ may allow us to estimate the position of the critical point in the QCD phase diagram. However, even if the series expansion continues to higher-orders, there is always the possibility to miss the critical point behavior due to truncation errors. An alternative approach is to use a black hole engineered holographic model, which displays a critical point at large densities and matches lattice susceptibilities at $\mu_B = 0$. Using the thermodynamic data from this black hole model, we obtain the freeze-out points extracted from the net-protons distribution measured at STAR and explore higher order fluctuations at the lowest energies at the beam energy scan to investigate signatures of the critical point.

1. Introduction

One of the most important goals of heavy ion physics is the mapping of the phases of strongly interacting matter at high temperatures, T, and densities. It is well established by lattice QCD calculations that at sufficiently high temperatures and zero baryonic chemical potential, μ_B, there is a crossover between baryonic matter and a deconfined state of quarks and gluons [1]. With increasing μ_B, the crossover is expected to end in a critical point (CEP) that separates the crossover from a first order phase transition. This prominent feature of the QCD phase diagram is inferred from taking into account the effects of nonzero quark masses on chiral models [2].

The Beam Energy Scan (BES) at RHIC is exploring high density regions of the QCD phase diagram looking for experimental signatures of the CEP. One promising signature involves comparing moments of the distribution of the net protons compared to the susceptibilities of the pressure [3], which are expected to show non-monotonic behavior in the kurtosis [4].

While lattice QCD is unable to perform full calculations at finite μ_B due to the Fermi sign problem, a small finite μ_B region can be reached employing different techniques such as Taylor expansions at $\mu_B = 0$ [5, 6] and analytic continuations from an imaginary chemical potential [7–10]. Even if the CEP were located at an accessible density, it is possible that the non-monotonicity is missed by the extrapolation scheme, due to the singular behavior of the thermodynamic quantities close to the CEP.

An alternative approach consists of studying the behavior of strongly interacting matter with a critical point using the holographic duality [11]. Ref. [12] showed how to construct black hole solutions of higher dimensional gravitational theories with thermodynamic properties similar to the QGP computed on the lattice at $\mu_B = 0$. The generalization of this type of model to nonzero

μ_B was done in [13, 14], where it was shown that these models can display a CEP at large baryon densities. These "black hole engineered" non-conformal models possess a nonzero bulk viscosity [15, 16], which plays an important role in hydrodynamics simulations [17–20], and can be used to compute baryonic susceptibilities and transport coefficients at nonzero μ_B [21, 22].

Recently, the parameters in the effective dilaton potential of this black hole model were revised in [23] to provide a better match to current lattice data at $\mu_B = 0$. The details of the corresponding extension to $\mu_B \neq 0$ will be published elsewhere [24]. In these proceedings, we use the preliminary data from this revised black hole model to compare its baryonic susceptibilities with the fluctuations of net-protons [25] and estimate the corresponding (T, μ_B) freeze-out line.

2. Baryon Number Fluctuations and the CEP

The baryon number susceptibility is defined by the following derivatives of the pressure, $\chi_n(T, \mu_B) = \frac{\partial^n}{\partial(\mu_B/T)^n}\left(\frac{P}{T^4}\right)$, and these quantities can be numerically calculated in the black hole model at $\mu_B \neq 0$. Susceptibilities provide essential information about the effective degrees of freedom of a system, and are related directly to the moments of the distribution, from which volume-independent susceptibility ratios can be formed

$$\begin{aligned} \text{mean}: \quad M &= \chi_1 & M/\sigma^2 &= \chi_1/\chi_2 \\ \text{variance}: \quad \sigma^2 &= \chi_2 & S\sigma &= \chi_3/\chi_2 \\ \text{skewness}: \quad S &= \chi_3/\chi_2^{3/2} & \kappa\sigma^2 &= \chi_4/\chi_2 \\ \text{kurtosis}: \quad \kappa &= \chi_4/\chi_2^2 & S\sigma^3/M &= \chi_3/\chi_1 \end{aligned} \tag{1}$$

In a heavy ion collision, the moments of the distribution are fixed at the chemical freeze-out such that comparisons to them may be used to extract T and μ_B at freeze-out [26–28]. Based on the singular behavior of thermodynamic variables close to the CEP in the theory of second order phase transitions, the susceptibilities scale with different powers of the (diverging) correlation length ξ close to the CEP. For instance, $\chi_2 \sim VT\xi^2$ (at mean field level) where V is the volume and, for a homogeneous system in equilibrium, one can show that the high order susceptibilities diverge with higher powers of ξ [29]. In practice, the divergence of ξ is limited by the system size and by finite time effects. In Ref. [4] it was argued that the characteristic behavior of the ratio $\kappa\sigma^2$ close to the critical point has a non-monotonic dependence as one approaches the CEP.

3. Results

In Fig. 1 (left) the susceptibility ratios calculated in the model are shown as a function of T for different values of μ_B. From these curves, we were able to extract the bands in T and μ_B for each collision energy $\sqrt{s} = 7.7 - 200$ GeV in Fig. 1 (right), by imposing that they reproduce the corresponding χ_1/χ_2 and χ_3/χ_2 from STAR. While some energies show a clear overlap, some of the lower energies do not, which may be due to other effects such as decays or acceptance cuts. In those cases, the freeze-out is obtained from the middle point of the overlapping area. For the rest of the energies, we choose from each area the point that gives the smallest distance of one area with respect to the other one. The freeze-out points, with their respective error bars, are extracted from the gap between those two points.

Fig. 2 shows an estimate for the location of the CEP found in the model, as well as our results for the (T, μ_B) freeze-out points. It is interesting to note that they are placed between the line corresponding to the inflection point of χ_2/T^2 (solid line) and the one corresponding to the minimum of c_s^2 (dashed line), computed using the revised black hole model at nonzero μ_B [24].

In Fig. 3, model calculations for χ_1/χ_2 and χ_3/χ_2 are compared to their corresponding experimental data points from STAR [25]. We also show in Fig. 3 the χ_4/χ_2 ratio computed

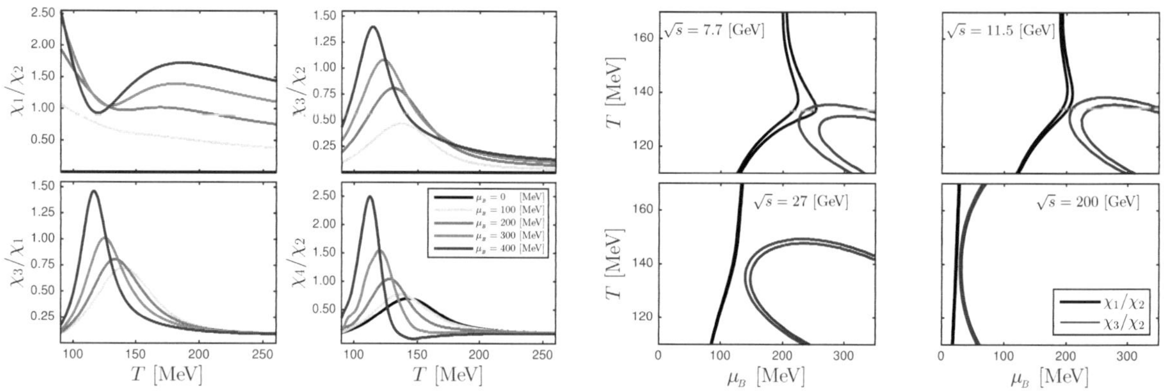

Figure 1. (Color online) Susceptibility ratios obtained from the black hole model (left panel) for $\mu_B = 0$ (black lines), $\mu_B = 100$ MeV (cyan lines), $\mu_B = 200$ MeV (red lines), $\mu_B = 300$ MeV (magenta lines), and $\mu_B = 400$ MeV (blue lines). Trajectories in the (T, μ_B) plane (right panel) that satisfy the STAR data [25] as a function of $\sqrt{s}$.

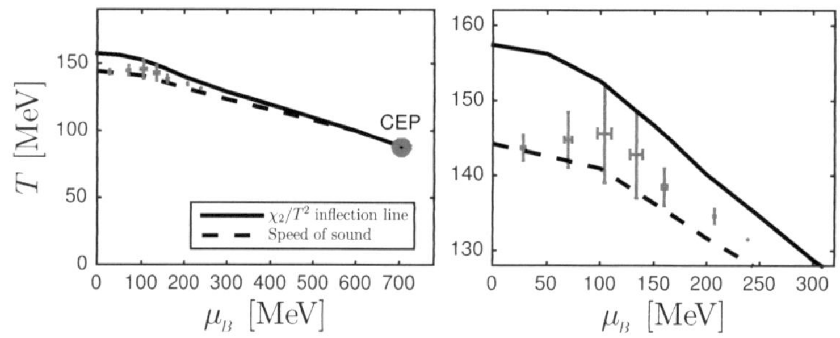

Figure 2. (Color online) Freeze-out line (red data points) extracted by comparing the model calculations to STAR data [25]. The preliminary estimate for the location of the black hole model's CEP is shown. The locations of the minimum of c_s^2 (dashed line) and the inflection point of χ_2/T^2 (solid line) of the black hole model [24] are also shown for a comparison.

at the extracted freeze-out (T, μ_B) values. One can see that this ratio increases with decreasing $\sqrt{s}$, following the trend of current experimental data at low $\sqrt{s}$, even though the CEP of the model only appears at much larger densities.

4. Conclusions

We used preliminary data from the black hole engineered holographic model [24] to extract the freeze-out line across collision energies at the BES. The computed χ_4/χ_2 ratio along the freeze-out line grows with decreasing $\sqrt{s}$, even though the freeze-out values are still far from the high density region where the CEP of the model is located [24]. Preliminary STAR data with higher p_T cuts ($0.4 < p_T < 2.0$ GeV) [30] show larger enhancement for χ_4/χ_2 at low $\sqrt{s}$ and we plan to examine this new data taking into account effects from acceptance cuts and decays in the future.

References

[1] Aoki Y, Endrodi G, Fodor Z, Katz S D and Szabo K K 2006 *Nature* **443** 675–678

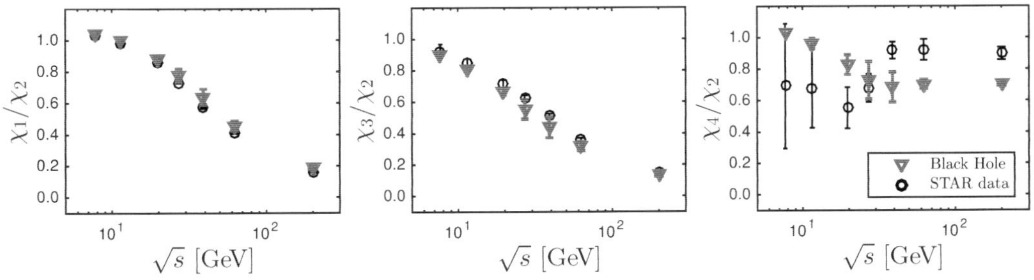

Figure 3. (Color online) Low order baryon susceptibility ratios, χ_1/χ_2 and χ_3/χ_2, used to obtain the freeze-out line compared to the net-proton distribution from STAR [25]. Also shown is the χ_4/χ_2 ratio from STAR [25] (black points) together with the black hole model calculations (red arrows) along the freeze-out line.

[2] Stephanov M A, Rajagopal K and Shuryak E V 1998 *Phys. Rev. Lett.* **81** 4816–4819
[3] Karsch F 2012 *Central Eur. J. Phys.* **10** 1234–1237
[4] Stephanov M A 2011 *Phys. Rev. Lett.* **107** 052301
[5] Allton C e a 2002 *Phys. Rev.* **D66** 074507 (*Preprint* hep-lat/0204010)
[6] Gavai R V and Gupta S 2008 *Phys. Rev.* **D78** 114503 (*Preprint* 0806.2233)
[7] Wu L K, Luo X Q and Chen H S 2007 *Phys. Rev.* **D76** 034505 (*Preprint* hep-lat/0611035)
[8] D'Elia M and Lombardo M P 2003 *Phys. Rev.* **D67** 014505 (*Preprint* hep-lat/0209146)
[9] de Forcrand P and Philipsen O 2002 *Nucl. Phys.* **B642** 290–306 (*Preprint* hep-lat/0205016)
[10] Gunther J, Bellwied R, Borsanyi S, Fodor Z, Katz S D, Pasztor A and Ratti C 2016 (*Preprint* 1607.02493)
[11] Maldacena J M 1999 *Int. J. Theor. Phys.* **38** 1113–1133 [Adv. Theor. Math. Phys.2,231(1998)]
[12] Gubser S S and Nellore A 2008 *Phys. Rev.* **D78** 086007 (*Preprint* 0804.0434)
[13] DeWolfe O, Gubser S S and Rosen C 2011 *Phys. Rev.* **D83** 086005 (*Preprint* 1012.1864)
[14] DeWolfe O, Gubser S S and Rosen C 2011 *Phys. Rev.* **D84** 126014 (*Preprint* 1108.2029)
[15] Gubser S S, Nellore A, Pufu S S and Rocha F D 2008 *Phys. Rev. Lett.* **101** 131601 (*Preprint* 0804.1950)
[16] Finazzo S I, Rougemont R, Marrochio H and Noronha J 2015 *JHEP* **02** 051 (*Preprint* 1412.2968)
[17] Noronha-Hostler J, Noronha J and Greiner C 2009 *Phys. Rev. Lett.* **103** 172302 (*Preprint* 0811.1571)
[18] Noronha-Hostler J, Denicol G S, Noronha J, Andrade R P G and Grassi F 2013 *Phys. Rev.* **C88** 044916
[19] Noronha-Hostler J, Noronha J and Grassi F 2014 *Phys. Rev.* **C90** 034907 (*Preprint* 1406.3333)
[20] Ryu S, Paquet J F, Shen C, Denicol G S, Schenke B, Jeon S and Gale C 2015 *Phys. Rev. Lett.* **115** 132301
[21] Rougemont R, Noronha J and Noronha-Hostler J 2015 *Phys. Rev. Lett.* **115** 202301 (*Preprint* 1507.06972)
[22] Rougemont R, Ficnar A, Finazzo S and Noronha J 2016 *JHEP* **04** 102 (*Preprint* 1507.06556)
[23] Finazzo S I, Critelli R, Rougemont R and Noronha J 2016 *Phys. Rev.* **D94** 054020 (*Preprint* 1605.06061)
[24] Finazzo S I, Critelli R, Rougemont R and Noronha J 2016, to appear
[25] Adamczyk L *et al.* (STAR) 2014 *Phys. Rev. Lett.* **112** 032302
[26] Borsanyi S, Fodor Z, Katz S D, Krieg S, Ratti C and Szabo K K 2014 *Phys. Rev. Lett.* **113** 052301
[27] Alba P, Alberico W, Bellwied R, Bluhm M, Mantovani Sarti V, Nahrgang M and Ratti C 2014 *Phys. Lett.* **B738** 305–310 (*Preprint* 1403.4903)
[28] Noronha-Hostler J, Bellwied R, Gunther J, Parotto P, Pasztor A, Vazquez I P and Ratti C 2016 (*Preprint* 1607.02527)
[29] Stephanov M A 2009 *Phys. Rev. Lett.* **102** 032301 (*Preprint* 0809.3450)
[30] Luo X (STAR) 2015 *PoS* **CPOD2014** 019 (*Preprint* 1503.02558)

IOP Conf. Series: Journal of Physics: Conf. Series **832** (2017) 012042 doi:10.1088/1742-6596/832/1/012042

Chiral magnetic effect in the anomalous transport model

Yifeng Sun[1], **Che Ming Ko**[1] **and Feng Li**[2]

[1]Cyclotron Institute and Department of Physics and Astronomy, Texas A&M University, 77843 College Station, USA

[2]Frankfurt Institute for Advanced Studies , Ruth-Moufang-Str. 1, 60438 Frankfurt, Germany

E-mail: sunyfphy@physics.tamu.edu

Abstract. We report our recent study of the effect of the magnetic field on the elliptic flows of quarks and antiquarks in relativistic heavy ion collisions using the anomalous transport model. We have found that the magnetic field can lead to a charge quadrupole moment in the transverse plane of a heavy ion collisions if the magnetic field lasts sufficiently long and the charge asymmetry of produced matter is nonzero. As the matter expands, the charge quadrupole moment leads to a splitting between the elliptic flows of quarks and antiquarks. However, the elliptic flow difference increases with the charge asymmetry, which is expected from the chiral magnetic wave formed in the quark-gluon plasma, only if the Lorentz force is neglected and the quark-antiquark scattering is dominated by the chirality changing channel.

The high temperature achieved in heavy ion collisions can lead to the restoration of chiral symmetry in the produced quark-gluon plasma. As a result, light quarks become almost massless and thus have definite chiralities. In the presence of a magnetic field, a novel gapless collective excitations called the chiral magnetic wave (CMW) emerges as a result of the interplay between the density waves for axial and vector charges [1, 2]. By solving the wave equation for the electric charge density in a schematic model, a splitting between the elliptic flows of positively and negatively charged particles has been found with a magnitude comparable to the experimental data if the magnetic field is sufficiently strong and long-lived [2, 3]. For a more quantitative study, the anomalous hydrodynamics and the chiral kinetic equation have been developed, which extend the normal hydrodynamics and kinetic equation for describing the time evolution of produced hot dense matter in a heavy ion collision to also include the effect of the chiral magnetic wave [4, 5, 6, 7, 8, 9, 10].

In the chiral kinetic equation, the equations of motion for a massless spin-1/2 particle in a magnetic field take the following form:

$$\frac{d\mathbf{r}}{dt} = \frac{\hat{\mathbf{p}} + Q(\hat{\mathbf{p}}\cdot\mathbf{b})\mathbf{B}}{1+Q\mathbf{B}\cdot\mathbf{b}} \quad \text{and} \quad \frac{d\mathbf{p}}{dt} = \frac{Q\hat{\mathbf{p}}\times\mathbf{B}}{1+Q\mathbf{B}\cdot\mathbf{b}}, \tag{1}$$

where $\mathbf{b} = \pm\frac{\mathbf{p}}{2p^3}$ with the plus and minus signs for particles of positive and negative helicities, respectively. The above two equations describe, respectively, the chiral kinetic motion (CKM) and the Lorentz force (LF) on a spin-1/2 particle.

For the scattering of massless quarks and antiquarks, the conservation of helicity requires that there is no chirality change in quark-quark and antiquark-antiquark scatterings. However,

IOP Conf. Series: Journal of Physics: Conf. Series **832** (2017) 012042 doi:10.1088/1742-6596/832/1/012042

quark and antiquark of same chirality can change their chiralities through the s-channel process. For simplicity, we have included only the chirality changing scattering (CCS) between a quark and an antiquark, and all cross sections are taken to be isotropic in the center of mass of the colliding particles.

To model heavy ion collisions, we take the initial condition from a blast wave model by using a Woods-Saxon form for the number density distribution. Furthermore, the ratio μ/T of the charge chemical potential to the temperature is taken to be uniform in space and the ratio of the axial charge chemical potential to temperature μ_5/T to be zero everywhere in order to exclude other effects that might mimic the effect due to the CMW. For simplicity, charges of quarks and antiquarks are taken to be $e/2$ and $-e/2$, respectively. For the magnetic field generated in heavy ion collisions, we assume that it is uniform along the y-axis, i.e., perpendicular to the reaction plane, and has the time dependence $eB = eB_0/[1 + (t/\tau)^2]$ [11] with $eB_0 = 7m_\pi^2$ and $\tau = 6$ fm/c. We take the initial temperature to be 300 MeV and use kinetic equation to describe the motions of massless quarks and antiquarks as well as a temperature-dependent cross section $\sigma = \sigma_0(T_0/T)^3$ for their scattering until the temperature of the system decreases to the chiral restoration temperature T_C = 150 MeV, at which quarks and antiquarks are converted to pions using the duality ansatz. The resulting pions are then described by the normal equation of motion without the Lorentz force and undergo scattering using empirically known cross sections [12] until the kinetic freeze-out temperature T_f = 120 MeV.

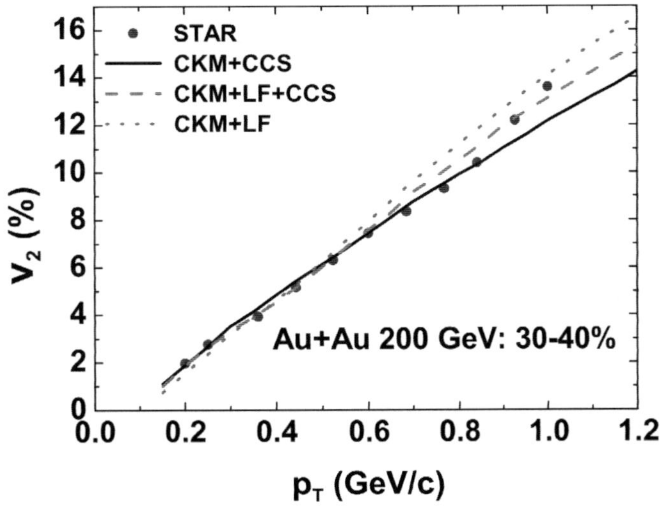

Figure 1. Elliptic flow of midrapidity pions as a function of their transverse momenta for the three cases of chiral kinetic motion with Lorentz force and chirality changing scattering (CKM+LF+CCS), without chirality changing scattering (CKM+LF), and without Lorentz force (CKM+CCS). Experimental data (solid circles) are from Ref. [13].

We first show in Fig. 1 the elliptic flow of midirapidity pions as a function of transverse momentum for the three cases of chiral kinetic motion with Lorentz force and chirality changing scattering (CKM+LF+CCS), without chirality changing scattering (CKM+LF), and without Lorentz force (CKM+CCS). With the cross section parameter σ_0 taking to be 13.7, 13.1 and 15.5 mb, respectively, for the three cases from fitting the experimentally measured integrated elliptic flow (0.036) of pions with transverse momenta in the range of $0.15 \leq p_T \leq 0.5$ GeV/c, all three results agree well with the experimental data.

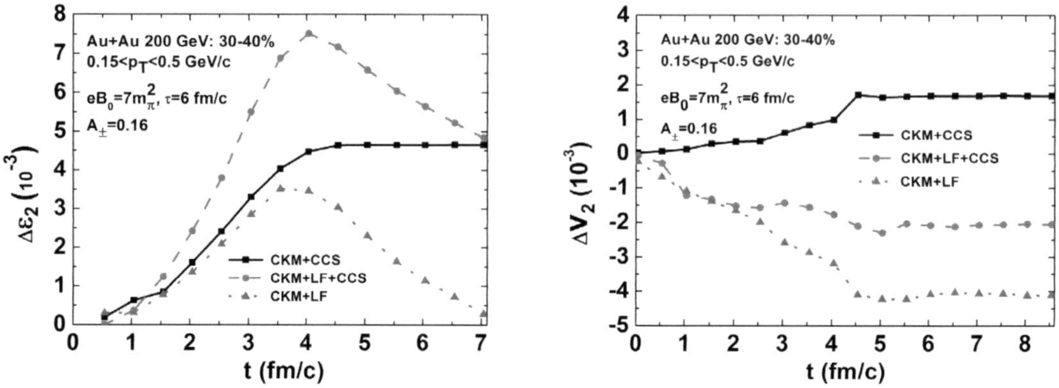

Figure 2. Eccentricity (left window) and elliptic flow (right window) differences between negatively and positively charged particles as a function of time for different scenarios of parton dynamics when the total charge asymmetry of the quark matter is $A_\pm = 0.16$.

Figure 2 shows the time evolution of the eccentricity (left window) and elliptic flow (right) differences between negatively and positively charged particles. The solid line shows that CKM+CCS can lead to larger eccentricity and elliptic flow for negatively charged particles than for positively charged particles when the charge asymmetry is nonzero. This result is due to the appearance of a axial charge dipole and charge quadrupole moments in the transverse plane, shown in the left and right windows of Fig. 3, respectively, as expected from the CMW. However, including the Lorentz force results in a larger elliptic flow for positively charged particles than for negatively charged particles, although it does not change the sign of the eccentricity difference.

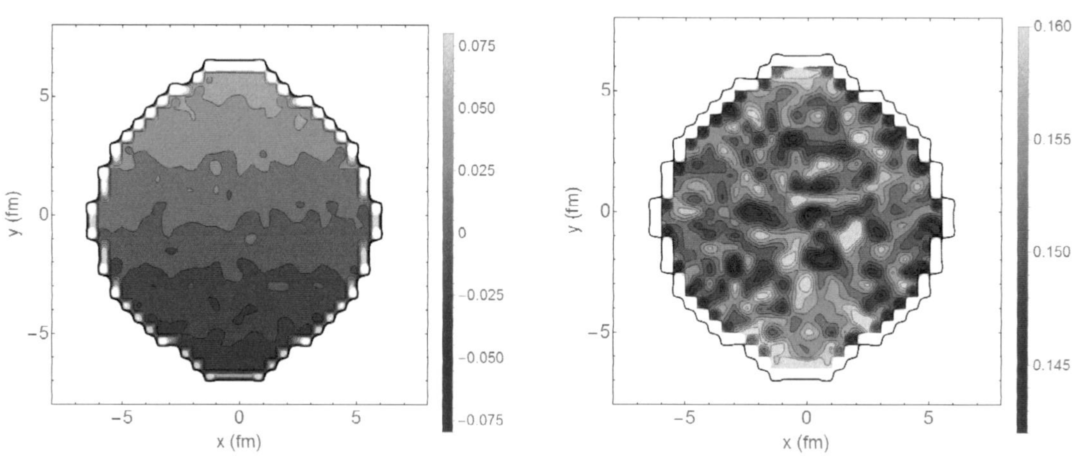

Figure 3. Axial charge chemical potential μ_5/T (left window) and charge chemical potential μ/T (right window) distributions in the transverse plane $z = 0$ at time $t = 5$ fm/c for events with charge asymmetry $A_\pm = 0.16$ for the case of including chiral kinetic motion and chirality changing quark-antiquark scattering but without the Lorentz force.

In Fig. 4, we show the charge asymmetry dependence of elliptic flow difference between negatively and positively charged particles. The black curve shows that CKM+CCS leads to a

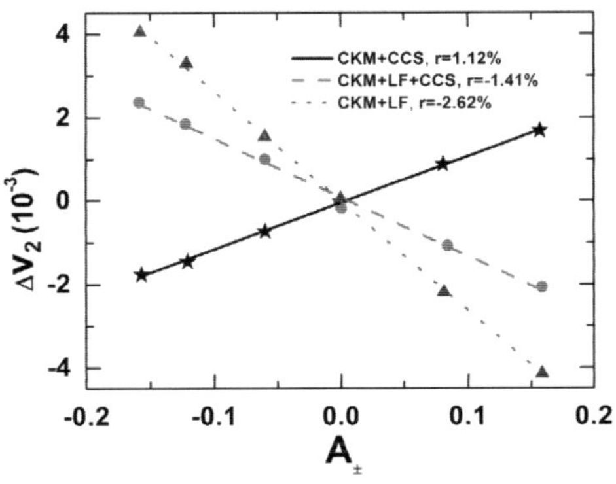

Figure 4. Elliptic flow difference Δv_2 as a function of charge asymmetry $A_\pm$ for different scenarios for the parton dynamics.

positive slope parameter as expected from the CMW [2] and consistent with the experimental data [3], although its value is smaller. The inclusion of the Lorentz force, on the other hand, changes the slope parameter to negative.

The anomalous transport model, which includes the kinetic equation of motion for massless quarks and antiquark and allows the change of chiralities in the scattering of quark and antiquark of same chirality, has been used to study the effect of CMW in heavy ion collisions. Our results clearly show that the elliptic flow difference between negatively and positively charged particles increases linearly with the charge asymmetry of produced matter, resulting in a positive slope parameter that is, however, smaller than the experimentally measured value. Allowing for a finite initial axial charge dipole moment that is proportional to the charge asymmetry may enhance the elliptic flow difference. On the other hand, the Lorentz force is found to enhance the elliptic flow of positively charged particles if the charge asymmetry is nonzero, thus leading to a negative slope parameter. Also, the justification for the existence of a long-lived magnetic field in relativistic heavy ion collisions remains missing.

Acknowledgments

This work was supported in part by the US Department of Energy under Contract No. DE-SC0015266 and the Welch Foundation under Grant No. A-1358.

References

[1] Kharzeev D E and Yee H U 2011 *Phys. Rev. D* **83** 085007
[2] Burnier Y, Kharzeev D E, Liao J and Yee H U 2011 *Phys. Rev. Lett.* **107** 052303
[3] Wang G 2013 *Nuc. Phys. A* **904-905** 248c
[4] Son D T and Surowka P 2009 *Phys. Rev. Lett.* **103** 191601
[5] Stephanov M A and Yin Y 2012 *Phys. Rev. Lett.* **109** 162001
[6] Son D T and Yamamoto N 2012 *Phys. Rev. Lett.* **109** 181602
[7] Son D T and Yamamoto N 2013 *Phys. Rev. D* **87** 085016
[8] Gao J H, Liang Z T, Pu S, Wang Q and Wang X N 2012 *Phys. Rev. Lett.* **109** 232301
[9] Chen J W, Pu S, Wang Q and Wang X N 2013 *Phys. Rev. Lett.* **110** 262301
[10] Manuel C and Torres-Rincon J M 2014 *Phys. Rev. D* **90** 076007
[11] Basar G, Kharzeev D E and Skokov V 2012 *Phys. Rev. Lett.* **109** 202303
[12] Li B A and Ko C M 1995 *Phys. Rev. C* **52** 2037
[13] Adams J, Aggarwal M M, Ahammed Z, *et al* 2005 *Phys. Rev. C* **72** 014904

IOP Conf. Series: Journal of Physics: Conf. Series **832** (2017) 012043 doi:10.1088/1742-6596/832/1/012043

Rapidity correlations test stochastic hydrodynamics

C Zin[1], S Gavin[1] and G Moschelli[2]

[1] Department of Physics and Astronomy, Wayne State University, Detroit, MI, 48202
[2] Lawrence Technological University, 21000 West Ten Mile Road, Southfield, MI 48075

E-mail: christopherzin@wayne.edu, sean.gavin@wayne.edu, gmoschell@ltu.edu

Abstract. We show that measurements of the rapidity dependence of transverse momentum correlations can be used to determine the characteristic time τ_π that dictates the rate of isotropization of the stress energy tensor, as well as the shear viscosity $\nu = \eta/sT$. We formulate methods for computing these correlations using second order dissipative hydrodynamics with noise. Current data are consistent with $\tau_\pi/\nu \sim 10$ but targeted measurements can improve this precision.

1. Introduction

From experiments at the Relativistic Heavy Ion Collider and the Large Hadron Collider we have seen that hydrodynamic models can be used to describe the expansions of a range of collision systems. In large nucleus-nucleus systems these descriptions were used as evidence of the production of a liquid-like quark gluon plasma. Smaller proton-proton and proton-nucleus collisions lack the temperature and density necessary to create quark gluon plasma so experiment was met with surprise when hydrodynamic models were successfully applied to these systems. To investigate the differences between these systems we turn our interest to the processes by which these systems thermalize.

Our focus is on the rapidity dependence of p_t correlations which give information about shear viscosity and the transport coefficient τ_π, the relaxation time for pressure isotropization. We approach the problem by using a stochastic description of hydrodynamics to develop differential equations for the momentum correlation function of the system. In doing so, we highlight the importance of including stochastic processes in the description of an evolving collision system. Our results, which are compared with experiment, show that momentum correlations are propagated through the system via second order diffusion. For a full discussion of the material in these proceedings, see [1].

2. Brownian Motion

In heavy ion collisions, the initial phase space distribution $f = dN/d^3p\,d^3x$ fluctuates from event to event due to impact parameter variations and the distributions of the colliding nuclei. Stochastic noise introduces important two-particle correlations to the system which we model after the familiar example of Brownian motion[1]. In Brownian motion, a heavy particle suspended in a fluid obeys the Langevin equation

$$\Delta v = -\gamma v(t)\Delta t + \Delta W \tag{1}$$

[1] A detailed discussion of Brownian motion and stochastic processes can be found in [2].

IOP Conf. Series: Journal of Physics: Conf. Series **832** (2017) 012043 doi:10.1088/1742-6596/832/1/012043

where $v(t)$ is the velocity of the heavy particle and $\Delta v = v(t+\Delta t) - v(t)$ is the change in velocity during the time step Δt. The first term on the right hand side of Eq. (1) represents the viscous drag force felt by the heavy particle, having strength γ. The second term is the contribution to Δv from stochastic noise which represents collisions of the heavy particle with the fluid medium.

One assumes that collisions are random in both magnitude and direction and that the variance of the noise is proportional to Δt. We represent this as $\langle \Delta W \rangle = 0$ and $\langle \Delta W^2 \rangle = \Gamma \Delta t$, where Γ determines the strength of the noise. Notice that as a result of this assumption ΔW becomes an infinitesimal of order $\Delta t/2$.

Taking the average of Eq. (1) and the limit $\Delta t \to 0$ gives a differential equation for $\langle v \rangle$

$$\Delta \langle v \rangle = -\gamma \langle v(t) \rangle \Delta t + \langle \Delta W \rangle \quad \Rightarrow \quad d\langle v \rangle / dt = -\gamma \langle v \rangle \tag{2}$$

which is free of noise. The effect of the noise is felt by $\langle v^2(t) \rangle$, however, as we can see by averaging the square of Eq. (1) and taking $\Delta t \to 0$:

$$\Delta \langle v^2 \rangle = 2\langle v \Delta v \rangle + \langle \Delta v \Delta v \rangle = -2\gamma \langle v^2 \rangle \Delta t + \Gamma \Delta t \quad \Rightarrow \quad d\langle v^2 \rangle / dt = -2\gamma \langle v^2 \rangle + \Gamma. \tag{3}$$

The first equality in Eq. (3) is known as the Itô product rule in stochastic calculus and is required so as to keep all terms of order Δt.

Our goal is to obtain a deterministic equation for the square of the mean velocity. To this end, notice that in equilibrium the time derivative in Eq. (3) must vanish, which can be seen by applying the equipartition theorem to $\langle v^2 \rangle_{eq}$. This leaves $\Gamma = 2\gamma \langle v^2 \rangle_{eq} = 2\gamma T/m$ where the second equality follows again through use of the equipartition theorem. Determining the strength of the noise in this manner is an application of the fluctuation-dissipation theorem. In the context of Brownian motion, the fluctuation-dissipation theorem says the energy of the Brownian particle which is dissipated via friction is added to the medium. In turn, the fluctuations in the medium increase, which is felt by the Brownian particle via more energetic collisions.

Notice, at this point, the importance of including stochastic noise in our calculation. Had we ignored the noise term in Eq. (3) we would have found that $\langle v^2(t) \rangle$ vanishes in equilibrium, a clear violation of the equipartition theorem. By including stochastic noise, through use of the Langevin equation, we have allowed our results to be consistent with well established physics.

Returning to Eq. (3), we define the variance of the velocity as $r = \langle v^2 \rangle - \langle v \rangle^2$ and the deviation of r from its equilibrium value as $\Delta r = r - r_{eq}$. We then find Δr satisfies $d\Delta r/dt = -2\gamma \Delta r$. This is a completely deterministic equation for Δr as the noise was absorbed into r_{eq} by application of the fluctuation-dissipation theorem. Also we emphasize that Δr relaxes with timescale $1/2\gamma$, half the value of the relaxation of $\langle v \rangle$ as seen in Eq. (2). These points are important for our main results later on.

3. Hydrodynamic Fluctuations

We turn now to heavy ion collisions. After a collision fluctuations in density and temperature cause the transverse velocity of neighboring system cells to vary along the beam axis. Viscosity works to diffuse fluctuations in the velocity throughout the system. This spread in fluctuations is dominated by shear hydrodynamic modes. Small fluctuations in velocity produce correspondingly small fluctuations in the momentum current $\mathbf{M} = w\mathbf{v}$ where $w = e + p$ is the enthalpy density with energy density e and pressure p. This momentum density satisfies the linear first order Navier-Stokes equation

$$\frac{\partial}{\partial t}\mathbf{M} + \nabla p = \frac{\zeta + \frac{1}{3}\eta}{w}\nabla(\nabla \cdot \mathbf{M}) + \frac{\eta}{w}\nabla^2 \mathbf{M} \tag{4}$$

where ζ and η are the bulk and shear viscosities. To separate out the shear modes we perform a Helmholtz decomposition on $\mathbf{M}$ by writing $\mathbf{M} = \mathbf{g}_l + \mathbf{g}$ where the longitudinal modes satisfy

$\nabla \times \mathbf{g}_l = 0$ and the shear modes satisfy $\nabla \cdot \mathbf{g} = 0$. Using this, we find the shear modes obey the diffusion equation $\partial \mathbf{g}/\partial t = \nu \nabla^2 \mathbf{g}$ for the kinematic viscosity $\nu = \eta/w$.

We apply our treatment of Brownian motion to the shear modes by writing this as a difference equation with stochastic noise: $\Delta g_i = \nu \nabla^2 g_i \Delta t + \Delta W_i$. Here, the noise satisfies $\langle \Delta W_i \rangle = 0$ and $\langle \Delta W_i(x_1) \Delta W_j(x_2) \rangle = \Gamma_{12}^{ij} \Delta t$. We then define the correlation function $r = \langle g_i(x_1) g_j(x_2) \rangle - \langle g_i(x_1) \rangle \langle g_j(x_2) \rangle$ and the deviation of r from its equilibrium value $\Delta r = r - r_{eq}$ where $r_{eq} = wT\delta_{ij}\delta(x_1 - x_2)$ is found through application of the fluctuation-dissipation theorem. This allows us to write down an explicit form for the noise $\Gamma_{12}^{ij} = -(\nabla_1^2 + \nabla_2^2) r_{eq}$. Notice that due to the δ-functions in r_{eq} there is no noise-noise correlation between the different components of $\mathbf{g}$ nor at different spatial positions $x_1 \neq x_2$.

We can now write an evolution equation for the momentum correlations of the system

$$\left[\frac{\partial}{\partial \tau} - \frac{\nu}{\tau^2}\left(\frac{\partial^2}{\partial \eta_1^2} + \frac{\partial^2}{\partial \eta_2^2}\right)\right] \Delta r = 0 \tag{5}$$

where τ is the proper time and η is the spatial rapidity. Thus, correlations of the shear modes spread through the system via first order diffusion. Unfortunately, first order diffusion is acausal so a δ-function spike in momentum will become a Gaussian, instantaneously spreading its tails to infinity. To remedy this issue we must repeat this procedure in second order hydrodynamics.

Starting with a second order Müller-Israel-Stewart equation we find that the shear modes satisfy a Maxwell-Cattaneo equation to linear order

$$\left[\tau_\pi \frac{\partial^2}{\partial t^2} + \frac{\partial}{\partial t}\right] \mathbf{g} = \nu \nabla^2 \mathbf{g}. \tag{6}$$

This equation introduces the transport coefficient τ_π, the relaxation time for isotropization of the stress-energy tensor. Applying our Brownian prescription to Eq. (6), we find a deterministic evolution equation for the correlations of the shear modes

$$\left[\frac{\tau_\pi^*}{2}\frac{\partial^2}{\partial \tau^2} + \frac{\partial}{\partial \tau} - \frac{\nu^*}{\tau^2}\left(\frac{\partial^2}{\partial \eta_1^2} + \frac{\partial^2}{\partial \eta_2^2}\right)\right] \Delta r = 0 \tag{7}$$

where the, now time dependent, coefficients are given by $\tau_\pi^* = \tau_\pi/(1+\kappa\tau_\pi/\tau)$, $\nu^* = \nu/(1+\kappa\tau_\pi/\tau)$ and $\kappa = (1 + d\ln(\tau_\pi/\eta T)/d\ln\tau)/2$. For simplicity we take these to be constant in this work.

Equation (7) has two important features we wish to emphasize. First, the combination of wave-like and diffusion-like behavior given by the first two terms renders the theory causal, correcting the problem of Eq. (5). Second, due to the factor of 1/2 in the first term, Δr relaxes at half the rate of the average, as seen in Eq. (6). This is the same type of behavior we saw in the case of Brownian motion.

For comparison to experiment we choose the transverse momentum covariance

$$\mathcal{C} = \langle N \rangle^{-2} \left\langle \sum_{j\neq i} p_{ti}\, p_{tj} \right\rangle - \langle p_t \rangle^2 \tag{8}$$

where $\langle N \rangle$ is the average multiplicity and $\langle p_t \rangle = \langle \sum_i p_{ti} \rangle / \langle N \rangle$ is the average transverse momentum. This observable has been used to study viscosity [3] and has a particularly simple connection to Δr given by $\langle N \rangle^2 \mathcal{C} = \int \Delta r \, dx_1 dx_2$. We study $\mathcal{C}$ in both the first and second order theory by using Eq. (5) and Eq. (7), with constant coefficients, to calculate Δr. Our results are shown in Fig. (1).

The experimental data has a Gaussian profile in the peripheral centrality bins. In more central bins the data develop a double bump structure for points close in relative rapidity. Our first order lines fit the Gaussian profile of the peripheral bins but maintain their Gaussian shape as we move to more central bins, losing agreement with the data. We see that as the second order line evolves to more central collisions the peak flattens to hit the broad shoulders developed by the data. Overall we find that the second order line better agrees with experiment.

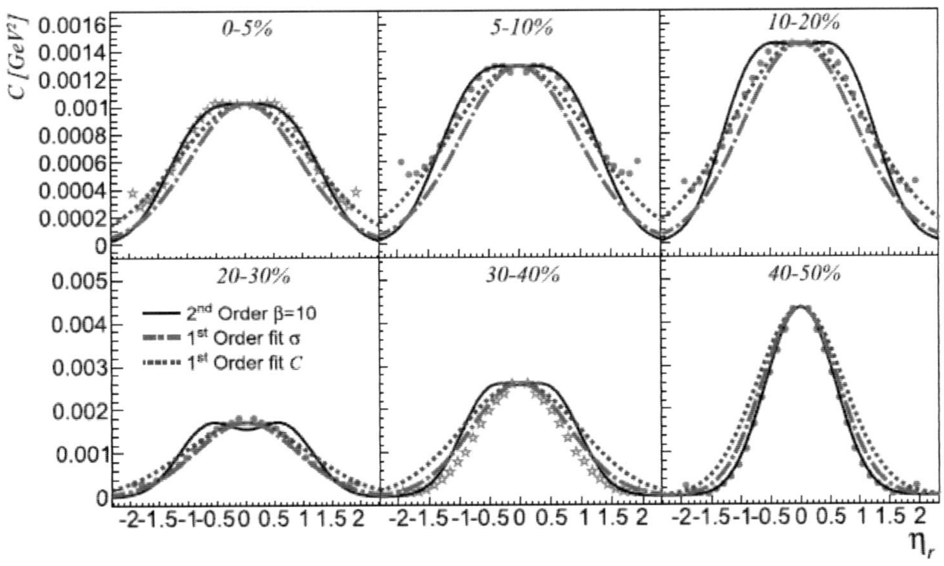

Figure 1. First and second order calculations of momentum correlations in relative rapidity $\eta_r = \eta_2 - \eta_1$. First order lines are fit to the rapidity width σ of Δr and directly to the shape of $\mathcal{C}$. Data are from [3] (open stars) and [4] (filled circles). Percentages indicate centrality bins.

4. Conclusions

Our goal was to develop equations for the evolution of the correlations of the shear momentum modes in a heavy ion collision system. We made an effort to study correlation functions relative to their values in equilibrium in order to account for the stochastic nature of particle collisions. Our main result, Eq. (7), describes the evolution of momentum correlations in second order hydrodynamics. We find that experimental data cannot be adequately explained using first order hydrodynamics. Second order hydrodynamics better explains the features shown by experiment. This introduces the relaxation time τ_π, which sets the rate for pressure isotropization of the system. Using our best fits to experiment we find $\tau_\pi/\nu \sim 10$, double the value predicted by kinetic theory. We feel this estimate can be improved by using more realistic parameters in Eq. (7) but save this for future work.

Acknowledgments

We thank Rajendra Pokharel for contributing to the early stages of this project as well as Monika Sharma and Claude Pruneau for discussing the STAR data. This work was supported in part by the U.S. NSF grant PHY-1207687.

References

[1] Gavin S, Moschelli G and Zin C 2016 *Phys. Rev.* **C94** 024921 (*Preprint* 1606.02692)
[2] Gardiner C 2004 *Handbook of Stochastic Methods for Physics, Chemistry and the Natural Sciences* (New York: Springer)
[3] Agakishiev H *et al.* (STAR) 2011 *Phys. Lett.* **B704** 467–473 (*Preprint* 1106.4334)
[4] Sharma M and Pruneau C (private communication, 2010)

IOP Conf. Series: Journal of Physics: Conf. Series **832** (2017) 012044 doi:10.1088/1742-6596/832/1/012044

Multiplicity and transverse momentum evolution of charge-dependent correlations in pp, p-Pb, and Pb-Pb collisions at the LHC

Jinjin Pan for the ALICE Collaboration
Wayne State University, 666 W. Hancock St, Detroit, MI 48201, USA
E-mail: jinjin.pan@cern.ch

Abstract. We report measurements of two-particle charge-dependent correlations in pp, p-Pb, and Pb-Pb collisions at $\sqrt{s_{\rm NN}}$ = 7, 5.02, and 2.76 TeV, respectively, as a function of pseudorapidity and azimuthal angle differences. These measurements, carried out using the balance function, probe the charge creation time and the development of collectivity in the produced systems. The balance function is studied as a function of the event multiplicity as well as the transverse momentum ($p_{\rm T}$) of charged particles detected in the range $|\eta| < 0.8$. In the low transverse momentum region, $0.2 < p_{\rm T} < 2.0$ GeV/c, the balance function becomes narrower in both $\Delta\eta$ and $\Delta\varphi$ directions in all three systems for events with higher multiplicity. For higher values of transverse momenta, the balance function becomes even narrower but exhibits no significant multiplicity dependence, indicating that the observed narrowing with increasing multiplicity at low $p_{\rm T}$ is a feature of bulk particle production.

1. Introduction

The charge-dependent part of two-particle correlations is studied with the balance function (BF) [1], defined by Formula (1), where $\langle N_{++}\rangle$, $\langle N_{+-}\rangle$, $\langle N_{-+}\rangle$ and $\langle N_{--}\rangle$ denote the average number of different charged particle pairs, while $\langle N_{+}\rangle$ and $\langle N_{-}\rangle$ for the average number of single charged particles. For each pair, the first (trigger) particle has a transverse momentum $p_{\rm T,trig}$, while the second (associated) charged particle has a transverse momentum $p_{\rm T,assoc}$. Particle production is often governed by conservation laws. Local charge conservation, in particular, ensures that each produced charged particle is balanced by an oppositely-charged partner, created at the same location in space and time, but transported in momentum space according to the collision dynamics and system evolution. By design, the BF reflects the distribution of balancing charges in momentum space. Theoretical studies have shown that the BF is a sensitive probe of both the pair creation time [1,2] and the collective nature of the system's evolution [3,4]. In particular, a reduced BF width is expected for heavy systems in which particle production dominantly occurs at low temperature following an isentropic expansion of Quark Gluon Plasma [1-4].

$$B(\Delta\eta,\Delta\varphi) = \frac{1}{2}\{\frac{\langle N_{+-}\rangle(\Delta\eta,\Delta\varphi) - \langle N_{++}\rangle(\Delta\eta,\Delta\varphi)}{\langle N_{+}\rangle} + \frac{\langle N_{-+}\rangle(\Delta\eta,\Delta\varphi) - \langle N_{--}\rangle(\Delta\eta,\Delta\varphi)}{\langle N_{-}\rangle}\} \quad (1)$$

In these proceedings, we present recent BF measurements in pp, p-Pb, and Pb-Pb collisions at $\sqrt{s_{\rm NN}}$ = 7, 5.02, and 2.76 TeV, respectively by the ALICE collaboration[5,6]. Results are presented as a function of produced particle multiplicity and transverse momentum (p_T) to investigate potential scaling properties, similarities, and differences between the three systems.

2. Balance Function Results

Figure 1 presents inclusive charge BFs as a function of $\Delta\eta$ and $\Delta\varphi$ for three multiplicity classes of Pb-Pb, p-Pb, and pp collisions. The BFs feature a prominent near-side peak at $|\Delta\varphi| < \pi/2$, which is observed to narrow with increasing multiplicity for all three collision systems. The peak amplitude also changes with multiplicity, with higher values corresponding to collisions with higher multiplicity. On the away-side, $\pi/2 < \Delta\varphi < 3\pi/2$, the BF has a larger magnitude for lower multiplicity events. In addition, a depletion in the correlation pattern around $(\Delta\eta,\Delta\varphi) = (0, 0)$ starts to emerge in mid-central (e.g. 30-40% multiplicity class) events in Pb-Pb collisions and becomes more pronounced in p-Pb and pp collisions with decreasing multiplicity. This depletion is in large part associated with the HBT effect which produces a relatively narrow peak in like-sign correlation whose width is expected to decrease with increasing source size.

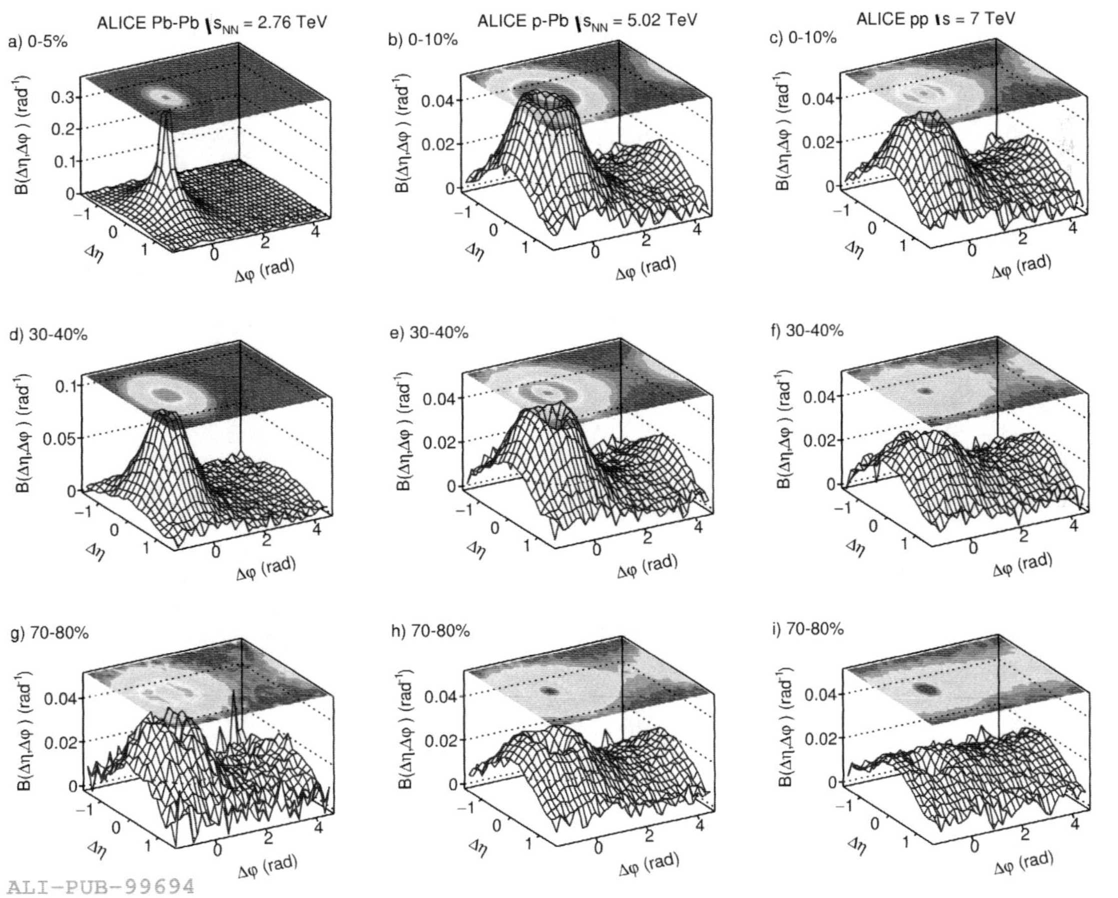

Figure 1. Balance function, $B(\Delta\eta,\Delta\varphi)$, for charged particles with $0.2 < p_{T,\mathrm{assoc}} < p_{T,\mathrm{trig}} < 2.0$ GeV/c, in Pb-Pb, p-Pb, and pp collisions at $\sqrt{s_{NN}}$ = 2.76, 5.02, and 7 TeV, respectively, for selected multiplicity classes.

Figure 2 presents projections of the BFs onto $\Delta\eta$ on the near-side in panels (a-c), away-side in panels (d-f), and onto $\Delta\varphi$ in panels (g-i). Statistical uncertainties are represented by

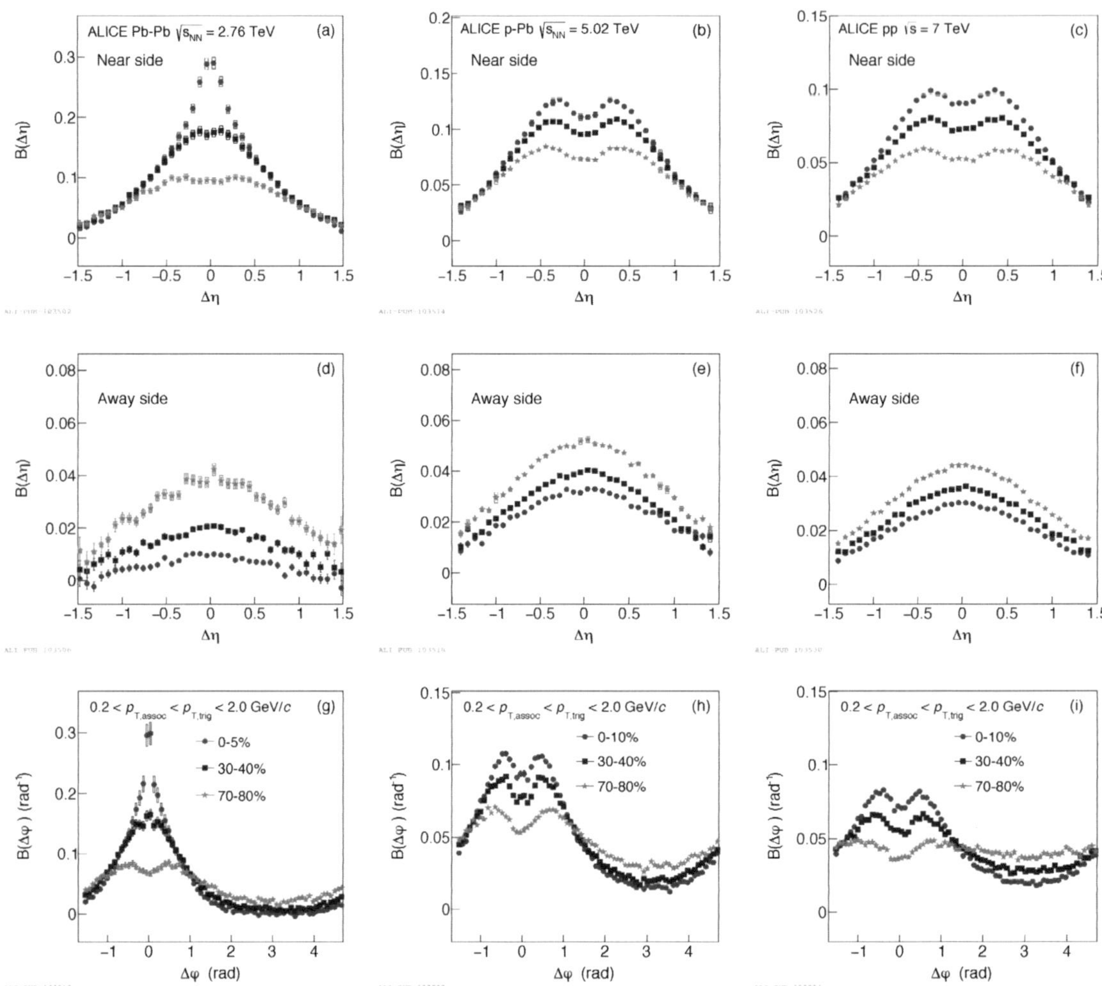

Figure 2. BF for charged particles with $0.2 < p_{\mathrm{T,assoc}} < p_{\mathrm{T,trig}} < 2.0$ GeV/c as a function of $\Delta\eta$ on the near-side (upper row) and away-side (middle row) and $\Delta\varphi$(lower row) in different multiplicity classes of Pb-Pb in panels a, d and g, p-Pb in panels b, e and h, and pp collisions in panels c, f and i at $\sqrt{s_{\mathrm{NN}}} = 2.76$, 5.02, and 7 TeV, respectively.

vertical error bars while systematic uncertainties, calculated as the quadratic sum of correlated and uncorrelated contributions, are shown as shaded boxes. One finds that the BF near-side peak shape and width exhibit a significant multiplicity dependence for all collision systems. In particular, the distribution narrows and the peak amplitude increases for higher multiplicity events. However, the magnitude of the BF on the away-side exhibits a different trend, with larger values measured for lower multiplicity events.

Figure 3 presents the charged-particle multiplicity and p_{T} dependence of the width of BF in $\Delta\eta$ (a) and $\Delta\varphi$ (b) for all three systems. The multiplicity is determined from the number of charged particles reconstructed in $|\eta| < 0.8$ and $p_{\mathrm{T}} > 0.2$ GeV/c. One observes that between the pp and the p-Pb systems, and for overlapping multiplicities in the low p_{T} region, the widths in $\Delta\eta$ and $\Delta\varphi$ have similar values. This could indicate that the charge-

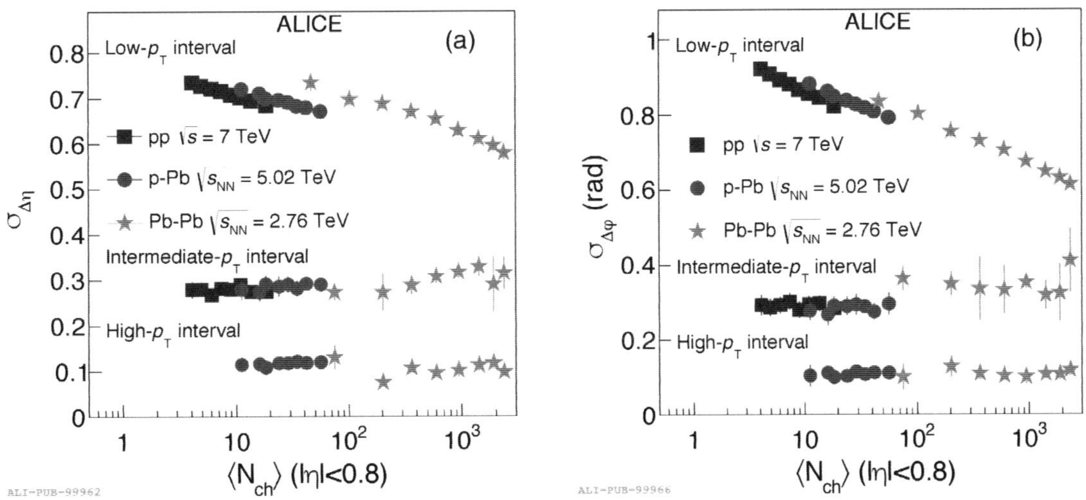

Figure 3. BF widths in $\Delta\eta$ (a) and in $\Delta\varphi$ (b) measured in pp, p-Pb, and Pb-Pb collision as a function of the charged-particle multiplicity, measured in $|\eta| < 0.8$ and $p_T > 0.2$ GeV/c. The low-, intermediate-, and high-p_T intervals correspond to $0.2 < p_{T,assoc} < p_{T,trig} < 2.0$ GeV/c, $2.0 < p_{T,assoc} < 3.0 < p_{T,trig} < 4.0$ GeV/c, and $3.0 < p_{T,assoc} < 8.0 < p_{T,trig} < 15.0$ GeV/c, respectively.

dependent correlations have similar origin in these two systems. On the other hand, widths observed in p-Pb and Pb-Pb at similar multiplicities show significant differences, which may be attributed to the strong radial flow profile in Pb-Pb collisions. These differences may also signal that a different mechanism drives the charge-dependent correlations in smaller systems. With increasing transverse momentum, the BFs become narrower and exhibit no significant multiplicity dependence for all systems. The agreement of $\sigma_{\Delta\eta}$ and $\sigma_{\Delta\varphi}$ for all multiplicities over all three systems clearly indicates that the dynamics responsible for the high-p_T charge-dependent correlations do not change significantly between pp, p-Pb, and Pb-Pb collisions.

3. Summary

These proceedings presented measurements, with the ALICE detector, of balance functions (BF) for charged particles in pp, p-Pb, and Pb-Pb collisions at the LHC. BFs were studied in ranges $0.2 < p_{T,assoc} < p_{T,trig} < 2.0$ GeV/c, $2.0 < p_{T,assoc} < 3.0 < p_{T,trig} < 4.0$ GeV/c, and $3.0 < p_{T,assoc} < 8.0 < p_{T,trig} < 15.0$ GeV/c. The $\Delta\eta$ and $\Delta\varphi$ widths of balance functions were found to decrease with increasing multiplicity in three systems only in the low p_T region (for $p_T < 2.0$ GeV/c). For higher p_T ranges, the BF widths dependence on produced particle multiplicities is drastically suppressed relative to that observed at low p_T.

4. References

[1] S.A. Bass, P. Danielewicz, S. Pratt 2000 *Phys. Rev. Lett.* **85** 2689
[2] S. Jeon, S. Pratt 2002 *Phys. Rev.* C **65** 044902
[3] P. Bozek 2005 *Phys. Lett.* B **609** 247
[4] S.A. Voloshin 2005 *Nucl. Phys.* A **749** 287
[5] ALICE Collaboration, B. Abelev et al. 2013 *Phys. Lett.* B **723** 267
[6] ALICE Collaboration, J. Adam et al. 2016 *Eur. Phys. J.* C **76** 86

IOP Conf. Series: Journal of Physics: Conf. Series **832** (2017) 012045 doi:10.1088/1742-6596/832/1/012045

Spinodal instability of baryon-rich quark matter

Feng Li[1] and Che Ming Ko[2]

[1] Frankfurt Institute for Advanced Studies, Ruth-Moufang Str. 1, Frankfurt 60438, Germany
[2] Cyclotron Institute and Department of Physics, Texas A&M University, College Station, Texas 77843-3366, USA

E-mail: fengli@fias.uni-frankfurt.de

Abstract. The spinodal instabilities of both confined and expanding baryon-rich quark matters are studied in a transport model derived from the Nambu-Jona-Lasino model. Appreciable higher-order density moments are seen as a result of the first-order phase transition in both cases. The skewness of the quark number event-by-event distribution in a small sub-volume of the system becomes appreciable for the confined quark matter. For the expanding quark matter, the density fluctuations lead to enhanced anisotropic flows and dilepton yield.

1. Introduction

The beam energy scan (BES) experiments [1, 2, 3] at the Relativistic Heavy Ion Collider (RHIC) are designed for the search of the critical end point (CEP) in the baryon-rich region of the QCD phase diagram. At CEP, the first-order phase transition between the hadronic matter and the quark-gluon-plasma (QGP) changes to a second-order one. To understand the properties of this phase transition, both schematic models and hydrodynamic approach [4, 5, 6] have been used to study how the spinodal instability of a QGP leads to the self-amplified deviation from its equilibrium state during the phase transition. Using the test particle method, the spinodal instabilities of both the confined and expanding quark matters have been studied in a transport model that is derived from the Nambu-Jona-Lasino (NJL) model [7]. By varying the coupling strength of the vector interaction G_V in the NJL model, different equations of state are obtained for a baryon-rich quark matter. A first-order phase transition occurs if $G_V = 0$ with the CEP at about $T_c \simeq 70$ MeV and $n_c \simeq 0.9$ fm^{-3} [8], and it is absent if $G_V = G_S$. The density fluctuations induced by the spinodal instability can be quantified by the scaled density moments $\langle\rho^N\rangle/\langle\rho\rangle^N$ [6], where $\langle\rho^N\rangle \equiv \int d^3\mathbf{r}\rho(\mathbf{r})^{N+1}/\int d^3\mathbf{r}\rho(\mathbf{r})$. They are all equal to one for a uniform density distribution but becomes larger if the density is nonuniform. Also, density fluctuations can affect the anisotropic flows and the dilepton yield in heavy ion collisions. Results on these observables for the two cases of with and without a first-order phase transition are compared in the following for both the confined and the expanding quark matter.

2. Quark matter in a box

Shown in the left window of Fig. 1 by the dotted, dashed, and solid lines are the time evolution of scaled density moments for $N = 2$, 4 and 6, respectively, during the first-order phase transition of a quark matter that is confined in a box with an initially uniform net quark density $n_q = 0.5$ fm^{-3} and temperature $T = 20$ MeV. They are obtained by averaged over 1,000 events. The scaled moments are seen to increase during the phase separation and reach their saturated values at

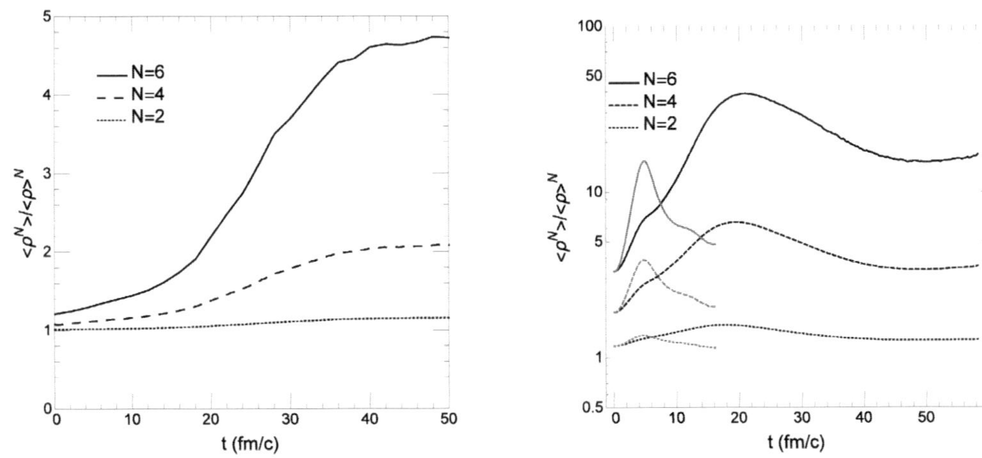

Figure 1. Time evolution of scaled density moments during a first-order phase transition for a quark matter confined in a box with an initially uniform net quark density $n_q = 0.5$ fm^{-3} and temperature $T = 20$ MeV (left window), and for an expanding quark matter with a spherical Wood-Saxon distribution of an initial net quark density $n_q = 1.5$ fm^{-3} and temperature $T_0 = 70$ MeV at the center (right window). Red lines in the right panel are for an expanding quark matter with same initial conditions but without a first-order phase transition.

about $t = 40$ fm/c, when the phase separation almost ends. Also, moments with larger N increase faster and saturate at larger values.

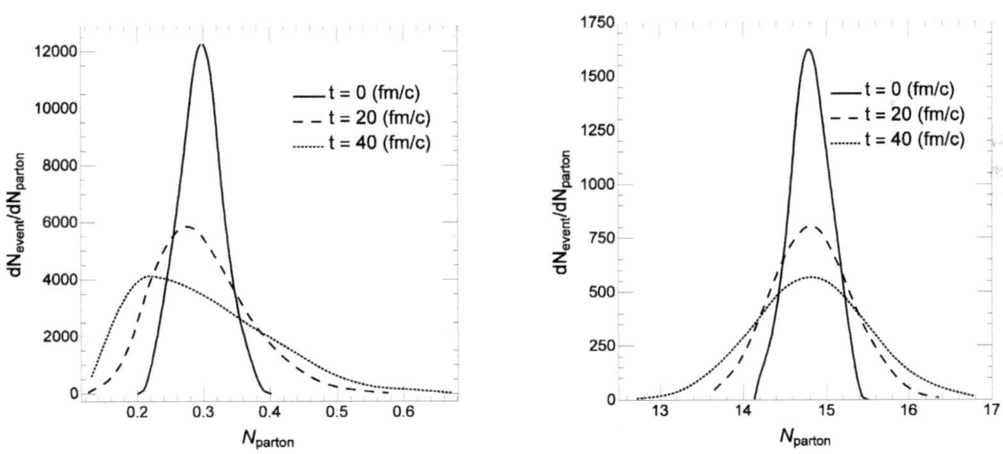

Figure 2. Time evolution of the event-by-event distribution of the number of quarks in a sub-volume of size 0.6 fm^3 (left window) and 30 fm^3 (right window) for a quark matter of temperature $T = 20$ MeV and average net quark density $n_q = 0.5$ fm^{-3} inside the spinodal region. The total number of events is 1000.

The number of quarks inside a sub-volume also varies drastically from event to event during a first-order phase transition as shown in Fig. 2 for the central cell from 1000 events at $t = 0$, 20, and 40 fm/c by the solid, dashed and dotted lines, respectively, for the two cases of sub-volume of size 0.6 fm^3 (left window) and 30 fm^3 (right window). The left panel of Fig. 2 clearly shows that the distribution for the small sub-volume becomes asymmetric as time increases, starting with an initial skewness of 0.11 and increasing to 0.60 at 20 fm/c and 0.75 at 40 fm/c, agreeing with the prediction in Ref. [9]. This feature is absent in the right panel of Fig. 2 for the larger sub-volume, where the distribution remains essentially symmetric with increasing time.

3. Expanding quark matter

For an expanding quark matter as in relativistic heavy ion collisions, the blast wave initial conditions of a spherical Wood-Saxon distribution with radius $R = 5$ fm, skin thickness $a = 0.5$ fm, and central density $\rho_0 = 1.5$ fm^3, and temperature $T_0 = 70$ MeV are used in Ref. [8] for two different equations of state with ($G_V = 0$) and without ($G_V = G_S$) a first-order phase transition. The phase trajectory of the central part of the quark matter is shown to traverse through the spinodal instablity region in the case of $G_V = 0$.

Shown in the right window of Fig. 1 by the dotted, dashed, and solid lines are the time evolution of the density moments for $N = 2$, 4, and 6, respectively. The black and red lines correspond to the two cases with and without a first-order phase transition, respectively. In both cases, the scaled density moments first increase and then decrease with time. Although the scaled density moments in the case without a first-order phase transition are larger than one, those in the case with a first-order phase transition are even larger, reflecting the effect due to density clumps that distribute randomly inside the expanding quark matter.

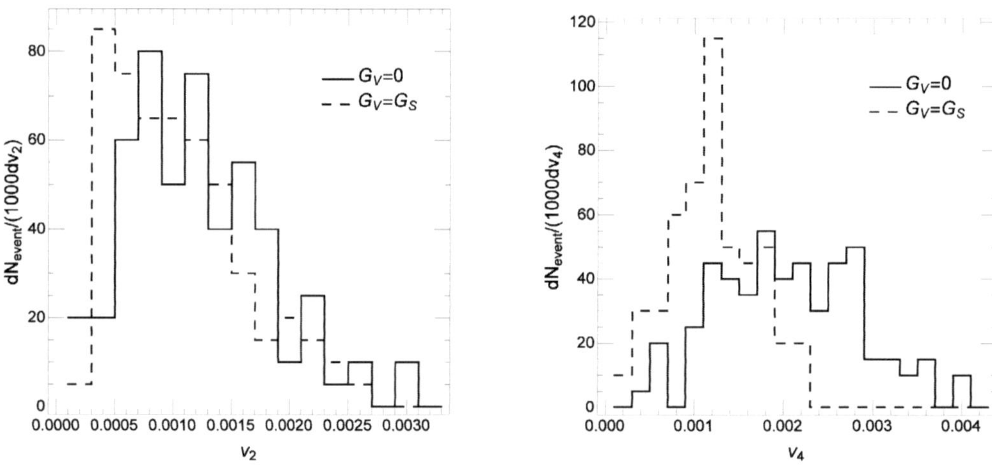

Figure 3. Event distributions of final anisotropic flows v_2 (left window) and v_4 (right window) for 100 events of an expanding quark matter with the blast wave initial conditions.

Since density fluctuations can lead to spatial anisotropy even in central heavy ion collisions, it can affect the anisotropic flows in the transverse plane of heavy ion collisions [11, 12]. The latter are characterized by the coefficients v_n in the Fourier expansion of the azimuthal angle distribution of particle transverse momenta and can be calculated by the two particle cumulant method [13], namely, $v_n\{2\} = \sqrt{\langle\cos(n\Delta\phi)\rangle}$ by averaging over all pairs of particles in an event. The event distributions of final $v_2\{2\}$ (left window) and $v_4\{2\}$ (tight window) of freeze-out partons for 100 events are shown in Fig. 3, with the solid and dashed lines for the cases with and without first order phase transition, respectively. Both distributions peak at a larger value for the case with a first-order phase transition, particularly the v_4, so enhanced anisotropic flows provide a plausible signal for the first-oder phase transition in the quark matter produced in relativistic heavy ion collisions.

Because the rate of the dilepton production through quark-antiquark annihilation is proportional to the product of their densities, it will also be enhanced as the density fluctuations

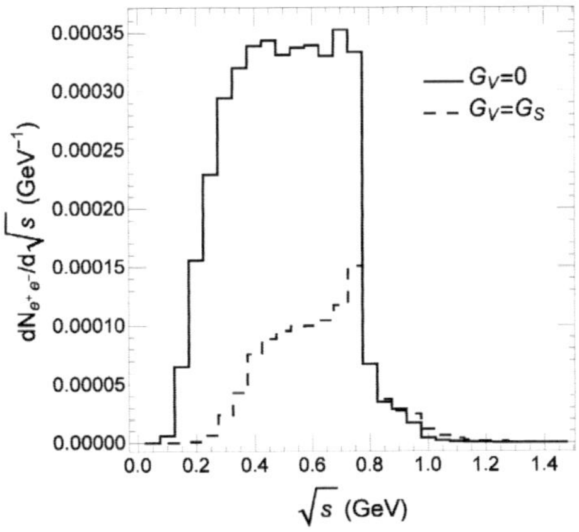

Figure 4. Dilepton yield as a function of the invariant mass $\sqrt{s}$ for the cases with (solid line) and without (dashed line) a first-order phase transition in an expanding quark matter with the blast wave initial conditions.

become large. The dilepton invariant mass spectrum obtained by averaging over 100 events are plotted in Fig. 4 by the solid and dashed lines for the cases with and without a first-order phase transition, respectively. As expected, more dileptions are produced from a quark matter with a first-order phase transition.

4. Summary

The spinodal instability of a baryon-rich quark matter has been studied in the transport model that is based on from the NJL model. Dense clumps are observed in both confined and expanding quark matters, and they lead to the enhancement of higher-order scaled density moments. The skewness of the quark number event-by-event distribution in a small sub-volume also increases during the first-order phase transition in the confined quark matter, but this feature disappears if the sub-volume is large. The density fluctuations further affect the anisotropy in the momentum space and enhance the dilepton production through quark-antiquark annihilation in the expanding quark matter.

Acknowledgments

This work was supported in part by the US Department of Energy under Contract No. DE-SC0015266 and the Welch Foundation under Grant No. A-1358.

References

[1] Adamczyk L *et al.* (STAR) 2014 *Phys. Rev. Lett.* **112** 032302
[2] Adamczyk L *et al.* (STAR) 2014 *Phys. Rev. Lett.* **113** 092301
[3] McDonald D (STAR) 2015 *EPJ Web Conf.* **95** 01009
[4] Steinheimer J and Randrup J 2012 *Phys. Rev. Lett.* **109**(21) 212301
[5] Herold C, Nahrgang M, Mishustin I and Bleicher M 2014 *Nucl. Phys. A* **925** 14 – 24 ISSN 0375-9474
[6] Steinheimer J and Randrup J 2013 *Phys. Rev. C* **87**(5) 054903
[7] Klevansky S P, Ogura A and Hufner J 1997 *Ann. Phys.* **261** 37–73
[8] Li F and Ko C M 2016 *Phys. Rev. C* **93**(3) 035205
[9] Stephanov M A 2009 *Phys. Rev. Lett.* **102**(3) 032301
[10] Randrup J 2010 *Phys. Rev. C* **82**(3) 034902
[11] Herold C, Nahrgang M, Mishustin I and Bleicher M 2014 *J. Phys. Conf. Ser.* **509** 012065
[12] Steinheimer J, Randrup J and Koch V 2014 *Phys. Rev. C* **89**(3) 034901
[13] Borghini N, Dinh P M and Ollitrault J Y 2001 *Phys. Rev. C* **64**(5) 054901

IOP Conf. Series: Journal of Physics: Conf. Series **832** (2017) 012046 doi:10.1088/1742-6596/832/1/012046

Hydrodynamic Overview at Hot Quarks 2016

Jacquelyn Noronha-Hostler

Department of Physics, University of Houston, Houston, TX 77204, USA

E-mail: jakinoronhahostler@gmail.com

Abstract. Event-by-event relativistic hydrodynamics has been extremely successful in describing flow observables in heavy-ion collisions. However, the initial state and viscosity simultaneously affect comparisons to data so a discussion of experimental observables that help to distinguish the two follows. Specific problems that arise in the hydrodynamical modeling at the Beam Energy Scan are also addressed.

1. Introduction

The Quark-Gluon Plasma (QGP) existed microseconds after the Big Bang where its size was significantly larger and its expansion significantly slower than the QGP experimentally measured in heavy ion collisions. The QGP produced in the laboratory is the hottest, smallest, and densest fluid known to mankind so the techniques used to describe its properties are still being developed. With the discovery that the QGP formed at RHIC and the LHC was a nearly perfect liquid, a "standard model" of heavy-ion collisions has emerged that includes fluctuating initial conditions, event-by-event relativistic viscous hydrodynamics, and a hadronic afterburner.

The main signatures of nearly perfect fluidity are the flow harmonics (Fourier coefficients of the particle spectra) that can be reproduced within relativistic hydrodynamical calculations with an extremely small shear viscosity to entropy density ratio, $\eta/s \sim 0.08$. Originally, it was understood that there should be a lower bound for $\eta/s \sim 1/4\pi$ by combining a quasiparticle description with the uncertainty principle [1]. The derivation of the KSS limit from strong coupling holography [2] initially gave support to such a bound, although, now it is known that there are holographic examples where η/s can be even smaller, e.g., [3]. It is expected that a minimum exists close to the crossover temperature [4] as one goes from a strongly interacting QGP phase into an eventually weakly interacting hadron gas phase [5, 6]. Further references for the temperature dependence of viscosity can be found in [7].

While a reasonable estimate for the range of values of η/s can be made using sophisticated Bayesian techniques [8, 9], its exact value is extremely dependent on the initial state formed immediately after the two heavy-ions collide. Over the years, a plethora of initial state models have been developed, each of which has a corresponding range of valid transport coefficients that allow for reasonable fit to experimental data. One of the most pressing issues remaining in heavy ions is finding observables either sensitive to only the initial state or only to transport coefficients, which will be discussed in detail in this proceedings.

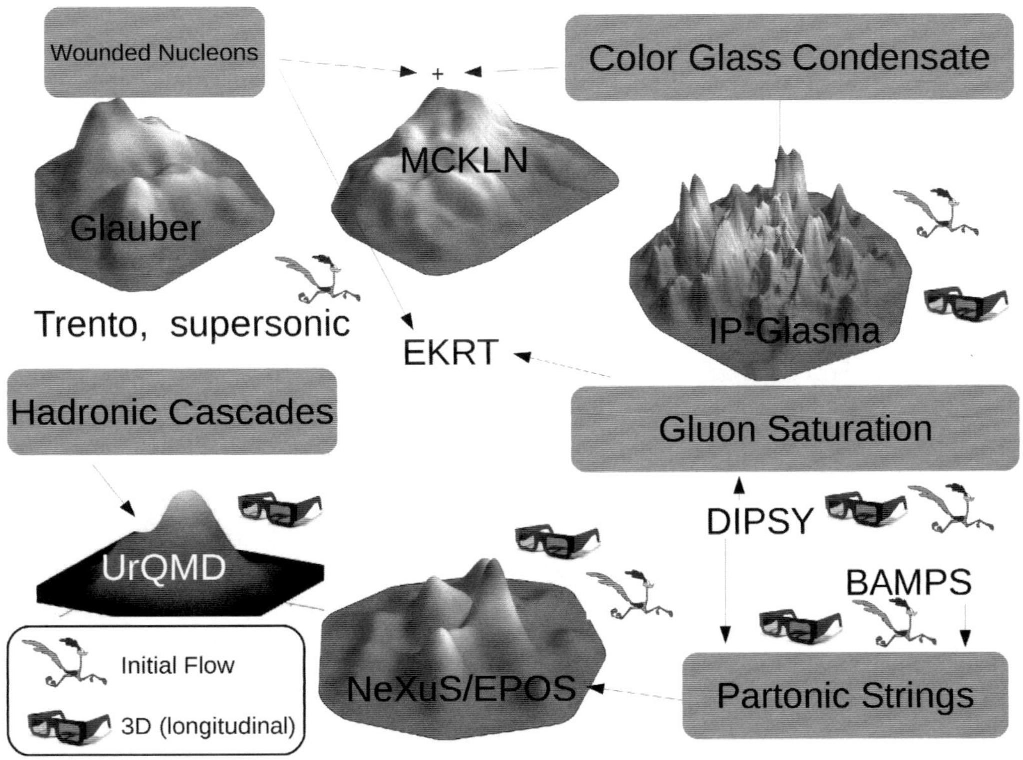

Figure 1. Initial conditions used in heavy ion collisions organized by their basic assumptions.

2. Collective Flow

When two heavy ions are smashed together, clear geometrical effects occur depending on if they hit head-on (central collisions), have a grazing collision (peripheral), or hit somewhere in between (mid-central). Experimentally, heavy ions are collided billions of times (each collision is an event) and each event produces a different number of particles that participated in the collision, known as multiplicity. The more central the collisions, the larger multiplicities are produced so the events are then sorted by their multiplicities into central classes where 0% centrality have the highest multiplicities and 100% centrality indicates the lowest possible multiplicities. Central collisions produce on average a circular shape in the transverse plane to the beam axis whereas an approximate almond shape is produced for mid-central collisions and beyond.

Due to quantum fluctuations in the initial position of protons and neutrons within each ion, a multitude of shapes can be produced [10]. Each highly inhomogeneous initial condition runs separately through hydrodynamics on an event-by-event basis. Experimentally the initial state cannot be measured directly, rather pressure gradients convert the initial geometrical shapes into a corresponding momentum space distribution anisotropy, measured via flow harmonics. To obtain the flow harmonics, one calculates the Fourier coefficients of the particle spectra (with special care to reproduce the exact way experimentalists measure flow harmonics [11] where multiplicity weighing and centrality rebinning should not be ignored [12, 13]).

A number of parameters that go into hydrodynamical modeling such as the initial time after which one assumes the system admits a hydrodynamic description, and also the switching temperature below which a hadronic transport is used. The initial time, τ_0, depends on the

IOP Conf. Series: Journal of Physics: Conf. Series **832** (2017) 012046 doi:10.1088/1742-6596/832/1/012046

collisional energy as well as the initial condition type and these issues were discussed in more detail in the Beam Energy Scan and anisotropic hydrodynamics talks at this conference. While the τ_0 can indicate an average initial temperature of the QGP, due to the lumpy nature of the medium, the initial temperatures is, in fact, a temperature field with strong variations on an event-by-event basis. The maximum switching temperature, T_{SW}, is constrained by the hadronization temperature indicated from Lattice QCD [14] to ensure that one switches to the correct degrees of freedom.

The flow harmonics themselves are most sensitive to the choice in the initial conditions as well as the transport coefficients (η/s, bulk viscosity to entropy density, ζ/s, and their corresponding relaxation times- τ_π and τ_Π, respectively). In the next two sections experimental observables that constrain the initial state and transport coefficients are discussed.

3. Separating the Initial State from Transport Coefficients

At LHC energies a number of initial conditions such as IP-Glasma [15], EKRT [16], and Trento (tuned to IP-Glasma) [17] manage to fit well the two and four particle flow cumulants as well as the distribution of v_n's. In Fig. 1 a schematic cartoon of the most well-known initial condition models categorized by their basic properties is shown. Additionally, extremely accurate predictions for the flow harmonics of LHC run 2 to the order of $\sim 5\%$ [18, 19] were made and later experimentally confirmed in [20]. However, the different models used in these predictions varied parameters like viscosity, freeze-out, and the inclusion of initial flow.

It has been well-established that an approximately linear relationship exists between the initial energy/entropy density eccentricities, ε_n, and the experimentally measured flow harmonics v_n's [21, 22, 23] and that sub-nucleon fluctuations do not appear to play a significant role in the calculation of the lowest order harmonics in large collision systems [24]. That being said, higher order flow harmonics are significantly more complicated and depend on a variety of eccentricities [22, 23]. Additionally, v_1 is especially complicated [23] and may depend on the full $T^{\mu\nu}$ initialization [25].

In order to more easily quantify the distribution of flow harmonics, multiparticle cumulants are used. For two particle correlations a rapidity gap is used to minimize non-flow effects whereas for 4+ particle correlations it is not possible. Cumulants of the v_2 distribution are directly connected to the moments of the distribution via $v_2\{4\}/v_2\{2\})^4 = 2 - \langle v_2^4\rangle/\langle v_2^2\rangle^2$, which can indicate the degree of which the system is fluctuating (see Fig. 2). As shown in Fig. 2, the kurtosis has a sharper peak to compensate for large fluctuations in the tails so the ratio of the fourth moment over the second indicates how much the tails are fluctuating versus the variance. If there are no fluctuations in the system the p_T-integrated $v_2\{4\}/v_2\{2\} \rightarrow 1$, whereas $v_2\{4\}/v_2\{2\} < 1$ and $v_2\{4\} \sim v_2\{6\} \sim \dots$ is a sign of the collective behavior measured in heavy ion collisions. Note, however, that the higher order cumulants are not exactly identical, small deviations can exist due to the skewness of the initial conditions [26]. Complications do exist for more peripheral collisions where deviations are seen between the linear mapping of the initial eccentricities and the final elliptical flow [16], which can be explained due to cubic response [27].

Recently, symmetric cumulants [28] that measure the correlation of different order flow harmonics on an event-by-event basis have been measured in PbPb collisions and $NSC(3,2)$ (which involves elliptic and triangular flow) appears to be almost entirely driven by the initial eccentricities. The symmetric cumulants are 4 particle correlations such that

$$NSC(m,n) = \langle v_m^2 v_n^2\rangle - \langle v_m^2\rangle\langle v_n^2\rangle / \langle v_m^2\rangle\langle v_n^2\rangle \tag{1}$$

where $m \neq n$. While it has been thought that $NSC(4,2)$ was sensitive to $\eta/s(T)$, the sensitivity diminishes after proper treatment of the multiplicity weighing and centrality binning in order to be consistent with experimental data [12].

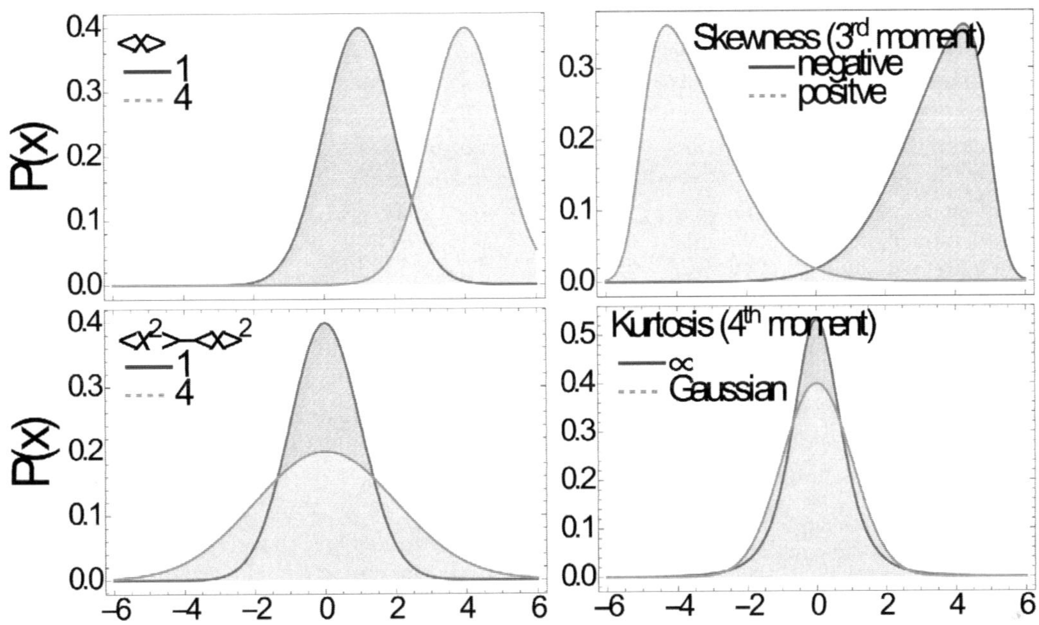

Figure 2. First moment (mean), second moment (variance), third moment (skewness), and fourth moment (kurtosis) of a distribution.

Another possibility that exists for extracting the properties of the initial state is via the correlations between soft and hard sectors of heavy ion collisions, i.e., soft-hard event engineering (SHEE) as described in [32] where it was found that the high p_T flow harmonics are directly linked to fluctuating initial eccentricities (due to the path length dependence of jet quenching). Moreover, it was recently found that viscosity plays little role in the high p_T flow harmonics [13] so it's especially interesting to exploit this relationship. Further studies may also be possible in the heavy flavor sector as well [33], as done for instance in [34].

Event plane correlations [31] were originally to correlate flow event plane angles of different orders. One of the simplest examples is the correlation between the second and fourth event planes:

$$\langle \cos 4\left(\Phi_2 - \Phi_4\right)\rangle\{\mathrm{SP}\} = \frac{\langle v_2^2 v_4 \cos 4(\Psi_2 - \Psi_4)\rangle}{\sqrt{\langle v_2^4\rangle\langle v_4^2\rangle}} \tag{2}$$

where a rapidity gap is implemented to reduce non-flow contributions. Interestingly enough, it appears that there is a direct link between the symmetric cumulants and the event plane correlations [30], though the latter are significantly more sensitive to viscous effects [16]. In fact, event plane correlations appear to be very sensitive to $\eta/s(T)$ where a small viscosity $\lesssim 0.2$ in the hadron resonance gas phase seems to be preferred (such a small η/s can occur due to a large number of massive, degenerate states close to the crossover transition [5, 6]). Additionally, $v_4\{4\}^4$ may indicate that a larger η/s is needed but further testing is still needed [35].

The extraction of transport coefficients is significantly more intricate due to the interplay between shear and bulk viscosity [36, 37, 38]. This is further complicated due to the large influence of the bulk viscous corrections at hadronization. In fact, different implementations of the bulk viscosity correction to the particle distribution at freeze-out can either increase the shear viscosity [9, 36, 37] or decrease it, as in [38]. Furthermore, second order transport coefficients may also become relevant [39, 40] especially in small collision systems, and the effects of many of these coefficients have not been systematically studied within the context of event-by-event hydrodynamics.

Additionally, $\langle p_T \rangle$ was recently used in [38] (and later in [9]) to help constrain the temperature dependence of the bulk viscosity. There is a very intricate connection between the temperature dependence of ζ/s and the equation of state (bulk viscosity is only nonzero in a non-conformal system, which indicates that there should be a peak in bulk viscosity at the crossover [41]) so this quantity is expected to play an even more significant role in the Beam Energy Scan.

4. Beam Energy Scan

So far in this paper we focused on LHC energies where the baryon chemical potential, μ_B, is approximately zero. With the Beam Energy Scan at RHIC one is able to probe matter that would correspond to other regions in the QCD Phase Diagram and, hopefully, also approach the critical end point. However, a significant number of questions remain at almost all level of hydrodynamical modeling. Initial calculations within kinetic theory/hadron resonance gas [42, 43, 44], which do not contain a critical point, show a decrease in η/s as one increases μ_B. However, if the dynamic universality class of the QCD critical point is that of model H [45], then one would expect the transport coefficients to diverge at the critical point. Assuming that the dynamic universality class is H and there is a diverging viscosity, then this would be a nice explanation for the turning off of triangular flow observed at low energies [46].

However, there may be a more simple explanation for the decrease in v_3 at low energies that connects to the decreased amount of time spent in the hydrodynamic phase as the beam energy decreases [47]. This would be more consistent either with no critical point (or perhaps one placed at much larger μ_B) or perhaps with a different dynamic universality class such as that of model B [48]. Significantly more research needs to be done to better understand both the location of the critical point as well as the dynamic properties of the expanding QGP in this regime and, hopefully, the results from the upcoming Beam Energy Scan II run at STAR will lead to unambiguous experimental verification of QCD critical behavior.

5. Outlook

Relativistic viscous hydrodynamics has become a powerful tool to explore the strongly interacting QCD matter formed in heavy ion collisions. Significant progress has been made in finding "orthogonal" measurements that can independently aid in extracting either the initial state of the system or its transport properties. A number of open-ended questions still remain such as if there is still perfect fluidity close to the QCD critical point as well as what special considerations need to be taken into account at finite baryon density. While not covered here, other important questions discussed at this conference remain, e.g., "is anisotropic hydrodynamics needed? [49] or can pre-equilibrium flow be a reasonable substitute?" [50, 51]. Additionally, the origin of flow harmonics (and their corresponding cumulants) in small systems is still up for debate [52].

Acknowledgements

JNH thanks the HotQuarks2016 organizers for inviting her to give the Hydrodynamic Overview of Heavy-Ion Collisions and she was supported by the National Science Foundation under grant no. PHY-1513864.

References

[1] P. Danielewicz and M. Gyulassy, Phys. Rev. D **31**, 53 (1985).
[2] P. Kovtun, D. T. Son and A. O. Starinets, JHEP **0310**, 064 (2003)
[3] M. Brigante, H. Liu, R. C. Myers, S. Shenker and S. Yaida, Phys. Rev. D **77**, 126006 (2008)
[4] Y. Aoki, G. Endrodi, Z. Fodor, S. D. Katz and K. K. Szabo, Nature **443**, 675 (2006)

[5] J. Noronha-Hostler, J. Noronha and C. Greiner, Phys. Rev. Lett. **103**, 172302 (2009)
[6] J. Noronha-Hostler, J. Noronha and C. Greiner, Phys. Rev. C **86**, 024913 (2012)
[7] J. Noronha-Hostler, arXiv:1512.06315 [nucl-th].
[8] J. E. Bernhard, P. W. Marcy, C. E. Coleman-Smith, S. Huzurbazar, R. L. Wolpert and S. A. Bass, Phys. Rev. C **91**, no. 5, 054910 (2015)
[9] J. E. Bernhard, J. S. Moreland, S. A. Bass, J. Liu and U. Heinz, Phys. Rev. C **94**, no. 2, 024907 (2016)
[10] B. Alver and G. Roland, Phys. Rev. C **81**, 054905 (2010) Erratum: [Phys. Rev. C **82**, 039903 (2010)]
[11] M. Luzum and J. Y. Ollitrault, Phys. Rev. C **87**, no. 4, 044907 (2013)
[12] F. G. Gardim, F. Grassi, M. Luzum and J. Noronha-Hostler, arXiv:1608.02982 [nucl-th].
[13] B. Betz, M. Gyulassy, M. Luzum, J. Noronha, J. Noronha-Hostler, I. Portillo and C. Ratti, arXiv:1609.05171 [nucl-th].
[14] S. Borsanyi, Z. Fodor, S. D. Katz, S. Krieg, C. Ratti and K. K. Szabo, Phys. Rev. Lett. **113**, 052301 (2014)
[15] C. Gale, S. Jeon, B. Schenke, P. Tribedy and R. Venugopalan, Phys. Rev. Lett. **110**, no. 1, 012302 (2013)
[16] H. Niemi, K. J. Eskola and R. Paatelainen, Phys. Rev. C **93**, no. 2, 024907 (2016)
[17] J. S. Moreland, J. E. Bernhard and S. A. Bass, Phys. Rev. C **92**, no. 1, 011901 (2015)
[18] H. Niemi, K. J. Eskola, R. Paatelainen and K. Tuominen, Phys. Rev. C **93**, no. 1, 014912 (2016)
[19] J. Noronha-Hostler, M. Luzum and J. Y. Ollitrault, Phys. Rev. C **93**, no. 3, 034912 (2016)
[20] J. Adam *et al.* [ALICE Collaboration], Phys. Rev. Lett. **116**, no. 13, 132302 (2016)
[21] D. Teaney and L. Yan, Phys. Rev. C **83**, 064904 (2011)
[22] F. G. Gardim, F. Grassi, M. Luzum and J. Y. Ollitrault, Phys. Rev. C **85**, 024908 (2012)
[23] F. G. Gardim, J. Noronha-Hostler, M. Luzum and F. Grassi, Phys. Rev. C **91**, no. 3, 034902 (2015)
[24] J. Noronha-Hostler, J. Noronha and M. Gyulassy, Phys. Rev. C **93**, no. 2, 024909 (2016)
[25] F. G. Gardim, F. Grassi, Y. Hama, M. Luzum and J. Y. Ollitrault, Phys. Rev. C **83**, 064901 (2011)
[26] G. Giacalone, L. Yan, J. Noronha-Hostler and J. Y. Ollitrault, arXiv:1608.01823 [nucl-th].
[27] J. Noronha-Hostler, L. Yan, F. G. Gardim and J. Y. Ollitrault, Phys. Rev. C **93**, no. 1, 014909 (2016)
[28] J. Adam *et al.* [ALICE Collaboration], arXiv:1604.07663 [nucl-ex].
[29] Y. Zhou, Adv. High Energy Phys. **2016**, 9365637 (2016) doi:10.1155/2016/9365637 [arXiv:1607.05613 [nucl-ex]].
[30] G. Giacalone, L. Yan, J. Noronha-Hostler and J. Y. Ollitrault, Phys. Rev. C **94**, no. 1, 014906 (2016)
[31] G. Aad *et al.* [ATLAS Collaboration], Phys. Rev. C **90**, no. 2, 024905 (2014)
[32] J. Noronha-Hostler, B. Betz, J. Noronha and M. Gyulassy, Phys. Rev. Lett. **116**, no. 25, 252301 (2016)
[33] M. Nahrgang, J. Aichelin, S. Bass, P. B. Gossiaux and K. Werner, Phys. Rev. C **91**, no. 1, 014904 (2015)
[34] C. A. G. Prado, J. Noronha-Hostler, M. R. Cosentino, M. G. Munhoz, J. Noronha and A. A. P. Suaide, arXiv:1609.06093 [nucl-th].
[35] G. Giacalone, L. Yan, J. Noronha-Hostler and J. Y. Ollitrault, arXiv:1608.06022 [nucl-th].
[36] J. Noronha-Hostler, G. S. Denicol, J. Noronha, R. P. G. Andrade and F. Grassi, Phys. Rev. C **88**, 044916 (2013)
[37] J. Noronha-Hostler, J. Noronha and F. Grassi, Phys. Rev. C **90**, no. 3, 034907 (2014)
[38] S. Ryu, J.-F. Paquet, C. Shen, G. S. Denicol, B. Schenke, S. Jeon and C. Gale, Phys. Rev. Lett. **115**, no. 13, 132301 (2015)
[39] E. Molnar, H. Niemi, G. S. Denicol and D. H. Rischke, Phys. Rev. D **89**, no. 7, 074010 (2014)
[40] S. I. Finazzo, R. Rougemont, H. Marrochio and J. Noronha, JHEP **1502**, 051 (2015)
[41] F. Karsch, D. Kharzeev and K. Tuchin, Phys. Lett. B **663**, 217 (2008)
[42] N. Demir and S. A. Bass, Phys. Rev. Lett. **102**, 172302 (2009)
[43] G. S. Denicol, C. Gale, S. Jeon and J. Noronha, Phys. Rev. C **88**, no. 6, 064901 (2013)
[44] G. P. Kadam and H. Mishra, Nucl. Phys. A **934**, 133 (2014)
[45] D. T. Son and M. A. Stephanov, Phys. Rev. D **70**, 056001 (2004)
[46] L. Adamczyk *et al.* [STAR Collaboration], Phys. Rev. Lett. **116**, no. 11, 112302 (2016)
[47] J. Auvinen and H. Petersen, Phys. Rev. C **88**, no. 6, 064908 (2013)
[48] R. Rougemont, J. Noronha and J. Noronha-Hostler, Phys. Rev. Lett. **115**, no. 20, 202301 (2015)
[49] M. Nopoush, R. Ryblewski and M. Strickland, Phys. Rev. C **90**, no. 1, 014908 (2014)
[50] U. W. Heinz and J. Liu, Nucl. Phys. A **956**, 549 (2016)
[51] L. Keegan, A. Kurkela, A. Mazeliauskas and D. Teaney, JHEP **1608**, 171 (2016)
[52] See P. Tribedy, HotQuarks 2016 proceedings to appear.

IOP Conf. Series: Journal of Physics: Conf. Series **832** (2017) 012047 doi:10.1088/1742-6596/832/1/012047

Equilibrium and non-equilibrium properties of the QGP in a magnetic field: a holographic approach

Renato Critelli

Instituto de Física, Universidade de São Paulo, Rua do Matão, 1371, Butantã, CEP 05508-090, São Paulo, SP, Brazil

E-mail: renato.critelli@usp.br

Abstract. This manuscript reviews recent theoretical progress on the understanding of the quark gluon plasma in a magnetic field that I presented on the conference Hot Quarks 2016, held at South Padre Island, Texas, USA, 12-17 September 2016. It is shown that, using a holographic bottom-up Einstein-Maxwell-Dilaton model, one can have a good quantitative agreement with Lattice result for QCD equation of state and Polyakov loop with nonzero magnetic field. I also present results for the anisotropic shear viscosity ratio $\eta_{\parallel}/\eta_{\perp}$, with the conclusion that $\eta_{\parallel} < \eta_{\perp}$ for non-zero magnetic field.

1. Introduction

The early stages of noncentral ultrarelativistic heavy ion collisions produce extremely large magnetic fields (see the review [1]). For comparison, the magnetic field B created at top LHC collision energies may reach values up to $B \sim 10^{19}$ G (or $eB \sim 15m_{\pi}^2 \sim 0.3$ GeV^2 in natural units), whilst the Earth's magnetic field has typical values of $B \sim 0.5$ G. Such extreme value of magnetic field has sparked great interest on its role on the dynamics of the strongly interacting quark gluon plasma (QGP) [2, 3] formed in heavy ion collisions.

A fundamental property of the QGP is its extremely small value of its shear viscosity over the entropy density ratio, $\eta/s \approx 0.095$, according to state-of-the-art hydrodynamic simulations [4] that match experimental data. Near the crossover region [5], i.e. the strongly coupled regime of QCD, weak coupling calculations are not applicable and one has to decide what effective model to use in order to study fundamental properties of the QGP. While lattice QCD is the non-perturbative method to compute the properties of the QGP in equilibrium, there are technical difficulties to extract real time observables from the lattice [6]. In this work, the gauge/gravity duality [7, 8, 9, 10] (or holography) is used as an effective many-body theory to describe thermal QCD near the crossover region. Besides the unique opportunity to study strongly coupled gauge theories both in and out of equilibrium, this holographic approach naturally describes nearly perfect liquids where $\eta/s = 1/(4\pi)$ [11, 12].

The holographic AdS/CFT duality introduced by Maldacena [7] relates type IIB string theory in an anti-de Sitter space (times a 5-sphere) with a supersymmetric conformal field theory in 4 dimensions, namely $\mathcal{N} = 4$ super Yang-Mills (SYM) theory. Although supersymmetry is broken when the temperature is introduced, QCD is definitely not conformal near the crossover region and, thus, in phenomenological applications to QGP physics it is necessary to go beyond conformal modeling [13]. In fact, the QCD trace anomaly has a peak around the crossover

IOP Conf. Series: Journal of Physics: Conf. Series **832** (2017) 012047 doi:10.1088/1742-6596/832/1/012047

critical temperature [14] and bulk viscosity [15, 16], which is absent in conformal plasmas, is needed for a consistent hydrodynamic description of the QGP [4, 17, 18, 19]. In order to make more contact with QCD phenomenology and to make quantitative predictions, we resort to a phenomenological bottom-up model introduced by Gubser and collaborators [20, 21, 22], sometimes called "black hole engineering" [13, 23], where black hole solutions of a holographic theory are constructed to have the same thermodynamic properties of the QGP as determined on the lattice. The holographic model in our case is defined by an Einstein-Maxwell-Dilaton (EMD) gravitational action in which the dilaton field (scalar field) is responsible for the breaking of conformal symmetry while the Maxwell field is necessary to include a magnetic field.

The EMD holographic model is currently being used to unveil important properties of the QGP near the crossover region, such as hydrodynamic transport coefficients [24], effects of baryonic chemical potential [25, 26, 27, 28], photon and dilepton production [29], and, for this proceedings in particular, magnetic field effects [30, 31, 23].

2. Gravitational setup and results

The gravitational bulk EMD Lagrangian $\mathcal{L}$ used to describe the strongly coupled QGP near the crossover region in the model is

$$16\pi G_5\,\mathcal{L} = R - \frac{(\partial_\mu\phi)^2}{2} - V(\phi) - \frac{f(\phi)F_{\mu\nu}^2}{4}, \tag{1}$$

where G_5 is the 5-dimensional Newton's constant, R is the Ricci scalar, ϕ is the dilaton field, and $F_{\mu\nu}$ is the Maxwell field. The dilaton potential $V(\phi)$, G_5, and the gauge-dilaton coupling $f(\phi)$ are free inputs that are fixed by matching the lattice results for the QCD equation of state and the magnetic susceptibility with (2+1) flavors at $B = 0$ [14, 32]. Therefore, once the inputs are fixed at $B = 0$, the results computed at non-zero magnetic field are genuine predictions of the holographic model.

Our results for the thermodynamics with non-zero magnetic field [30, 31], which are equilibrium properties of QCD and, therefore, can be compared with lattice results, are presented in Fig. 1. One can see that our calculations are in very good agreement with the lattice results [14, 38]. Evidently, the results with zero magnetic field are not predictions of the holographic model since they were engineered to match the results of lattice QCD. Such a quantitative agreement in equilibrium is the minimal requirement that must be fulfilled in phenomenological QGP applications before non-equilibrium calculations can be performed within the holographic model.

Regarding the holographic Polyakov loop [33, 34, 35], we followed previous calculations of this quantity [36, 37] and extended them to the more realistic non-conformal modeling pursued here. We show the results for this non-local observable in Fig. 2 with a comparison to lattice results [39, 40]. With the Polyakov loop at hand, we were able to compute the holographic heavy quark entropy [37], which is also found to be in remarkable agreement with lattice results [41].

Regarding the non-equilibrium properties of the QGP, such as transport coefficients, we calculated how the magnetic field may change the value of η/s. However, one has to keep in mind that the magnetic field breaks spatial isotropy and, thus, there are two shear viscosity coefficients, $\eta_\parallel$ (the shear coefficient parallel to the direction of B) and $\eta_\perp$ (shear coefficient in the directions orthogonal to the B direction) [42, 43, 31]. Fig. 3 shows the result for the ratio $\eta_\parallel/\eta_\perp$ as function of magnetic field and temperature in our model that describes QGP lattice data near the crossover transition.

3. Conclusions

Holography can be used to describe the effects of strong magnetic fields on the non-conformal QGP near the crossover transition both in and out-of equilibrium. In the near future, we intend to investigate the non-equilibrium aspects of the model considered here undergoing a Bjorken expansion.

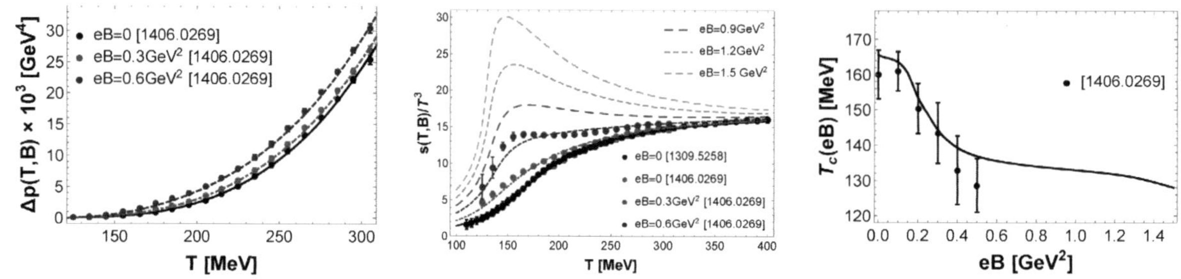

Figure 1. (Color online) Holographic predictions for QCD thermodynamics with nonzero magnetic field near the crossover region compared with lattice results (represented by the dots). *Left:* Pressure difference. *Center:* Entropy density. *Right:* (Pseudo) Critical temperature obtained from the inflection point of s/T^3. Lattice result taken from [14, 38].

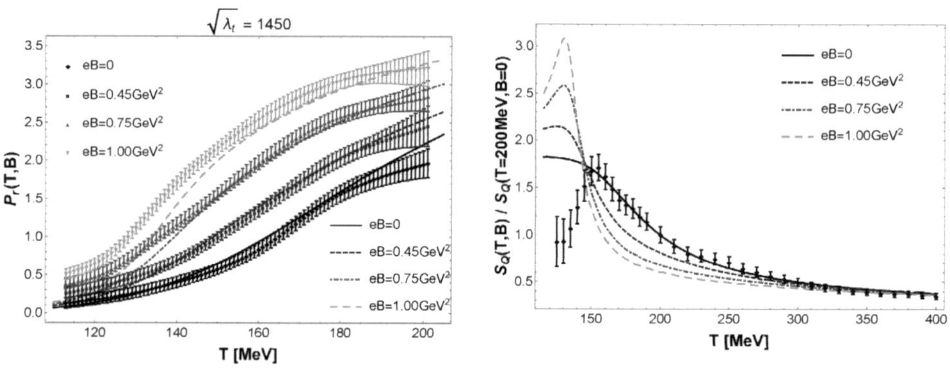

Figure 2. (Color online) *Left:* Holographic Polyakov loop compared with lattice result (represented by the dots) [39, 40]. *Right:* Holographic heavy quark entropy compared to lattice results from Ref. [41].

Acknowledgements
I thank J. Noronha for comments on this manuscript and the São Paulo Research Foundation (FAPESP), under FAPESP grant number 2016/09263-2, for financial support.

References
[1] D. Kharzeev, K. Landsteiner, A. Schmitt and H. -U. Yee, Lect. Notes Phys. **871**, 1 (2013).
[2] M. Gyulassy and L. McLerran, Nucl. Phys. A **750**, 30 (2005).
[3] U. Heinz and R. Snellings, Ann. Rev. Nucl. Part. Sci. **63**, 123 (2013).
[4] S. Ryu, J.-F. Paquet, C. Shen, G. S. Denicol, B. Schenke, S. Jeon and C. Gale, Phys. Rev. Lett. **115**, no. 13, 132301 (2015).
[5] Y. Aoki, G. Endrodi, Z. Fodor, S. D. Katz and K. K. Szabo, Nature **443**, 675 (2006).
[6] H. B. Meyer, Eur. Phys. J. A **47**, 86 (2011).
[7] J. Maldacena, Adv. Theor. Math. Phys. **2**, 231 (1998).
[8] S. S. Gubser, I. R. Klebanov and A. M. Polyakov, Phys. Lett. B **428**, 105 (1998),
[9] E. Witten, Adv. Theor. Math. Phys. **2**, 253 (1998).

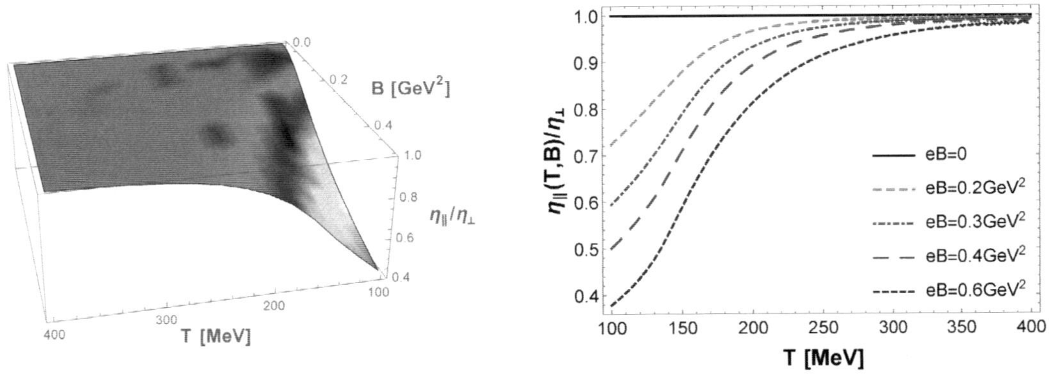

Figure 3. (Color online) *Left:* $\eta_{\parallel}/\eta_{\perp}$ as function of the magnetic field B and the temperature T. *Right:* Anisotropy of shear viscosity as function of the T for selected values of B.

[10] E. Witten, Adv. Theor. Math. Phys. **2**, 505 (1998).
[11] P. Kovtun, D. T. Son and A. O. Starinets, Phys. Rev. Lett. **94**, 111601 (2005).
[12] A. Buchel and J. T. Liu, Phys. Rev. Lett. **93**, 090602 (2004).
[13] R. Rougemont, arXiv:1610.06124 [nucl-th].
[14] S. Borsanyi, Z. Fodor, C. Hoelbling, S. D. Katz, S. Krieg and K. K. Szabo, Phys. Lett. B **730**, 99 (2014).
[15] F. Karsch, D. Kharzeev and K. Tuchin, Phys. Lett. B **663**, 217 (2008).
[16] J. Noronha-Hostler, J. Noronha and C. Greiner, Phys. Rev. Lett. **103**, 172302 (2009).
[17] J. Noronha-Hostler, G. S. Denicol, J. Noronha, R. P. G. Andrade and F. Grassi, Phys. Rev. C **88**, 044916 (2013).
[18] J. Noronha-Hostler, J. Noronha and F. Grassi, Phys. Rev. C **90**, no. 3, 034907 (2014).
[19] J. E. Bernhard, J. S. Moreland, S. A. Bass, J. Liu and U. Heinz, Phys. Rev. C **94**, no. 2, 024907 (2016).
[20] S. S. Gubser, S. S. Pufu and F. D. Rocha, JHEP **0808**, 085 (2008).
[21] S. S. Gubser and A. Nellore, Phys. Rev. D **78**, 086007 (2008).
[22] S. S. Gubser, A. Nellore, S. S. Pufu and F. D. Rocha, Phys. Rev. Lett. **101**, 131601 (2008).
[23] R. Critelli, R. Rougemont, S. I. Finazzo and J. Noronha, arXiv:1606.09484 [hep-ph].
[24] S. I. Finazzo, R. Rougemont, H. Marrochio and J. Noronha, JHEP **1502**, 051 (2015).
[25] O. DeWolfe, S. S. Gubser and C. Rosen, Phys. Rev. D **83**, 086005 (2011).
[26] O. DeWolfe, S. S. Gubser and C. Rosen, Phys. Rev. D **84**, 126014 (2011).
[27] R. Rougemont, A. Ficnar, S. Finazzo and J. Noronha, JHEP **1604**, 102 (2016).
[28] R. Rougemont, J. Noronha and J. Noronha-Hostler, Phys. Rev. Lett. **115**, no. 20, 202301 (2015).
[29] S. I. Finazzo and R. Rougemont, Phys. Rev. D **93**, no. 3, 034017 (2016).
[30] R. Rougemont, R. Critelli and J. Noronha, Phys. Rev. D **93**, no. 4, 045013 (2016).
[31] S. I. Finazzo, R. Critelli, R. Rougemont and J. Noronha, Phys. Rev. D **94**, no. 5, 054020 (2016).
[32] C. Bonati, M. D'Elia, M. Mariti, F. Negro and F. Sanfilippo, Phys. Rev. D **89**, no. 5, 054506 (2014).
[33] J. M. Maldacena, Phys. Rev. Lett. **80**, 4859 (1998).
[34] S. J. Rey, S. Theisen and J. T. Yee, Nucl. Phys. B **527**, 171 (1998).
[35] A. Brandhuber, N. Itzhaki, J. Sonnenschein and S. Yankielowicz, Phys. Lett. B **434**, 36 (1998).
[36] J. Noronha, Phys. Rev. D **81**, 045011 (2010).
[37] J. Noronha, Phys. Rev. D **82**, 065016 (2010).
[38] G. S. Bali, F. Bruckmann, G. Endrdi, S. D. Katz and A. Schfer, JHEP **1408**, 177 (2014).
[39] F. Bruckmann, G. Endrodi and T. G. Kovacs, JHEP **1304**, 112 (2013).
[40] G. Endrodi, JHEP **07** (2015) 173.
[41] A. Bazavov, N. Brambilla, H.-T. Ding, P. Petreczky, H.-P. Schadler, A. Vairo and J. H. Weber, Phys. Rev. D **93**, no. 11, 114502 (2016).
[42] X. G. Huang, A. Sedrakian and D. H. Rischke, Annals Phys. **326**, 3075 (2011).
[43] R. Critelli, S. I. Finazzo, M. Zaniboni and J. Noronha, Phys. Rev. D **90**, no. 6, 066006 (2014).

IOP Conf. Series: Journal of Physics: Conf. Series **832** (2017) 012048 doi:10.1088/1742-6596/832/1/012048

Holographic black hole engineering at finite baryon chemical potential

Romulo Rougemont

Instituto de Física, Universidade de São Paulo, Rua do Matão, 1371, Butantã, CEP 05508-090, São Paulo, SP, Brazil

E-mail: romulo@if.usp.br

Abstract. This is a contribution for the Proceedings of the Conference Hot Quarks 2016, held at South Padre Island, Texas, USA, 12-17 September 2016. I briefly review some thermodynamic and baryon transport results obtained from a bottom-up Einstein-Maxwell-Dilaton holographic model engineered to describe the physics of the quark-gluon plasma at finite temperature and baryon density. The results for the equation of state, baryon susceptibilities, and the curvature of the crossover band are in quantitative agreement with the corresponding lattice QCD results with $2+1$ flavors and physical quark masses. Baryon diffusion is predicted to be suppressed by increasing the baryon chemical potential.

1. Introduction

Ultrarelativistic heavy ion collisions [1, 2, 3, 4, 5] constitute the experimental arena where many properties of deconfined QCD matter associated to the formation of a quark-gluon plasma (QGP) [6, 7, 8] are being currently uncovered at the Relativistic Heavy Ion Collider (RHIC) and at the Large Hadron Collider (LHC). One of the most striking features of the QGP produced in such collisions is the strongly coupled nature of the medium close to the hadronization crossover [9, 10]. This fact may be attested by the very small value of the ratio between the shear viscosity (η) and the entropy density (s) of the fluid, $\eta/s \approx 0.095$, obtained in hydrodynamic simulations of the spacetime evolution of the QGP simultaneously matching experimental data for different physical observables [11]. Such a small ratio is incompatible with perturbative QCD calculations of η/s [12, 13], but is remarkably close to the value $\eta/s = 1/4\pi$ valid for any isotropic and translationally invariant gauge/gravity dual [14, 15, 16, 17] with two derivatives for the metric field in the gravity action [18, 19]. However, this fairly general and robust property of holographic plasmas implies that, in order to attempt to obtain phenomenologically reliable insights for the physics of the QGP, one must take into account the behavior of other physical observables besides η/s, since many different gauge/gravity duals describing very different strongly correlated quantum fluids share the same value of η/s.

A largely employed "toy model" for the QGP in holographic studies is the $\mathcal{N} = 4$ super Yang-Mills (SYM) plasma [20], which is one of the simplest and best understood holographic systems at finite temperature. However, a close inspection of its properties reveals that it is a fairly inadequate "approximation" of the physics of the real-world QGP. In fact, the SYM plasma is conformal, while the QGP is strongly non-conformal in the crossover region, which is the region where, in principle, holographic plasmas could be useful for real-world phenomenology. Indeed, it is in the crossover region where the QGP is strongly coupled, while at very large temperatures it becomes weakly interacting. Quantum fluids in a weakly coupled regime cannot be described by any holographic system defined in the classical gravity limit of the AdS/CFT correspondence, since gauge/gravity duals are known to lack asymptotic freedom, displaying a strongly coupled instead of a trivial ultraviolet fixed point. As a consequence of the aforementioned issues, the equation of state (EoS) of the SYM plasma is completely different from lattice QCD results [21]. The most striking difference is imprinted in the trace anomaly of the energy-momentum tensor, which displays a marked peak at the crossover in QCD, while being identically zero for the SYM plasma. Moreover, in what regards transport properties of the QGP, it has been recently pointed out by hydrodynamic simulations [11, 22, 23, 24] that a nonzero bulk viscosity (ζ) with a peak for the ratio ζ/s at the crossover seems to be required in order to simultaneously describe different experimental data from heavy ion collisions. This is completely different from the vanishing bulk viscosity of the SYM plasma. Therefore, in order to apply holography to obtain useful results for the real-world QGP, one needs to resort to a different kind of gauge/gravity dual. In this regard, it was originally proposed by Steven Gubser and his collaborators in Refs. [25, 26] a non-conformal Einstein-Maxwell-Dilaton (EMD) model which is able to match the lattice QCD EoS at finite temperature (T) and zero baryon chemical potential (μ_B). In Refs. [27, 28, 29, 30] this construction was further refined and the holographic EoS as well as many transport coefficients were calculated in the (T, μ_B) plane. In Refs. [31, 32, 33] an anisotropic version of the EMD model was proposed to describe the physics of the QGP in the plane of temperature and magnetic field (at zero chemical potential), with the magnetized EoS and many transport properties being computed. These EMD constructions were found to be in quantitative agreement with a large set of observables calculated on the lattice, both at finite μ_B [21, 34, 35, 36, 37] and finite magnetic field [38, 39, 40, 41].

2. Thermodynamics and baryon transport in the (T, μ_B) plane

The EMD model of Refs. [28, 29] constitutes a promising tool to make new predictions about the behavior of physical observables which are very difficult to access via first principle lattice QCD calculations, like transport coefficients. This may be attested by the amount of simultaneous agreement with lattice QCD results for different physical observables, both at zero and nonzero μ_B, together with the *sine qua non* condition for a *bona fide* description of the strongly coupled QGP benchmarked by its nearly perfect fluidity. This property is encoded in a small value of η/s, which is naturally embedded in any gauge/gravity dual.

In Refs. [28, 29] the EoS and baryon susceptibilities were calculated in the holographic EMD model and compared to lattice QCD results from Refs. [21, 34, 35], as shown in Fig. 1. Furthermore, in Ref. [28], it was also obtained the curvature of the crossover band in the EMD model, $\kappa_{\mathrm{EMD}} \approx 0.013$, which is in quantitative agreement with the lattice results $\kappa^{(\mathrm{I})}_{\mathrm{latt.}} = 0.0135(20)$ from Ref. [36] and $\kappa^{(\mathrm{II})}_{\mathrm{latt.}} = 0.0149(21)$ from Ref. [37].

In Ref. [29] the second order baryon susceptibility, the baryon DC conductivity and the baryon diffusion constant, which controls the fluid response to inhomogeneities in the baryon charge density, were calculated as functions of T and μ_B in the EMD model, as displayed in Fig. 2. The diffusion of baryon charge is predicted to be suppressed as one increases the baryon density, which is compatible with the existence of a critical point in the EMD model for $\mu_B > 400$ MeV, as the baryon diffusion is expected to vanish at the critical point [42]. The precise location of the critical point of this holographic EMD model will be determined elsewhere.

Acknowledgements

I thank J. Noronha for useful comments on this manuscript and also acknowledge financial support by the São Paulo Research Foundation (FAPESP) under the grant no. 2013/04036-0.

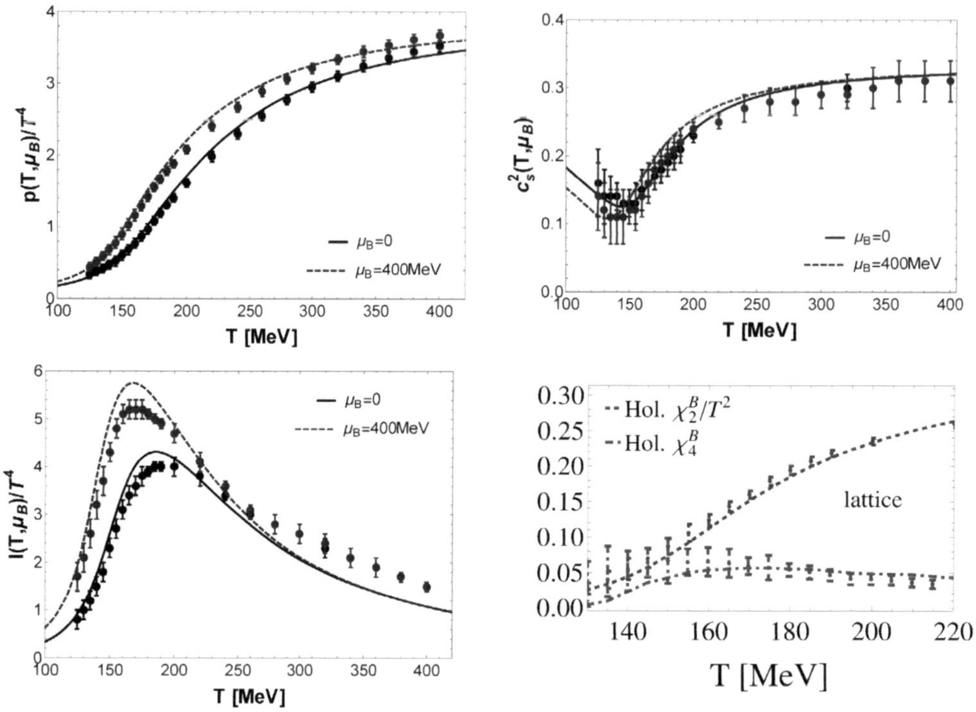

Figure 1. T dependence of pressure (top left), squared speed of sound (top right), and trace anomaly (bottom left) for different μ_B. Second and fourth order baryon susceptibilities at $\mu_B = 0$ (bottom right).

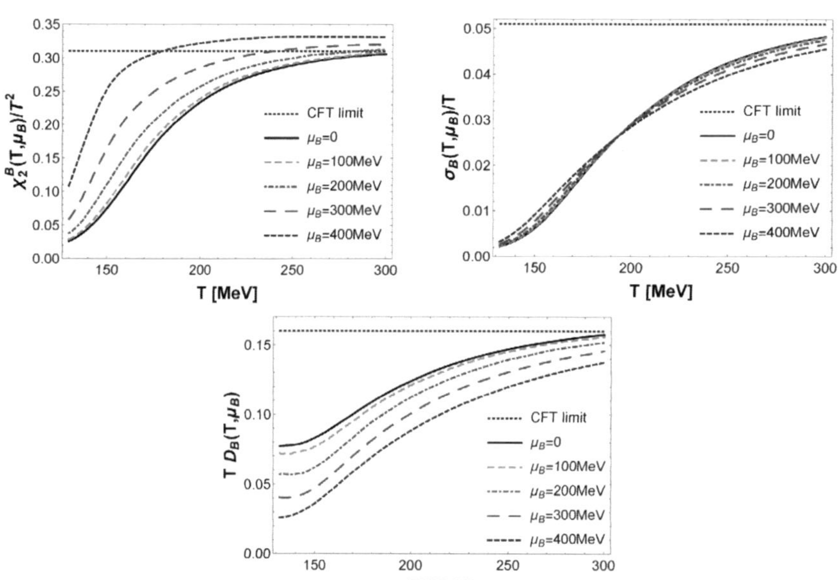

Figure 2. T dependence of baryon susceptibility (top left), baryon DC conductivity (top right), and baryon diffusion constant (bottom) for different μ_B. The corresponding conformal limits reached at $T \gg \mu_B$ are also displayed as straight lines in these plots.

IOP Conf. Series: Journal of Physics: Conf. Series **832** (2017) 012048 doi:10.1088/1742-6596/832/1/012048

References

[1] BRAHMS Collaboration, I. Arsene *et al.*, Nucl. Phys. A **757**, 1 (2005), [arXiv:nucl-ex/0410020].
[2] PHENIX Collaboration, K. Adcox *et al.*, Nucl. Phys. A **757**, 184 (2005), [arXiv:nucl-ex/0410003].
[3] PHOBOS Collaboration, B. B. Back *et al.*, Nucl. Phys. A **757**, 28 (2005), [arXiv:nucl-ex/0410022].
[4] STAR Collaboration, J. Adams *et al.*, Nucl. Phys. A **757**, 102 (2005), [arXiv:nucl-ex/0501009].
[5] ATLAS Collaboration, G. Aad *et al.*, JHEP **1311**, 183 (2013), [arXiv:1305.2942 [hep-ex]].
[6] M. Gyulassy and L. McLerran, Nucl. Phys. A **750**, 30 (2005), [arXiv:nucl-th/0405013].
[7] U. Heinz and R. Snellings, Ann. Rev. Nucl. Part. Sci. **63**, 123 (2013), [arXiv:1301.2826 [nucl-th]].
[8] E. Shuryak, [arXiv:1412.8393 [hep-ph]].
[9] Y. Aoki, G. Endrodi, Z. Fodor, S. D. Katz and K. K. Szabo, Nature **443**, 675 (2006) [hep-lat/0611014].
[10] S. Borsanyi *et al.*, Nature **539**, no. 7627, 69 (2016) [arXiv:1606.07494 [hep-lat]].
[11] S. Ryu, J.-F. Paquet, C. Shen, G. S. Denicol, B. Schenke, S. Jeon and C. Gale, Phys. Rev. Lett. **115**, no. 13, 132301 (2015) [arXiv:1502.01675 [nucl-th]].
[12] P. B. Arnold, G. D. Moore and L. G. Yaffe, JHEP **0011**, 001 (2000) [hep-ph/0010177].
[13] P. B. Arnold, G. D. Moore and L. G. Yaffe, JHEP **0305**, 051 (2003) [hep-ph/0302165].
[14] J. Maldacena, Adv. Theor. Math. Phys. **2**, 231 (1998), [arXiv:hep-th/9711200].
[15] S. S. Gubser, I. R. Klebanov and A. M. Polyakov, Phys. Lett. B **428**, 105 (1998), [arXiv:hep-th/9802109].
[16] E. Witten, Adv. Theor. Math. Phys. **2**, 253 (1998), [arXiv:hep-th/9802150].
[17] E. Witten, Adv. Theor. Math. Phys. **2**, 505 (1998), [hep-th/9803131].
[18] A. Buchel and J. T. Liu, Phys. Rev. Lett. **93**, 090602 (2004) [hep-th/0311175].
[19] P. Kovtun, D. T. Son and A. O. Starinets, Phys. Rev. Lett. **94**, 111601 (2005) [hep-th/0405231].
[20] J. Casalderrey-Solana, H. Liu, D. Mateos, K. Rajagopal and U. A. Wiedemann, [arXiv:1101.0618 [hep-th]].
[21] S. Borsanyi, G. Endrodi, Z. Fodor, S. D. Katz, S. Krieg, C. Ratti and K. K. Szabo, JHEP **1208**, 053 (2012) [arXiv:1204.6710 [hep-lat]].
[22] J. Noronha-Hostler, G. S. Denicol, J. Noronha, R. P. G. Andrade and F. Grassi, Phys. Rev. C **88**, 044916 (2013) [arXiv:1305.1981 [nucl-th]].
[23] J. Noronha-Hostler, J. Noronha and F. Grassi, Phys. Rev. C **90**, no. 3, 034907 (2014) [arXiv:1406.3333 [nucl-th]].
[24] J. E. Bernhard, J. S. Moreland, S. A. Bass, J. Liu and U. Heinz, Phys. Rev. C **94**, no. 2, 024907 (2016) [arXiv:1605.03954 [nucl-th]].
[25] S. S. Gubser and A. Nellore, Phys. Rev. D **78**, 086007 (2008) [arXiv:0804.0434 [hep-th]].
[26] O. DeWolfe, S. S. Gubser and C. Rosen, Phys. Rev. D **83**, 086005 (2011) [arXiv:1012.1864 [hep-th]].
[27] S. I. Finazzo, R. Rougemont, H. Marrochio and J. Noronha, JHEP **1502**, 051 (2015) [arXiv:1412.2968 [hep-ph]].
[28] R. Rougemont, A. Ficnar, S. Finazzo and J. Noronha, JHEP **1604**, 102 (2016) [arXiv:1507.06556 [hep-th]].
[29] R. Rougemont, J. Noronha and J. Noronha-Hostler, Phys. Rev. Lett. **115**, no. 20, 202301 (2015) [arXiv:1507.06972 [hep-ph]].
[30] S. I. Finazzo and R. Rougemont, Phys. Rev. D **93**, no. 3, 034017 (2016) [arXiv:1510.03321 [hep-ph]].
[31] R. Rougemont, R. Critelli and J. Noronha, Phys. Rev. D **93**, no. 4, 045013 (2016) [arXiv:1505.07894 [hep-th]].
[32] S. I. Finazzo, R. Critelli, R. Rougemont and J. Noronha, Phys. Rev. D **94**, no. 5, 054020 (2016) [arXiv:1605.06061 [hep-ph]].
[33] R. Critelli, R. Rougemont, S. I. Finazzo and J. Noronha, arXiv:1606.09484 [hep-ph].
[34] S. Borsanyi, Z. Fodor, S. D. Katz, S. Krieg, C. Ratti and K. Szabo, JHEP **1201**, 138 (2012) [arXiv:1112.4416 [hep-lat]].
[35] R. Bellwied, S. Borsanyi, Z. Fodor, S. D. Katz, A. Pasztor, C. Ratti and K. K. Szabo, Phys. Rev. D **92**, no. 11, 114505 (2015) [arXiv:1507.04627 [hep-lat]].
[36] C. Bonati, M. D'Elia, M. Mariti, M. Mesiti, F. Negro and F. Sanfilippo, Phys. Rev. D **92**, no. 5, 054503 (2015) [arXiv:1507.03571 [hep-lat]].
[37] R. Bellwied, S. Borsanyi, Z. Fodor, J. Gunther, S. D. Katz, C. Ratti and K. K. Szabo, Phys. Lett. B **751**, 559 (2015) [arXiv:1507.07510 [hep-lat]].
[38] G. S. Bali, F. Bruckmann, G. Endrodi, S. D. Katz and A. Schfer, JHEP **1408**, 177 (2014) [arXiv:1406.0269 [hep-lat]].
[39] F. Bruckmann, G. Endrodi and T. G. Kovacs, JHEP **1304**, 112 (2013) [arXiv:1303.3972 [hep-lat]].
[40] G. Endrodi, JHEP **07** (2015) 173, [arXiv:1504.08280 [hep-lat]].
[41] A. Bazavov, N. Brambilla, H.-T. Ding, P. Petreczky, H.-P. Schadler, A. Vairo and J. H. Weber, Phys. Rev. D **93**, no. 11, 114502 (2016) [arXiv:1603.06637 [hep-lat]].
[42] D. T. Son and M. A. Stephanov, Phys. Rev. D **70**, 056001 (2004) [hep-ph/0401052].

IOP Conf. Series: Journal of Physics: Conf. Series **832** (2017) 012049 doi:10.1088/1742-6596/832/1/012049

Anisotropic flow and related phenomena in Pb-Pb collisions at $\sqrt{s_{\rm NN}} = 5.02$ TeV with ALICE

Jacopo Margutti (for the ALICE Collaboration)
Nikhef, Science Park 105, 1098 XG Amsterdam, Netherlands
Utrech University, Domplein 29, 3512 JE Utrecht, Netherlands
E-mail: jacopo.margutti@cern.ch

Abstract. The ALICE experiment is designed and optimised to study the properties of the Quark-Gluon Plasma (QGP), a new state of matter, which is expected to be created at the high energy densities reached at the LHC. One of the key observables used to characterise the transport properties and the equation of state of the QGP is the azimuthal anisotropy in particle production, which is usually called anisotropic flow. In this presentation, we report the first measurements of anisotropic flow in Pb–Pb collisions at $\sqrt{s_{\rm NN}} = 5.02$ TeV, the highest energy ever achieved in heavy–ion collisions, and compare them with both theoretical predictions and experimental measurements at lower energies. This provides a unique opportunity to test the validity of the hydrodynamic paradigm at new energies and to further constrain key transport parameters of the QGP, such as the shear viscosity over entropy ratio.

1. Introduction

One key observable in the study of heavy–ion collisions has been the azimuthal anisotropies of particle production in momentum space with respect to the reaction plane, i.e. the plane spanned by the impact parameter and the beam axis. This phenomenon, usually referred to as anisotropic flow, has been most commonly interpreted at higher energies as the result of the hydrodynamic behaviour of the QGP, thus providing one of the most solid evidences of the non–trivial collective dynamics of such system. Anisotropic flow is also sensitive to the transport parameters (e.g. viscosities) and equation of state of the system, both in its QGP phase and after hadronisation, which then makes it a suitable probe of such proprieties. By using a general Fourier series decomposition of the azimuthal distribution of produced particles:

$$\frac{dN}{d\varphi} \propto 1 + \sum_{n=1}^{+\infty} v_n \cos\left[n(\varphi - \Psi_n)\right],$$

anisotropic flow is quantified with the Fourier coefficients v_n and corresponding symmetry planes Ψ_n [1]. While the 2nd harmonic (v_2), usually referred to as elliptic flow, is mostly determined by the approximately ellipsoidal shape of the overlap region in a non–central heavy–ion collision, the higher ones ($v_3, v_4, v_5 \ldots$) come from fluctuations in such shape. In these proceedings, we report the first measurements of anisotropic flow of charged particles in Pb–Pb collisions at $\sqrt{s_{\rm NN}} = 5.02$ TeV [3], obtained from two- and multi–particle cumulants, using the approach proposed in [2].

IOP Conf. Series: Journal of Physics: Conf. Series **832** (2017) 012049 doi:10.1088/1742-6596/832/1/012049

2. Analysis
The data used for this measurement have been recorded with the ALICE detector in November 2015, during the Run 2 of the LHC. Only one low luminosity run (with trigger rate of 27 Hz), being least affected by pile-up and distortions from space charge in the main tracking detector, the Time Projection Chamber (TPC), is used for this analysis. A sample of 140 k minimum bias Pb–Pb collisions passed the selection criteria. All charged tracks in the p_{T} range $0.2 < p_{\mathrm{T}} < 5$ GeV/c and pseudorapidity $|\eta| < 0.8$ are used in this analysis. These tracks are reconstructed using combined information from the Inner Tracking System (ITS) and the TPC. The systematic uncertainty related to the non-uniform reconstruction efficiency is found to be at the level of 1%. A variation of the track selection criteria is used to estimate the corresponding systematic uncertanty, which is found to be 0.5% at most. The flow coefficients from tracks that are reconstructed from TPC space points alone are compared to coefficients extracted from particles that used both TPC clusters and ITS hits and are found to agree within 2%. This difference is included in the total systematic uncertainty. Other systematic uncertanties that are studied relate to the centrality determination, the polarity of the magnetic field of the ALICE detector and the position of the reconstructed primary vertex, which are found to be negligible. The total systematic uncertanty is evaluated adding in quadrature the aforementioned contributions.

3. Results
Fig. 1 (a) shows the centrality dependence of the anisotropic flow coefficients v_2, v_3 and v_4 from two- and multi–particle cumulants, for 2.76 and 5.02 TeV Pb–Pb collisions, integrated in the p_{T} range $0.2 < p_{\mathrm{T}} < 5$ GeV/c. v_2 increases from central to peripheral collisions, reaches a maximum in the 40-50% centrality class and then decreases. v_3 and v_4, on the contrary show a much milder centrality dependence. These features do not vary with collision energy. The difference between two- and multi–particle cumulants are attributed to the opposite contribution (positive and negative, respectively) of flow fluctuations to these observables. Concerning the energy dependence, the flow coefficients are all found to increase going from 2.76 to 5.02 TeV, as shown in fig. 1 (b) and (c), without significant differences across centralities, at least in the centrality range 0-50%. The measurements are found to be compatible with theoretical predictions [4, 5]. In particular, compared to the different predictions from [4], the measurements seem to indicate that the shear viscosity over entropy ratio (η/s) does not vary significantly between collision energies. The two parametrisations of $\eta/s(T)$ that are found to be compatible with the data point to small or non existent temperature dependence of η/s in the QGP phase:

$$0 < \frac{d\eta/s}{dT} < 0.15\ [100\ \mathrm{MeV}^{-1}] \quad \text{for } T > 150\ \mathrm{MeV}.$$

This is consistent with other recent findings [6].

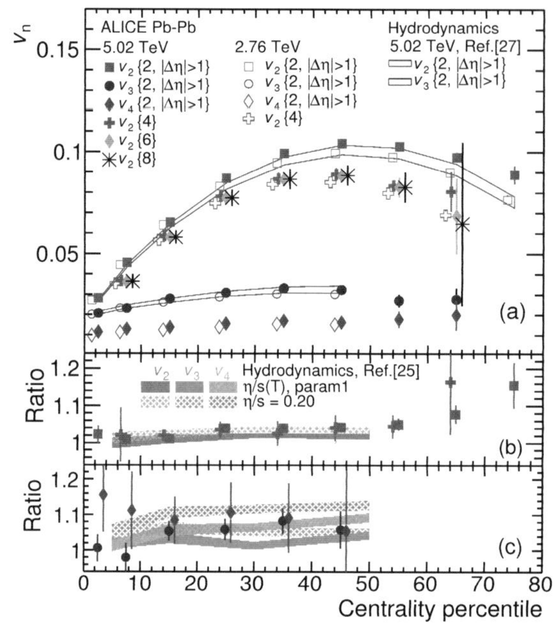

Figure 1. (a) $p_{\rm T}$-integrated anisotropic flow coefficients v_n as a function of event centrality, for 2.76 and 5.02 TeV Pb–Pb collisions. (b), (c) Ratios of v_n for these two energies. Various predictions from hydrodynamical models are also presented [4, 5].

In fig. 2 we report v_2, v_3 and v_4 as a function of $p_{\rm T}$, for 2.76 and 5.02 TeV Pb–Pb collisions, in the centrality classes 0-5% and 30-40%. Comparing measurements at different collision energies, the $p_{\rm T}$-differential flow coefficients are found to be compatible, indicating that the increase observed in the $p_{\rm T}$-integrated ones can be attributed to an increase of mean transverse momentum $\langle p_{\rm T}\rangle$, due to an increase in radial flow.

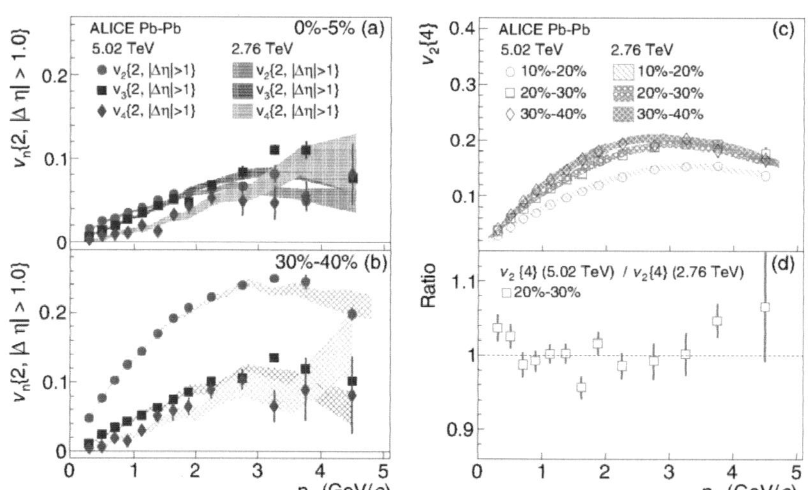

Figure 2. Anisotropic flow coefficients v_n as a function of $p_{\rm T}$ in the centrality classes 0-5% (a) and 30-40% (b) from two–particle cumulants. $v_2\{4\}(p_{\rm T})$ is shown in panel (c) and the ratio between the measurements at 2.76 and 5.02 TeV in (d).

Finally, in fig. 3 we show the fully p_T integrated v_2 measured in the 20–30% centrality range and compare it with results at lower energies and similar centralities. This measurements was performed combining two methods, one based on the fit of the p_T spectra and $v_2\{4\}(p_T)$, the other on the direct calculation of $v_2\{4\}$ from tracklets in the ITS, which has an acceptance of $p_T > 50$ MeV/c. The fully p_T integrated v_2 is observed to increase by $4.89 \pm 2.27\%$ between 2.76 and 5.02 TeV.

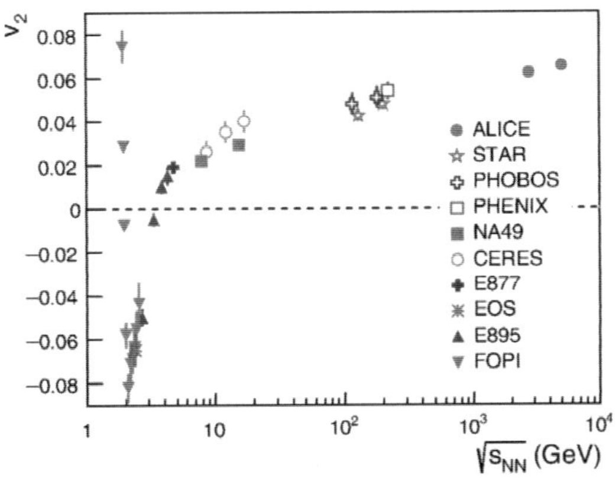

Figure 3. Fully p_T-integrated elliptic flow $v_2\{4\}$ in the 20–30% centrality range at 5.02TeV compared with v_2 measurements at lower energies with similar centralities.

4. Summary

In these proceedings we present the first measurements of anisotropic flow in Pb–Pb collisions at $\sqrt{s_{NN}} = 5.02$ TeV. Compared to measurements at 2.76 TeV, an increase is observed in the p_T–integrated flow coefficients, while no signifcant differences are observed in the p_T–differential ones. Therefore, the increase observed in p_T–integrated flow is attributed to an increase in average p_T. The data are found to be compatible with theoretical predictions. This constitutes an important test for the hydrodynamical models with which we most commonly describe ultra–relativistic heavy–ion collisions. We also have indications that the temperature dependence of the shear viscosity over entropy ratio is relatively small ($0 < d(\eta/s)/dT < 0.15$ [100 MeV^{-1}] for $T > 150$ MeV.).

References

[1] S. Voloshin and Y. Zhang, Z. Phys. C 70, 665-672 (1996)
[2] A. Bilandzic, R. Snellings and S. Voloshin, Phys. Rev. C 83, 044913 (2011)
[3] J. Adam *et al.* [ALICE Collaboration], Phys. Rev. Lett. 116, 132302 (2016)
[4] H. Niemi *et al.*, Phys. Rev. C 93, 014912 (2016)
[5] J. Noronha-Hostler *et al.*, Phys. Rev. C 93, 034912 (2016)
[6] J.E. Bernhard *et al.*, Phys. Rev. C 94, 024907 (2016)

IOP Conf. Series: Journal of Physics: Conf. Series **832** (2017) 012050 doi:10.1088/1742-6596/832/1/012050

Transient oscillating modes in a relativistic expanding gas

Dennis Bazow

Department of Physics, The Ohio State University, Columbus, OH 43210

Abstract. We show that a transient effective theory for rapid anisotropically expanding systems based on the Boltzmann equation exhibits transient oscillatory behavior.

1. Introduction and setup

The many-body dynamics of a weakly interacting, dilute system where two-particle correlations can be neglected is governed by the Boltzmann equation,

$$k^\mu \partial_\mu f(x,k) = C[f](x,k). \tag{1}$$

The complete description of the microscopic dynamics involves an infinite number of nonhydrodynamic modes contained in the linearized collision term $C[f]$; the virtual infinite complexity is reduced by systematically truncating these. (In general the collision term is non-linear, but in this work we will only considered the linearized collision term.) Hydrodynamics is an effective theory for the evolution of the conserved macroscopic quantities where only the so-called hydrodynamic modes, those that satisfy the dispersion relation $\lim_{\boldsymbol{k}\to 0}\omega_n(\boldsymbol{k}) = 0$, are included. It has always been previously thought that the nonhydrodynamic modes of the Boltzmann equation are exponentially damped on microscopic time scales [1]. This is in stark contrast to strongly coupled systems [2, 3] where the nonhydrodynamic modes are damped in an oscillatory manner.

In this proceedings we show that the former only holds when all local momentum anisotropies are treated perturbatively by expanding the Boltzmann equation around its local equilibrium value. This assumption is relaxed by considering a system that expands stronger in the longitudinal rather than transverse directions, such as the quark-gluon plasma created during relativistic heavy-ion collisions [4], where these local momentum anisotropies can become large and must be treated nonperturbatively. Here we do so by expanding the distribution function around a local "quasi-equilibrium" state $f_{\boldsymbol{k}} = f_{\mathrm{a},\boldsymbol{k}} + \delta\tilde{f}_{\boldsymbol{k}}$ by introducing a non-hydrodynamic degree of freedom $\xi(x)$ describing the deviation from momentum isotropy [5]:

$$f_{\mathrm{a},\boldsymbol{k}} = f_{\mathrm{eq}}\left(\beta_{\mathrm{a}}\sqrt{k_\perp^2+(1+\xi)k_z^2}\right). \tag{2}$$

$\beta_{\mathrm{a}}(x)$ is a temperature-like parameter that is varied on macroscopic time scales related to energy conservation while the anisotropy parameter ξ is not constrained by conservation laws, but controlled by the expansion of the system and by microscopic time scales associated with the

Boltzmann collision term. In a procedure that closely follows [6], but now resums and truncates the moment hierarchy according to a modified power counting scheme based on the Knudsen and *residual* inverse Reynolds numbers from which the largest contributions arising from the local momentum anisotropy have been eliminated, we derive a new type of transient relativistic fluid dynamics in which the slowest non-hydrodynamic mode (associated with the local momentum anisotropy) turns out to exhibit transient oscillations. It should not be surprising that the physical phenomena emerging from this situation is analogous to that of a plasma placed in an external field; here ξ plays the role, in this loose analogy, of a fictitious external (gravitational) field that causes momentum space inhomogeneities.

2. Main ingredients

For a gas of massless particles the only conserved current is the energy-momentum tensor which is decomposed for anisotropic systems as [7]

$$T^{\mu\nu} \equiv \mathcal{E} u^\mu u^\nu - \mathcal{P}_\perp \Delta^{\mu\nu} + (\mathcal{P}_L - \mathcal{P}_\perp) z^\mu z^\nu + \tilde{\pi}^{\mu\nu} \,, \tag{3}$$

where $\mathcal{E}$ is the LRF energy density, $\mathcal{P}_{L,\perp}$ are the longitudinal and transverse pressures, and $\tilde{\pi}^{\mu\nu}$ is the residual shear stress tensor arising from $\delta\tilde{f}$ effects. u^μ is the fluid velocity, z^μ is the z unit vector in the LRF, and $\Delta^{\mu\nu} \equiv g^{\mu\nu} - u^\mu u^\nu$. We now replace the Boltzmann equation by a hierarchy of moment equations. The solution to the Boltzmann equation is now equivalent to solving the dynamical equations for the moments $\tilde{\rho}_r^{\mu\nu}$ [6, 8]:

$$\dot{\tilde{\rho}}_r^{\mu\nu}(x) \equiv \Delta^{\mu\nu}_{\alpha\beta} D \tilde{\rho}_r^{\alpha\beta}(x) \equiv \Delta^{\mu\nu}_{\alpha\beta} D \int dK (u{\cdot}k)^r k^{\langle\alpha} k^{\beta\rangle} \delta\tilde{f}_{\boldsymbol{k}} \,, \tag{4}$$

where $D \equiv u{\cdot}\partial$ is the time derivative in the local rest frame (LRF). The angle brackets around any two Lorentz indices denote a tensor that is transverse to u and traceless. Linearizing the collilsion term and using the Boltzmann equation (1) in the form

$$\delta\dot{\tilde{f}}_{\boldsymbol{k}} = -\dot{f}_{\mathrm{a},\boldsymbol{k}} - \frac{1}{k{\cdot}u}\Big(k{\cdot}\nabla\big(f_{\mathrm{a},\boldsymbol{k}} + \delta\tilde{f}_{\boldsymbol{k}}\big) - C_{\boldsymbol{k}}[f]\Big) \,, \tag{5}$$

results in

$$\begin{aligned}
\Delta^{\mu\nu}_{\alpha\beta}\dot{\tilde{\rho}}_r^{\alpha\beta} + \sum_{n=0}^{N_2} (\mathcal{A}^{(2)}_{rn})^{\mu\nu}_{\alpha\beta} \tilde{\rho}_n^{\alpha\beta} &= \mathcal{L}_r^{\mu\nu}\dot{\xi} + \mathcal{M}_r^{\mu\nu\lambda}\dot{z}_\lambda + (\alpha^{(2)}_{\theta r})^{\mu\nu}\theta + (\alpha^{(2)}_{\sigma r})^{\mu\nu\lambda\rho}\sigma_{\lambda\rho} + (\alpha^{(2)}_{\omega r})^{\mu\nu\lambda\rho}\omega_{\lambda\rho} \\
&+ \frac{\mathcal{B}_r^{\mu\nu}}{\mathcal{J}_{2,0,-1}} \tilde{\pi}^{\alpha\beta}\sigma_{\alpha\beta} - \frac{2}{7}(2r{+}5)\tilde{\rho}_r^{\lambda\langle\mu}\sigma^{\nu\rangle}_\lambda + 2\tilde{\rho}_r^{\lambda\langle\mu}\omega^{\nu\rangle}{}_\lambda - \frac{1}{3}(r{+}4)\tilde{\rho}_r^{\mu\nu}\theta.
\end{aligned} \tag{6}$$

Here $\sigma_{\mu\nu} \equiv \nabla_{\langle\mu} u_{\nu\rangle}$ is the velocity shear tensor, $\omega_{\mu\nu} \equiv (\nabla_\mu u_\nu - \nabla_\nu u_\mu)/2$ is the vorticity tensor, and $\theta \equiv \partial{\cdot}u$ the scalar expansion rate. The matrix $\mathcal{A}^{(2)}$ contains all of the microscopic information of the linearized Boltzmann equation. Appearance of time derivatives (in the LRF) of the anisotropy parameter ξ and z^μ is due to the fact that $\dot{f}_{\mathrm{a}} \supset \dot{\beta}_a$, $\dot{u}^\mu$, $\dot{\xi}$, $\dot{z}^\mu$. $\dot{\beta}_a$ and $\dot{u}^\mu$ are replaced with the conservation laws in terms of only spatial (in the LRF) gradients of the local thermodynamic fields. However, the anisotropy parameter ξ does not evolve on purely macroscopic time scales and thus is not constrained by the conservation laws. We refer the reader to Ref. [9] for further details.

In order to close the system of equations (6), we need to identify and separate the microscopic times scales by determining the eigenmodes of $\mathcal{A}^{(2)}$ by introducing the matrix $\Omega^{(2)}$ that diagonalizes $\mathcal{A}^{(2)}$. This process results in an asymptotic "Navier-Stokes" limit for all microscopic eigenmodes except for the slowest one ($r = 0$):

$$\tilde{\rho}_i^{\mu\nu} - (\Omega_{i0}^{(2)})_{\alpha\beta}^{\mu\nu}\tilde{\pi}^{\alpha\beta} + \hat{\ell}_i^{\mu\nu}\dot{\xi} + \hat{m}_i^{\mu\nu\lambda}\dot{z}_\lambda + (\hat{\eta}_{\theta i}^{(2)})^{\mu\nu}\theta + (\hat{\eta}_{\sigma i}^{(2)})^{\mu\nu\lambda\rho}\sigma_{\lambda\rho} + (\hat{\eta}_{\omega i}^{(2)})^{\mu\nu\lambda\rho}\omega_{\lambda\rho} + \text{h. o. terms.} \tag{7}$$

The tensors $\hat{\ell}$, $\hat{m}$, $\hat{\eta}_\theta^{(2)}$, $\hat{\eta}_\sigma^{(2)}$ and $\hat{\eta}_\omega^{(2)}$ are built from u^μ, z^μ, $\Delta^{\mu\nu}$ and the microscopic relaxation scales. The fact that all higher order tensor moments are expressed in terms of the lowest-order one that appears in $T^{\mu\nu}$, $\tilde{\rho}_0^{\mu\nu} \equiv \tilde{\pi}^{\mu\nu}$, Eq. (6) can be completely decoupled from higher order moments. This is achieved by multiplying Eq. (6) by the microscopic relaxation times $(\tau_{nr}^{(2)})_{\alpha\beta}^{\mu\nu} = [(\mathcal{A}^{(2)})_{nr}^{-1}]_{\alpha\beta}^{\mu\nu}$ and then using Eq. (7). But the asymptotic "Navier-Stokes" limit (7) involves time derivatives (in the LRF) of the anisotropy parameter. This is the main ingredient. Since $\xi \sim \mathcal{O}(1)$ in our power counting scheme and does not evolve according to the conservation laws it cannot be neglected as a higher order term thus leading to second order comoving time derivatives $\ddot{\xi}$. (When $\dot{\tilde{\rho}}_r^{\mu\nu}$ in Eq. (6) is approximated according to (7) the $\hat{\ell}_i^{\mu\nu}\dot{\xi}$ term generates $\ddot{\xi}$ terms.) The result is a set of equations that take the following form:

$$\begin{aligned}(\tau_\pi^{(n)})_{\alpha\beta}^{\mu\nu}\dot{\tilde{\pi}}^{\alpha\beta} &+ (\Omega_{n0}^{(2)})_{\alpha\beta}^{\mu\nu}\tilde{\pi}^{\alpha\beta} + (\tau_\xi^{(n)})^{\mu\nu}\dot{\xi} + (\lambda_z^{(n)})^{\mu\nu\lambda}\dot{z}_\lambda + (\mathcal{D}_2^{(n)})^{\mu\nu} + (\mathcal{D}_1^{(n)})^{\mu\nu} \\ &= (\eta_\theta^{(2,n)})^{\mu\nu}\theta + (\eta_\sigma^{(2,n)})^{\mu\nu\alpha\beta}\sigma_{\alpha\beta} + (\eta_\omega^{(2,n)})^{\mu\nu\alpha\beta}\omega_{\alpha\beta} + (\mathcal{J}^{(n)})^{\mu\nu} + (\mathcal{K}^{(n)})^{\mu\nu}\,. \end{aligned} \tag{8}$$

See Ref. [9] for the explicit expressions for the coefficients. The main result is that the tensors $(\mathcal{D}_2^{(n)})^{\mu\nu}$ contain second-order time derivatives $\sim \ddot{\xi}$. Fig. 1 shows the result of the dispersion relations in the long wavelength limit $k \to 0$ for the eigenmodes of the linearized equations of motion (i.e. linear perturbations around a static background). The equal and opposite real and the negative imaginary parts of the frequencies of the nonhydrodynamic modes indicate that the momentum anisotropy parameter ξ undergoes transient (damped) oscillations.

3. Conclusions

We showed here that transient oscillations of the slowest nonhydrodynamic modes arise generically in the weakly coupled Boltzmann equation from large deviations from local momentum isotropy in rapidly expanding anisotropically expanding systems. If the distribution function is instead expanded around its local equilibrium value when deriving the macroscopic equations of motion from the Boltzmann equation, all nonhydrodynamic modes are found to be exponentially damped [1].

References

[1] G. S. Denicol, J. Noronha, H. Niemi and D. H. Rischke, Phys. Rev. D **83**, 074019 (2011).
[2] A. O. Starinets, Phys. Rev. D **66**, 124013 (2002).
[3] J. Noronha and G. S. Denicol, arXiv:1104.2415 [hep-th].
[4] M. Strickland, Acta Phys. Polon. B **45**, 2355 (2014).
[5] P. Romatschke and M. Strickland, Phys. Rev. D **68**, 036004 (2003).
[6] G. S. Denicol, H. Niemi, E. Molnar and D. H. Rischke, Phys. Rev. D **85**, 114047 (2012).
[7] D. Bazow, U. Heinz and M. Strickland, Phys. Rev. C **90**, 054910 (2014).
[8] G. S. Denicol, T. Koide and D. H. Rischke, Phys. Rev. Lett. **105**, 162501 (2010).
[9] D. Bazow, M. Martinez and U. W. Heinz, Phys. Rev. D **93**, 034002 (2016)

IOP Conf. Series: Journal of Physics: Conf. Series **832** (2017) 012050 doi:10.1088/1742-6596/832/1/012050

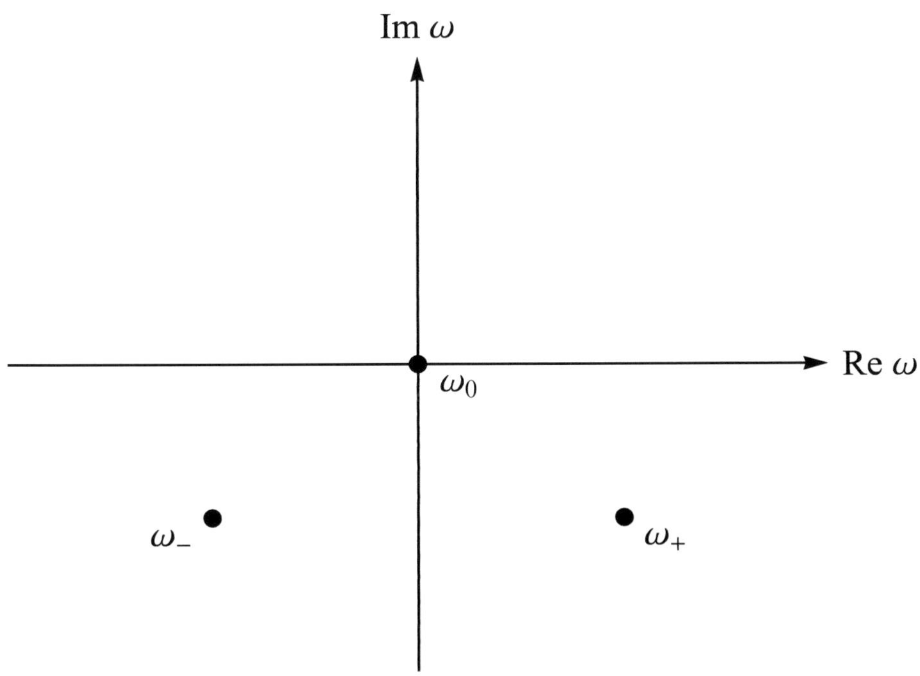

Figure 1. Shear channel eigenmodes in the zero wavenumber limit. Shown are the two degenerate hydrodynamic modes at $\omega_0 = 0$ and the first two non-hydrodynamic modes at $\omega_\pm$ that exhibit transient oscillatory behavior.

IOP Conf. Series: Journal of Physics: Conf. Series **832** (2017) 012051 doi:10.1088/1742-6596/832/1/012051

Quasiparticle anisotropic hydrodynamics

Mubarak Alqahtani and Michael Strickland
Kent State University, Kent, OH 44242 USA
E-mail: malqaht3@kent.edu (presenter of the talk)

Abstract. We study an azimuthally-symmetric boost-invariant quark-gluon plasma using quasiparticle anisotropic hydrodynamics including the effects of both shear and bulk viscosities. We compare results obtained using the quasiparticle method with the standard anisotropic hydrodynamics and viscous hydrodynamics. We consider the predictions of the three methods for the differential particle spectra and mean transverse momentum. We find that the three methods agree for small shear viscosity to entropy density ratio, η/s, but show differences at large η/s. Additionally, we find that the standard anisotropic hydrodynamics method shows suppressed production at low transverse-momentum compared to the other two methods, and the bulk-viscous correction can drive the primordial particle spectra negative at large p_T in viscous hydrodynamics.

1. Introduction

Ultrarelativistic heavy-ion collision experiments at RHIC and LHC study the quark-gluon plasma (QGP) using high-energy nuclear collisions. Relativistic hydrodynamics has been quite successful in describing the collective behavior seen in these experiments. Over the years, many approaches have been developed to model the physics of the QGP such as ideal hydrodynamics [1] and viscous hydrodynamics [2, 3, 4]. Viscous hydrodynamic studies indicate that the QGP is a momentum-space anisotropic plasma. In order to more accurately account for potentially large momentum-space anisotropies, anisotropic hydrodynamics was developed [5, 6, 7, 8, 9, 10, 11]. In recent years, there have been developments in the anisotropic hydrodynamics program which focused on making phenomenological predictions for heavy-ion physics [12, 13, 14]. For a recent review on anisotropic hydrodynamics, see Ref. [15].

A major conceptual difficulty for anisotropic hydrodynamics is how to impose a realistic equation of state (EoS) on a system which is potentially far from equilibrium. In quasiparticle hydrodynamics (aHydroQP), one imposes a realistic EoS which takes into account the non-conformality of QGP by introducing a single finite-temperature quasiparticle mass which is fit to lattice data. We compare aHydroQP with the standard anisotropic hydrodynamics (aHydro) where approximate conformality is assumed. Additionally, we compare both aHydro formulations with second-order viscous hydrodynamics (vHydro) [16, 17]. We then present comparisons of the primordial particle spectra and the average transverse momentum for pions, kaons, and protons predicted by these three methods [13].

2. Anisotropic hydrodynamics

In non-conformal anisotropic hydrodynamics, the anisotropy tensor is defined as [10]

$$\Xi^{\mu\nu} = u^\mu u^\nu + \xi^{\mu\nu} - \Delta^{\mu\nu}\Phi \,, \tag{1}$$

where u^μ is the fluid four-velocity, $\xi^{\mu\nu}$ is a symmetric and traceless anisotropy tensor, and Φ is associated with the bulk degree of freedom. The one-particle distribution function is assumed to be of the form

$$f(x,p) = f_{\rm iso}\left(\frac{1}{\lambda}\sqrt{p_\mu \Xi^{\mu\nu} p_\nu}\right), \tag{2}$$

where λ can be identified with the temperature only in the isotropic equilibrium limit. By expanding the argument of the square root in Eq. (2), one can write

$$f(x,p) = f_{\rm eq}\left(\frac{1}{\lambda}\sqrt{\sum_i \frac{p_i^2}{\alpha_i^2} + m^2}\right), \tag{3}$$

where $i \in \{x,y,z\}$ and the ellipticity parameters α_i are $\alpha_i \equiv (1 + \xi_i + \Phi)^{-1/2}$.

3. Equation of state

In quasiparticle hydrodynamics, we implement a realistic equation of state by assuming that we have an ensemble of massive quasiparticles with a single temperature-dependent mass. However, in order to not violate thermodynamical consistency, one must introduce a background field contribution to the energy-momentum tensor [12, 18]

$$T^{\mu\nu}_{\rm eq} = T^{\mu\nu}_{\rm kinetic,eq} + g^{\mu\nu} B_{\rm eq}, \tag{4}$$

where $B_{\rm eq}$ can be found by integrating

$$\frac{dB_{\rm eq}}{dT} = -4\pi \tilde{N} m^2 T K_1(\hat{m}_{\rm eq}) \frac{dm}{dT}, \tag{5}$$

with $m(T)$ obtained from this thermodynamic identity

$$\mathcal{E}_{\rm eq} + \mathcal{P}_{\rm eq} = T\mathcal{S}_{\rm eq} = 4\pi \tilde{N} T^4 \hat{m}^3_{\rm eq} K_3(\hat{m}_{\rm eq}), \tag{6}$$

where $\mathcal{E}_{\rm eq}$ and $\mathcal{P}_{\rm eq}$ are provided from lattice QCD calculations [19].

4. Boltzmann equation

The dynamical equations necessary to describe the system can be obtained by taking moments of Boltzmann equation [12, 18]

$$p^\mu \partial_\mu f + \frac{1}{2}\partial_i m^2 \partial^i_{(p)} f = -\mathcal{C}[f], \tag{7}$$

where $C[f]$ is the collisional kernel containing all interactions involved. For this work, we specialize to the relaxation time approximation (RTA) for the collisional kernel.

5. Anisotropic freeze-out

The freeze-out process occurs at the late times, leading to the particle production observed experimentally. In anisotropic hydrodynamics, we perform "anisotropic Cooper-Frye freeze-out" using the distribution function specified in Eq. (2), which ensures that the distribution function is positive definite in all regions in phase space [13, 14]. In viscous hydrodynamics, however, one linearizes around the equilibrium distribution function to take into account the dissipative correction and the positivity of the distribution function is no longer guaranteed [20, 21].

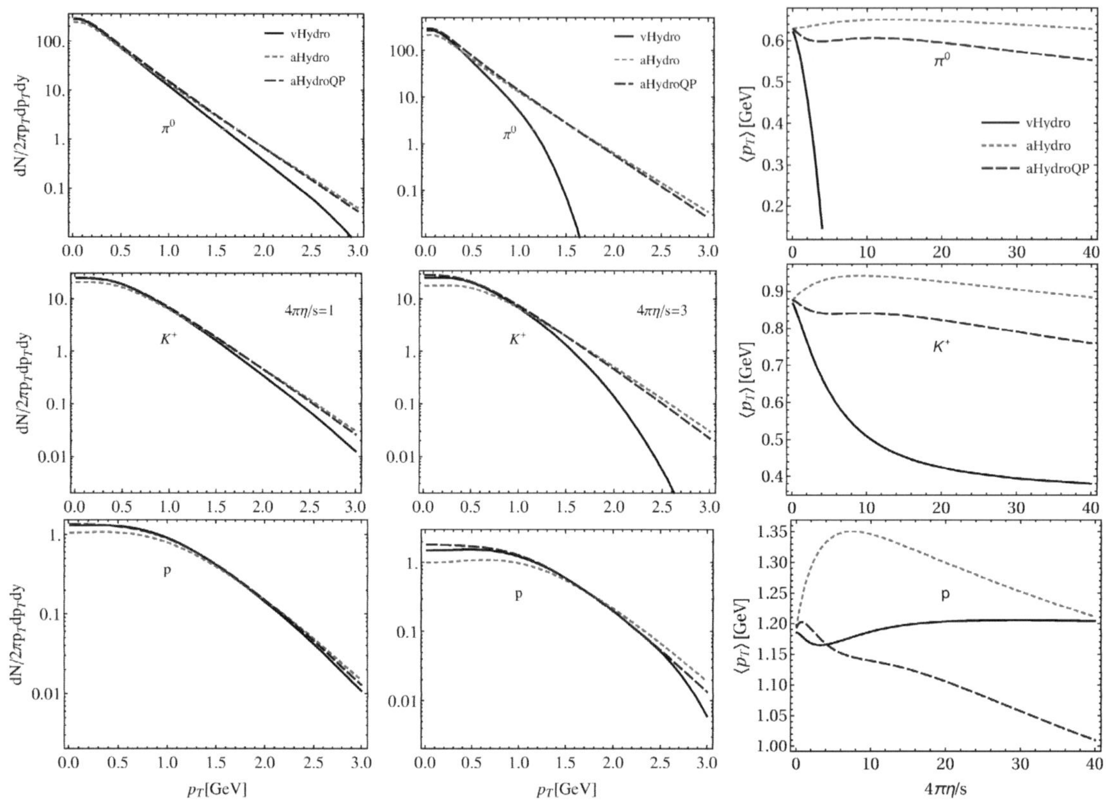

Figure 1. The left column shows the spectra of neutral pions, kaons (K^+), and protons as a function of transverse momentum p_T with $4\pi\eta/s = 1$. The middle column is the same as the left column with $4\pi\eta/s = 3$. The right column shows a comparison of the neutral pion, kaon (K^+), and proton average transverse momentum as a function of $4\pi\eta/s$. In all figures, the solid black line is the result from vHydro, the red dashed line is the result from aHydro, and the blue long-dashed line is the result from aHydroQP. Figures taken from Ref. [13].

6. Results

We now turn to the numerical results. We present comparisons of the results obtained using anisotropic hydrodynamics (both aHydroQP and standard aHydro) and second-order viscous hydrodynamics. For all results presented here, we consider a central collision with an optical Glauber profile. We assume an initial temperature of $T_0 = 600$ MeV at $\tau_0 = 0.25$ fm/c and take the freeze-out temperature to be $T_{\rm eff} = T_{\rm FO} = 150$ MeV in all cases shown. In Fig. 1 (left and middle columns), we present our results for the primordial pion, kaon, and proton spectra obtained for $4\pi\eta/s = 1$ and 3, respectively. For $4\pi\eta/s = 1$, both aHydro approaches are in good agreement over the entire p_T range with the largest differences seen at low momentum. The vHydro result, however, shows a significant downward curvature in the pion spectrum resulting in many fewer high-p_T pions. For larger η/s the vHydro primordial particle spectra become unphysical at lower momenta. For example, for $4\pi\eta/s = 3$, the middle column, the differential pion spectrum goes negative at $p_T \sim 1.6$ GeV.

In Fig. 1 (right column), we present the mean transverse momentum for pions, kaons, and protons as a function of $4\pi\eta/s$. We see that both aHydro approaches show a weak dependence of the pion and kaon $\langle p_T\rangle$ on the value of η/s. On the other hand, vHydro predicts a much more steep decrease in $\langle p_T\rangle$ for the pions and kaons. We also see in vHydro that $\langle p_T\rangle$ for pions becomes negative for $4\pi\eta/s \sim 5$. This rapid decrease can be understood from the negativity of the p_T-differential pion spectra at high p_T where the integrand is more sensitive to the high-p_T part of the spectra. Finally, we note that the proton spectra and hence $\langle p_T \rangle$ for protons is sensitive to the way in which the EoS is implemented when comparing the two aHydro approaches.

7. Conclusions and outlook

We compared three different viscous hydrodynamics approaches: aHydroQP, aHydro, and vHydro. We presented the predictions of these models for the primordial particle spectra and average transverse momentum. We found that they agree well for small shear viscosity to entropy density, η/s with the standard aHydro method showing suppressed production at low transverse-momentum compared to the other two methods. Finally, we showed that, when using standard viscous hydrodynamics, the bulk-viscous correction can drive the primordial particle spectra negative at large p_T, which is not seen with either aHydro method.

Looking to the future, we will try to extend the aHydroQP method to 3+1d since this approach takes into account the non-conformality of the system from the beginning. However, this will be numerically intensive compared with the standard aHydro. Until a 3+1d code is available, 1+1d results will be a reference point to show the differences between the two approaches used to impose a realistic EoS in aHydro.

Acknowledgments

We thank M. Nopoush for his collaboration. M. Strickland was supported by the U.S. Department of Energy under Award No. DE-SC0013470. M. Alqahtani was supported by a PhD fellowship from the University of Dammam.

References

[1] Huovinen P, Kolb P F, Heinz U W, Ruuskanen P V and Voloshin S A 2001 *Phys. Lett.* **B503** 58–64 (*Preprint* hep-ph/0101136)
[2] Romatschke P and Romatschke U 2007 *Phys. Rev. Lett.* **99** 172301 (*Preprint* 0706.1522)
[3] Ryu S, Paquet J F, Shen C, Denicol G S, Schenke B, Jeon S and Gale C 2015 *Phys. Rev. Lett.* **115** 132301 (*Preprint* 1502.01675)
[4] Niemi H, Denicol G S, Huovinen P, Molnar E and Rischke D H 2011 *Phys. Rev. Lett.* **106** 212302 (*Preprint* 1101.2442)
[5] Florkowski W and Ryblewski R 2011 *Phys. Rev.* **C83** 034907 (*Preprint* 1007.0130)
[6] Martinez M and Strickland M 2010 *Nucl. Phys.* **A848** 183–197 (*Preprint* 1007.0889)
[7] Martinez M, Ryblewski R and Strickland M 2012 *Phys. Rev.* **C85** 064913 (*Preprint* 1204.1473)
[8] Ryblewski R and Florkowski W 2012 *Phys. Rev.* **C85** 064901 (*Preprint* 1204.2624)
[9] Bazow D, Heinz U W and Strickland M 2014 *Phys. Rev.* **C90** 054910 (*Preprint* 1311.6720)
[10] Nopoush M, Ryblewski R and Strickland M 2014 *Phys. Rev.* **C90** 014908 (*Preprint* 1405.1355)
[11] Nopoush M, Ryblewski R and Strickland M 2015 *Phys. Rev.* **D91** 045007 (*Preprint* 1410.6790)
[12] Alqahtani M, Nopoush M and Strickland M 2015 *Phys. Rev.* **C92** 054910 (*Preprint* 1509.02913)
[13] Alqahtani M, Nopoush M and Strickland M 2016 (*Preprint* 1605.02101)
[14] Nopoush M, Strickland M, Ryblewski R, Bazow D, Heinz U and Martinez M 2015 *Phys. Rev.* **C92** 044912 (*Preprint* 1506.05278)
[15] Strickland M 2014 *Acta Phys. Polon.* **B45** 2355–2394 (*Preprint* 1410.5786)
[16] Denicol G S, Niemi H, Molnar E and Rischke D H 2012 *Phys. Rev.* **D85** 114047 [Erratum: Phys. Rev.D91,no.3,039902(2015)] (*Preprint* 1202.4551)
[17] Denicol G S, Jeon S and Gale C 2014 *Phys. Rev.* **C90** 024912 (*Preprint* 1403.0962)
[18] Romatschke P 2012 *Phys. Rev.* **D85** 065012 (*Preprint* 1108.5561)
[19] Borsanyi S, Endrodi G, Fodor Z, Jakovac A, Katz S D, Krieg S, Ratti C and Szabo K K 2010 *JHEP* **11** 077 (*Preprint* 1007.2580)
[20] Rose J B, Paquet J F, Denicol G S, Luzum M, Schenke B, Jeon S and Gale C 2014 *Nucl. Phys.* **A931** 926–930 (*Preprint* 1408.0024)
[21] Bozek P 2010 *Phys. Rev.* **C81** 034909 (*Preprint* 0911.2397)

IOP Conf. Series: Journal of Physics: Conf. Series **832** (2017) 012052 doi:10.1088/1742-6596/832/1/012052

Effective dynamical coupling of hydrodynamics and transport for heavy-ion collisions

Dmytro Oliinychenko[1,4] **and Hannah Petersen**[1,2,3]

[1] Frankfurt Institute for Advanced Studies, Ruth-Moufang-Straße 1, D-60438 Frankfurt am Main, Germany

[2] Institut für Theoretische Physik, Goethe-Universität, D-60438 Frankfurt am Main, Germany

[3] GSI Helmholtzzentrum für Schwerionenforschung GmbH, Planckstr. 1, 64291 Darmstadt, Germany

[4] Bogolyubov Institute for Theoretical Physics, Kiev 03680, Ukraine

E-mail: oliiny@fias.uni-frankfurt.de, petersen@fias.uni-frankfurt.de

Abstract. Present hydrodynamics-based simulations of heavy-ion collisions neglect the feedback from the frozen-out particles flying back into the hydrodynamical region. This causes an artefact called "negative Cooper-Frye contributions", which is negligible for high collision energies, but becomes significant for lower RHIC BES energies and for event-by-event simulations. To avoid negative Cooper-Frye contributions, while still preserving hydrodynamical behavior, we propose a pure hadronic transport approach with forced thermalization in the regions of high energy density. It is demonstrated that this approach exhibits enhancement of strangeness and mean transverse momenta compared to conventional transport – an effect typical for hydrodynamical approaches.

1. Introduction and Motivation

A major goal of heavy-ion collision experiments is the study of the QCD phase diagram and the search for its critical point. High energy heavy-ion collisions create a high-temperature and low-baryon density system, which undergoes a cross-over during cooling, but not a phase transition. A conjecture supported by many phenomenological models is that at lower collision energies, where a higher baryon density is reached, the system undergoes a first order phase transition. Testing this conjecture and finding the critical point, where the phase transition turns into a cross-over, motivates the Beam Energy Scan (BES) experiments at the Relativistic Heavy Ion Collider (RHIC) in Brookhaven (USA), as well as the construction of the Nuclotron-based Ion Collider fAcility (NICA) in Dubna (Russia) and the Facility for Antiproton and Ion Research (FAIR) in Darmstadt (Germany). At BES the beam energy was sequentially decreased from $\sqrt{s_{NN}} = 200$ GeV per nucleon pair down to 7.7 GeV. Also a fixed-target mode operation of STAR detector at RHIC would allow going down to $\sqrt{s_{NN}} \approx 3$ GeV [1]. Planned energy ranges per nucleon pair in heavy-ion collisions at NICA and FAIR are $\sqrt{s_{NN}} = (4-11)$ GeV [3] and $\sqrt{s_{NN}} = (2.3-5.3)$ GeV and eventually $(2.3-9.3)$ GeV [2]. This illustrates a substantial interest of the heavy-ion community in low-energy collisions.

In these proceedings we point out that traditional assumptions behind the hydrodynamical simulations of heavy ion collisions break at low collision energies. This motivates us to suggest an approach, which allows to relax these assumptions.

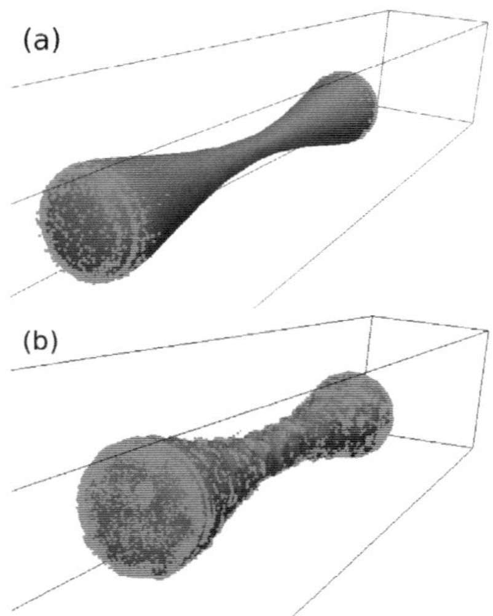

Figure 1. Demonstration of the lumpiness effect on "negative crossings". Hypersurfaces of constant Landau rest frame energy density 0.3 GeV/fm^3 for Au+Au collisions at 160 A GeV are shown at fixed time. Both are obtained from 1000 coarse-grained UrQMD events, but in (a) additional smoothing is applied (see text). Points mark particles crossing the hypersurface from outside to inside.

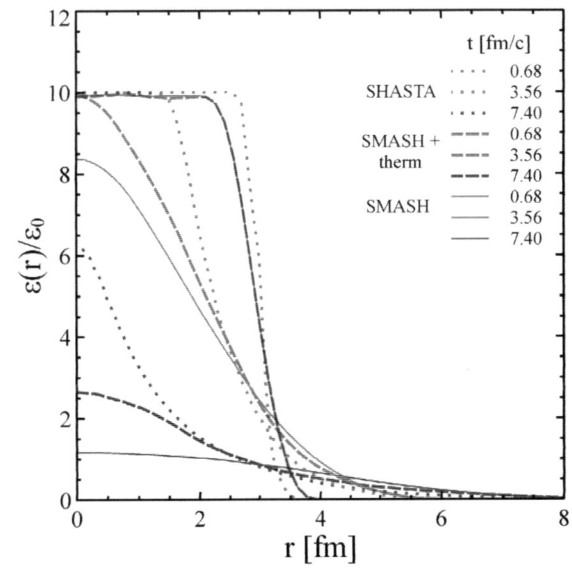

Figure 2. The time evolution of an expanding sphere is compared for ideal hydrodynamics (SHASTA, dotted lines), hadron cascade (SMASH, solid lines) and the same hadron cascade enhanced by forced thermalization where rest frame energy density exceeds 0.3 GeV/fm^3 (SMASH+therm, dashed lines).

2. Negative contributions on the freeze-out hypersurface

Relativistic hydrodynamics is very successful in the description of the dynamics of heavy-ion collisions and explains a plethora of measured particle spectra and flows (v_2, v_3, etc.) at RHIC and LHC. Due to this success of hydrodynamics it is stated that the hot and dense matter created in heavy-ion collisions at RHIC and LHC has a very small kinematic viscosity. All the current hydrodynamics-based simulations adopt the following approximations: the hydrodynamic equations are solved in the whole forward light cone, including regions where hydrodynamics is not applicable; the particlization hypersurface is chosen a posteriori from hydrodynamics; frozen-out particles flying again into the hydrodynamical region cannot cause feedback to hydrodynamics. Particle spectra produced from the hypersurface element $d\sigma_\mu$ are typically computed using the Cooper-Frye formula:

$$E\frac{d^3N}{d^3p} = f(p)\, p^\mu d\sigma_\mu \tag{1}$$

According to this formula, for space-like $d\sigma_\mu$, there exist momenta for which the number of produced particles is negative. This negative number corresponds to particles that fly back into the hydrodynamical region. In a simulation, where hydrodynamical equations for the dense region are coupled with kinetic equations for the sparse region, these negative contributions will be properly taken into account. However, currently such simulations only exist for non-relativistic hydrodynamics [4].

Are the above mentioned approximations adequate? In other words: are negative Cooper-Frye contributions negligible? A study dedicated to this question [5] found that they may be small at high collision energies, but become larger for the lower ones reaching 10-15% for pions at midrapidity at energies of FAIR, NICA and RHIC BES. Here we would like to mention one more feature that has not been discussed in detail in [5]: in event-by-event simulations, where the particlization hypersurface may be lumpy, negative contributions can be increased dramatically. This effect of a lumpy hypersurface is illustrated by Fig. 1. It shows two surfaces of constant Landau rest frame energy density, $e_c = 0.3$ GeV/fm^3, at fixed time. They were obtained from coarse-grained UrQMD simulation of Au+Au collision at 160 A GeV. Both were constructed from 1000 events, but for one simulation the energy density in the cells was smoothed using running average ($\epsilon = \frac{1}{27}\sum_{i=1}^{27}\epsilon_i$) where the sum is running over the cell neighbours. Particles crossing hypersurfaces from the outside to the inside (the "negative crossings") for all events are shown with dots. One can see that lumps increase the amount of "negative crossings" substantially. The lumpiness effect requires further investigation, because unlike in this coarse-grained transport study, in the hydrodynamic simulation the lumpy initial state may be smoothed by the hydrodynamical evolution. For viscous hydrodynamics the freeze-out hypersurface is smoother than for ideal hydrodynamics given the same initial state, so the effect is smaller. The larger the viscosity, the stronger the smoothing and the smaller the effect.

3. Forced thermalization at high density regions

To relax the approximations used by hydrodynamical approaches, but preserve their advantages, we have constructed and tested the following conceptually simple approach. In the hadronic transport calculation we have performed forced thermalization in the regions where the rest frame Landau energy density ϵ is larger than some ϵ_c. In this way we reach hydrodynamical behavior in the region where hydrodynamics is applicable. The kinetic Boltzmann equation is solved in the rest of the space, and it automatically turns out that the surface connecting both regions is found dynamically. This was implemented using the SMASH hadron cascade [6]. In the forced thermalization procedure we required the exact conservation of energy, momentum and quantum numbers. Its implementation and testing are described in detail in [7].

In Fig. 2 we show the comparison of our approach, pure SMASH transport and SHASTA ideal hydrodynamics [8] within a simple setup: a sphere of radius $R_0 = 3$ fm is initialized with constant energy density $10\epsilon_0$, where ϵ_0 is a nuclear ground state energy density, and allowed to evolve. Fig. 2 demonstrates that our approach exhibits intermediate behavior between hydrodynamics and transport. The difference between our approach and hydrodynamics emerges due to the fact that at the boundary of the hydrodynamical region hydrodynamics assumes a small Knudsen number, but in our approach in the vicinity of the boundary particles may simply fly out.

Finally, we apply our approach to Au+Au collision at $\sqrt{s_{NN}} = 3$ GeV and observe several expected effects characteristic for hydrodynamics. The main results are shown in Fig. 3. Firstly, we find that multiplicities of strange particles are larger in our approach than in a pure SMASH transport and tend to be close to the multiplicities from the UrQMD hybrid approach [9] initialized and treated in the same way. This is the effect of the forced thermalization: in pure transport the amount of strange particles is below the equilibrium values. Secondly, we observe that the average transverse momentum in our approach is larger for different particles species than in transport. This is happening due to the isotropization: more momentum is transferred from the collision axis to the midrapidity region. The increase of $\langle p_t\rangle$ seems to be independent on the time between thermalizations $\Delta t_{\rm th}$. Also, the effect is especially pronounced for K^-, which are produced only in the secondary strangeness exchange reactions.

To sum up, we have performed the first tests of a new approach, where in a pure hadronic transport forced thermalization is applied in the dense regions. This approach interpolates between transport and hydrodynamics, avoiding the traditional problem of negative Cooper-Frye contributions. The plan for the future besides additional tests is to understand bulk observables at low and intermediate collision energies.

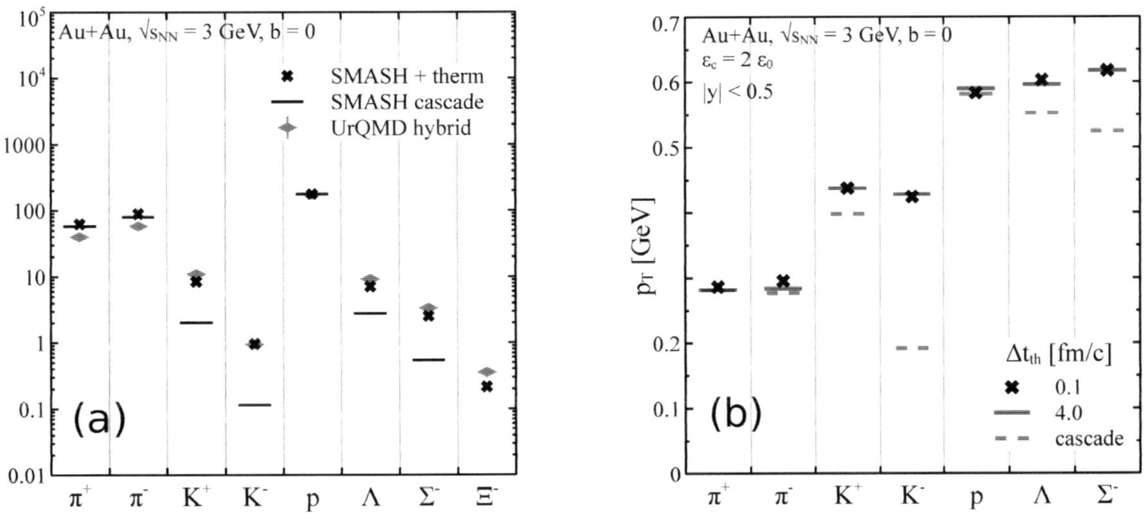

Figure 3. Central Au+Au collision at $\sqrt{s} = 3$ GeV. Panel (a): Final multiplicities are compared for SMASH, SMASH with thermalization, and UrQMD hybrid. Panel (b): Mean transverse momentum $p_T = \sqrt{p_x^2 + p_y^2}$ at midrapidity for the calculation with forced thermalization is compared to the cascade.

Acknowledgments
The authors thank L. G. Pang, H. Niemi and J. Steinheimer for fruitful discussions. Computational resources have been provided by the Center for Scientific Computing (CSC) at the Goethe-University of Frankfurt. The authors acknowledge funding of a Helmholtz Young Investigator Group VH-NG-822 from the Helmholtz Association and GSI. This work was supported by the Helmholtz International Center for the Facility for Antiproton and Ion Research (HIC for FAIR) within the framework of the Landes-Offensive zur Entwicklung Wissenschaftlich-Ökonomischer Exzellenz (LOEWE) program launched by the State of Hesse. D. O. acknowledges support by the Deutsche Telekom Stiftung and Helmholtz Graduate School for Hadron and Ion Research (HGS-HIRe).

References

[1] G. Odyniec, EPJ Web Conf. **95** (2015) 03027.
[2] J. M. Heuser [CBM Collaboration], Acta Phys. Polon. Supp. **9** (2016) 221.
[3] V. Kekelidze *et al.*, Nucl. Phys. A **956** (2016) 846.
[4] S. Tiwari, J. Comp. Phys. **144** (1998) 710-726.
[5] D. Oliinychenko, P. Huovinen and H. Petersen, Phys. Rev. C **91** (2015) no.2, 024906
[6] J. Weil *et al.*, arXiv:1606.06642 [nucl-th].
[7] D. Oliinychenko and H. Petersen, arXiv:1609.01087 [nucl-th].
[8] D. H. Rischke, S. Bernard and J. A. Maruhn, Nucl. Phys. A **595** (1995) 346
[9] H. Petersen, J. Steinheimer, G. Burau, M. Bleicher and H. Stocker, Phys. Rev. C **78** (2008) 044901

IOP Conf. Series: Journal of Physics: Conf. Series **832** (2017) 012053 doi:10.1088/1742-6596/832/1/012053

Momentum anisotropy at freeze out

S. Feld, N. Borghini, C. Lang

Universität Bielefeld, Fakultät für Physik, Bielefeld, Germany

E-mail: s.feld@physik.uni-bielefeld.de

Abstract. The transition from a hydrodynamical modeling to a particle-based approach is a crucial element of the description of high-energy heavy-ion collisions. Assuming this "freeze out" happens instantaneously at each point of the expanding medium, we show that the local phase-space distribution of the emitted particles is asymmetric in momentum space. This suggests the use of anisotropic hydrodynamics for the last stages of the fluid evolution. We discuss how observables depend on the amount of momentum-space anisotropy at freeze out and how smaller or larger anisotropies allow for different values of the freeze-out temperature.

1. The need for a kinetic freeze out

Hydrodynamics has emerged as being very successful at describing a large amount of the evolution of the fireball created in a ultrarelativistic heavy ion collision.

However, hydrodynamics is an effective field theory which assumes the medium to be continuous. An appropriate measure for this continuity is the Knudsen number Kn, defined as the ratio between the mean free path and some macroscopic length scale like the system size. For hydrodynamics to be applicable, this number has to be small. Even if at the beginning of the evolution the condition $\mathrm{Kn} \ll 1$ is fulfilled, yet as the fireball expands, it becomes more dilute and the Knudsen number grows. Obviously, at the very end detectors detect individual particles, which have long ceased to interact with each other. Going backwards in time to the point where interactions became rare, this amounts to a very large mean free path (and Kn), in which case kinetic theory provides an appropriate modeling of the particles evolution.

Thus, the description of the fireball evolution needs to include a prescription for switching from the hydrodynamic approach at $\mathrm{Kn} \ll 1$ to the transport models used at $\mathrm{Kn} \gg 1$. This prescription, which for consistency should conserve all relevant quantum numbers of the fireball as well as its local energy and momentum densities, is the kinetic freeze out. Note that in the simplified application we shall present hereafter, we consider the direct conversion from a fluid to free particles, omitting the stage consisting of interacting particles described by a kinetic theory.

2. The Cooper–Frye formula

The basic idea of the Cooper–Frye freeze out is that the medium description changes suddenly from a fluid-dynamical to a kinetic one as the medium is passing through a hypersurface Σ [1]. From this point on, the particles decouple from the fluid and, assuming the emitted particles are then free, their momenta are frozen. Σ is a closed three-dimensional surface in space-time, since eventually the whole fluid has to freeze out.

IOP Conf. Series: Journal of Physics: Conf. Series **832** (2017) 012053 doi:10.1088/1742-6596/832/1/012053

Introducing the on-shell phase-space distribution $f(x^\mu, \vec{p})$, the Lorentz-invariant particle spectrum, which basically expresses the number of particles per momentum interval, is given by the Cooper–Frye integral

$$E_p \frac{d^3N}{d^3\vec{p}} = \frac{1}{(2\pi)^3} \int_\Sigma f(x^\mu, \vec{p})\, p^\mu d^3\sigma_\mu(x^\mu), \tag{1}$$

where $d\sigma_\mu(x^\mu)$ is a unit vector standing perpendicular to the hypersurface Σ. For the particle distribution function $f(x^\mu, \vec{p})$ at freeze out entering the Cooper–Frye formula, popular choices are the Maxwell–Jüttner distribution, or its quantum generalizations to fermions and bosons, possibly with terms accounting for dissipative effects in the fluid.

An important issue is now to characterize the hypersurface Σ. It is often assumed to correspond to the space-time points where some property of the fluid (e.g. its temperature T, or its particle-number or energy density) reaches some "critical" value, or when a given proper time is reached.

One problem here is that there are regions of Σ where $d\sigma_\mu d\sigma^\mu < 0$, which likely leads to negative contributions to the particle spectrum (1). However the idea of the particle spectrum is just to count particles, not quantum numbers, so that negative contributions are unexpected (for a proposal on how to deal with this problem, see e.g. Ref. [2]).

A second issue with the Cooper–Frye freeze out is the dependence of the computed observables on the initially chosen freeze out parameter, which characterizes Σ. Since the hypersurface is not associated with a phase transition, there is no reason why an observable should depend on the point where a theoretical physicist changes her/his "tool" to describe the medium. Accordingly, one should rather expect that nature performs a smooth transition between the two asymptotic models valid at small and for large Knudsen numbers.

Because of its simplicity, the Cooper–Frye recipe is however much too attractive to be discarded. Our idea is, instead of just gluing together the two models (hydrodynamic and kinetic theory), to try to first tune them, so that they fit together in a better way. The desired result should be a much weaker (or almost no) dependence of observables on the point where we glued together the two models.

3. Anisotropic Cooper-Frye freeze out

3.1. Motivation

In our opinion, one way of achieving this "improved gluing" is to resort to an explicitly anisotropic distribution at freeze out [3]. A first hint towards that direction comes from a series of studies, starting with Ref. [4], of non-relativistic hypersonic flows which expand through a nozzle into vacuum. The conclusion of these studies is that the frozen-out particle distribution can be characterized by two effective temperatures, which stand on the one hand for the "thermal" motion perpendicular to the streamlines ($T_\perp$) and on the other hand for the motion parallel to the streamlines ($T_\parallel$). We are currently working on a generalization of these studies to the relativistic heavy ion collision case.

A second motivation for invoking a momentum-asymmetric phase-space distribution is that there is already a transition between models in the description of heavy ion collisions, namely from the pre-equilibrium model(s) to hydrodynamics, which can be improved by the use of anisotropic hydrodynamics, see e.g. Ref. [5].

3.2. Implementation

The basic idea of the anisotropic freeze out [3] is to implement an anisotropy in the phase-space distribution along the radial direction. As ansatz for the functional form of $f(x^\mu, \vec{p})$, we used

IOP Conf. Series: Journal of Physics: Conf. Series **832** (2017) 012053 doi:10.1088/1742-6596/832/1/012053

the Romatschke–Strickland distribution [6], which reads in the local rest frame:

$$f_{aniso}^{LRF}(x^\mu, \vec{p}, \Lambda, \xi) \propto exp\left(-\frac{\sqrt{(p \cdot u)^2 + \xi(x^\mu)(p^r)^2}}{\Lambda(x^\mu)}\right), \tag{2}$$

where $\xi(x^\mu)$ is the anisotropy parameter and $\Lambda(x^\mu)$ the "anisotropic temperature", which is going to be the temperature in the case of isotropy. To implement this distribution function into the Cooper–Frye formula, one must boost it into the laboratory frame.

As a consequence of using the Romatschke–Strickland distribution, we get an additional parameter, which gives us the possibility to tune the anisotropy in the radial direction. In principle this parameter has only the lower boundary -1, since we need the distribution function to be real valued. In order to generate a higher effective temperature/pressure in the radial direction, we need to choose ξ to be less than 0.

Because we are up to now just interested in how the technique of the anisotropic freeze out works, we inserted a blast-wave like fluid velocity profile of the form $u^\phi = u^\eta = 0$, $u^r = \bar{u}_{max}\,\frac{r}{R}\,(1 + 2\sum_n V_n \cos(n\phi))$, where we chose the parameters to be $R = 10$ fm, $\tau_{\text{fo}} = 7.5$ fm/c, $V_2 = V_3 = 0.05$ and we investigated the emission of pions. Because of the lack of a microscopical model we ignored a possible position-space dependence of the anisotropic temperature Λ and the anisotropy parameter ξ. Now that we have all ingredients (1) and (2) together, we are able to compute observables for the anisotropic freeze out.

4. Observables

4.1. Particle spectra

The first observable which we present is the transverse particle spectrum. We plotted this for various values for the anisotropy parameter in Figure 1. As one can see, the smaller the anisotropy parameter is chosen, the more particle one will detect with higher momenta. This behavior is to be expected, since in anisotropic hydrodynamics one obtains a higher pressure in a direction if one chooses the corresponding anisotropy parameter to be negative. Such a higher pressure will then push the particles to higher momenta in the out-direction, which results in the deviations of the ideal momentum spectrum. As a cross-check we plotted the spectrum for positive ξ-values as well.

In Figure 2 we now "optimized" (by eye) the pairs of anisotropy parameter ξ and anisotropic temperature Λ. As one can see there, we are able to reproduce the same particle spectrum while varying the "temperature" over a range of 30 MeV. This behavior is a first hint that the additional parameter ξ works the way it should, relaxing the need for a well-defined value of the freeze-out temperature

4.2. Anisotropic flow coefficients v_n

The second observables which we now present are the anisotropic flow coefficients v_n. These coefficients arise through a Fourier series of the particle spectrum in the momentum angle ϕ_p. We computed the coefficients up to v_5. They all follow the same trend, so here we are only showing the elliptic flow coefficient v_2.

In Figure 3 we plotted v_2 as a function of momentum for different values of the anisotropy. As one can see, v_2 decreases when the anisotropy decrease. This trend reflects the fact that the negative ξ induces a higher effective temperature in the out-direction. This higher effective temperature results in a higher thermal motion along this direction, which then overcomes the effect of collective motion that is measured by the flow coefficient v_2.

Now we followed the same procedure as before and optimized the pairs of anisotropic temperature Λ and anisotropy parameter ξ, so that we are able to reproduce the same observable although we vary the value of the temperature at freeze out. The results are plotted in Figure 4. Again we are able to reproduce nearly the same v_2 for an freeze out temperature interval of nearly 30 MeV.

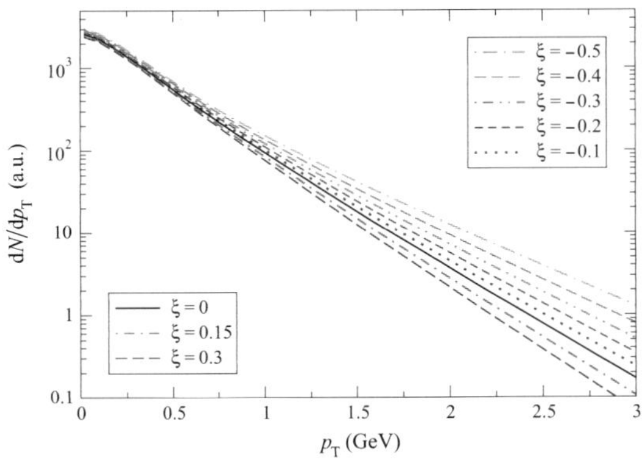

Figure 1. Transverse particle spectrum for various anisotropies ξ.

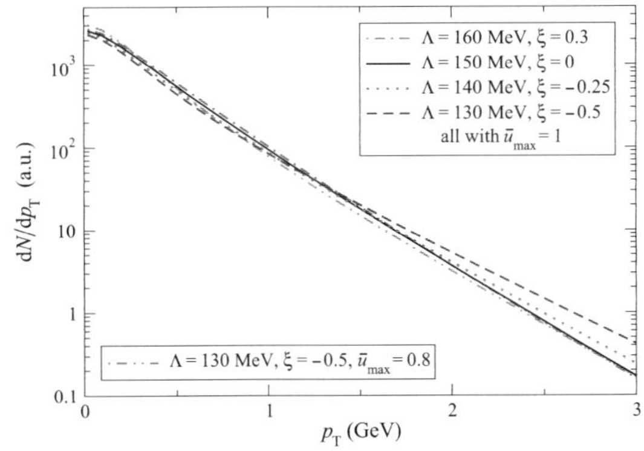

Figure 2. Transverse particle spectrum for optimized ξ, Λ pairs.

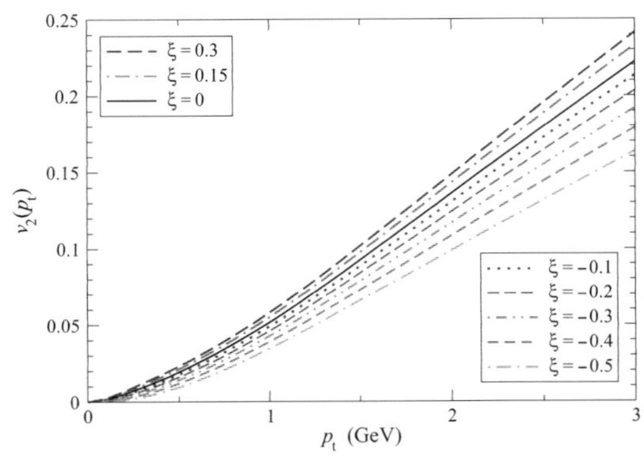

Figure 3. Elliptic flow coefficient v_2 for various ξ.

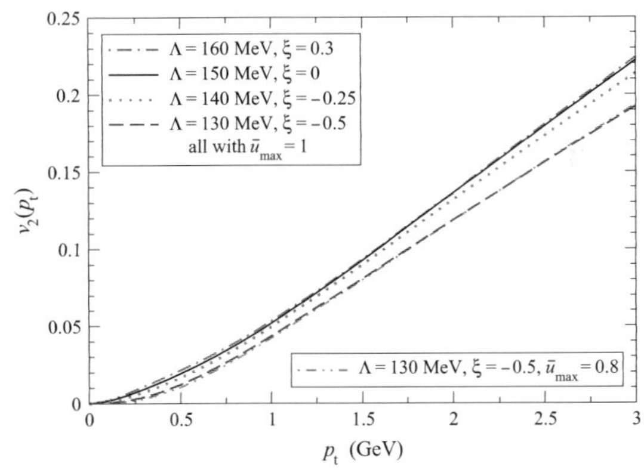

Figure 4. Elliptic flow coefficient v_2 for various pairs of ξ and Λ.

5. Conclusion

We showed that at the cost of an additional parameter, namely the anisotropy parameter ξ we are able to establish a interval of freeze out temperatures, which all give the same observables. So we get rid of the strong dependence of observables on the parameter where we glued together the two asymptotic models, hydrodynamics and kinetic theory. As we showed for decreasing freeze out temperatures Λ we have to choose more negative anisotropy parameters ξ. This reflects the fact that we are able to tune the late times hydrodynamics in such a way, that it crosses over to the (trivial) kinetic theory in a smoother manner.

References

[1] Cooper F and Frye G 1974 *Phys. Rev.* D **10** 186
[2] Oliinychenko D, Huovinen P and Petersen H 2015 *J. Phys. Conf. Ser.* **599** 012017 (*Preprint* 1412.2020)
[3] Borghini N, Feld S and Lang C 2015 *Eur. Phys. J.* C **75** 275 (*Preprint* 1412.5199)
[4] Hamel B and Willis R 1966 *Phys. Fluids* **9** 829
[5] Strickland M 2014 *Acta Phys. Polon.* B **45** 2355–2394 (*Preprint* 1410.5786)
[6] Romatschke P and Strickland M 2003 *Phys. Rev. D* **68** 036004 (*Preprint* hep-ph/0304092)

IOP Conf. Series: Journal of Physics: Conf. Series **832** (2017) 012054 doi:10.1088/1742-6596/832/1/012054

Phenomenological predictions of 3+1d anisotropic hydrodynamics

Mohammad Nopoush and Michael Strickland

Department of Physics, Kent State University, Kent, OH 44242 United States

Radoslaw Ryblewski

Institute of Nuclear Physics, Polish Academy of Sciences, PL-31342 Krakow, Poland

E-mail: mnopoush@kent.edu

Abstract. We make phenomenological predictions for particle spectra and elliptic flow in heavy-ion collisions using 3+1d anisotropic hydrodynamics (aHydro) including the effects of both shear and bulk viscosities. The dynamical equations necessary are derived by taking moments of the Boltzmann equation allowing for three distinct (diagonal) momentum-space anisotropy parameters. The formulation is based on relaxation-time approximation for the collisional kernel and a lattice-QCD-based equation of state. Evolving the system to late times, we calculate particle production using THERMINATOR 2, modified to account for an ellipsoidal distribution function. We obtain particle spectra for different particle species such as pions, kaons, and protons, and elliptic flow v_2 as a function of centrality, transverse momentum, and rapidity. In our model, we have four free parameters, i.e. freeze-out temperature, initial central energy density, initial momentum-space anisotropies, and shear viscosity to entropy density ratio. Using a multidimensional fit to LHC experimental data, we make a preliminary extraction of these parameters. We find reasonable agreement between 3+1d aHydro and available experimental data for $\eta/s \sim 0.23$.

1. Introduction

In recent years, hydrodynamics has enabled us to enhance our understanding of the space-time evolution of the quark-gluon plasma (QGP) created in ultra-relativistic heavy-ion collisions. Different schemes of hydrodynamics such as ideal hydrodynamics and later viscous hydrodynamics (vHydro) have been able to describe the evolution of QGP quite well [1]. However, the existence of large momentum-space anisotropies at early times and near the transverse edges of QGP is numerically challenging and can cause standard vHydro schemes to break down. Anisotropic hydrodynamics (aHydro) addresses this issue by taking into account the inherent momentum-space anisotropy of the QGP at leading order in a reorganized hydrodynamic expansion [2–11]. Importantly, aHydro has been shown to agree better with exact solution of the Boltzmann equation than vHydro models [12–17]. In this paper, we study the phenomenology of the QGP using a 3+1d aHydro formalism. Evolving the system to late times, we calculate the resulting identified particle spectra and elliptic flow generated using a fixed energy density freeze-out prescription. Our preliminary results show a reasonable agreement with the LHC experimental data for 2.76 TeV Pb+Pb collisions.

2. Setup and background

Leading order anisotropic hydrodynamics (aHydro) is based on an anisotropic distribution function of the form [6, 18]

$$f(x,p) \equiv f_{\text{iso}}\Big(\frac{\sqrt{p^\mu \Xi_{\mu\nu} p^\nu}}{\lambda}\Big)\,, \tag{1}$$

with λ being a temperature-like scale and $\Xi_{\mu\nu} \equiv u_\mu u_\nu + \xi_{\mu\nu} - \Phi\Delta_{\mu\nu}$ being the anisotropy tensor, which parametrizes the anisotropic form of distribution function. In this relation, u_μ is fluid four-velocity, $\xi_{\mu\nu}$ is a symmetric traceless anisotropy tensor which obeys $u_\mu \xi^{\mu\nu} = 0$, Φ is the bulk degree of freedom, and $\Delta_{\mu\nu} \equiv g_{\mu\nu} - u_\mu u_\nu$ is the transverse projection operator. Herein, the tensor $\xi_{\mu\nu}$ is taken to be diagonal, i.e. $\xi_{\mu\nu} = \text{diag}(0, \boldsymbol{\xi})$, where $\boldsymbol{\xi}$ are anisotropy parameters in different spatial directions. This form contains information about both shear and bulk corrections at leading order [6]. In what follows, we assume the distribution to be of Boltzmann form, i.e. $f_{\text{iso}}(p) \equiv \exp(-p)$.

3. Dynamical Equations

In order to derive the dynamical equations, one needs to set up the basis vectors in the lab frame (LF). To do so, one needs to perform a set of tensor transformations including a boost along and a rotation around the longitudinal beam direction, followed by a transverse boost which connects the local rest frame to the LF [2, 18]. The dynamical equations are obtained by taking moments of the Boltzmann equation

$$p^\mu \partial_\mu f = -\mathcal{C}[f]\,, \tag{2}$$

assuming the anisotropic form for the distribution function (1) and then taking the relevant projections using the LF basis vectors. The zeroth and the first moments lead to evolution equations for the particle four-current J^μ and the energy-momentum tensor $T^{\mu\nu}$. Taking higher moments, one obtains the dynamical equations for hydrodynamics quantities of higher rank, i.e. the second moment of the Boltzmann equations provides information about microscopic dissipation. The result is a closed set of dynamical equations,

$$\partial_\mu J^\mu = (n_{\text{eq}} - n)/\tau_{\text{eq}}\,, \quad \partial_\mu T^{\mu\nu} = 0\,, \quad \partial_\mu \mathcal{I}^{\mu\nu\lambda} = u_\mu(\mathcal{I}^{\mu\nu\lambda}_{\text{eq}} - \mathcal{I}^{\mu\nu\lambda})/\tau_{\text{eq}}\,, \tag{3}$$

which is obtained based on the relaxation-time approximation for the collisional kernel. In the equations above, J^μ, $T^{\mu\nu}$, and $\mathcal{I}^{\mu\nu\lambda}$ are the first, second, and third moments of the distribution function, respectively. Note that the nth-moment of distribution function is defined by $N_{\text{dof}} \int d^3p/((2\pi)^3 E) p^{\mu_1} \cdots p^{\mu_n} f(x,p)$, with N_{dof} being number of degrees of freedom. Herein, we obtain four equations from the first moment and three from the second moment. In addition, the conservation of energy provides us with an extra equation which allows us to compute the local effective temperature from the non-equilibrium energy density (Landau-matching).

4. aHydro equation of state

Implementing a realistic equation of state (EoS) in the framework of aHydro is conceptually challenging since the EoS is a relation between the energy density and pressure in isotropic equilibrium. We have devised two approaches to deal with this issue. In the quasiparticle approach, we assume that the system is comprised of quasiparticles with a temperature-dependent mass which is fit to the lattice QCD data [19, 20]. In the second approach, called the standard approach, we obtain the necessary dynamical equations in the conformal limit ($m \to 0$). In this limit, the components of $T^{\mu\nu}$ multiplicatively factorize and one can then connect the isotropic parts of energy density and pressures using a realistic EoS [21]. In this work, we have used the standard approach. The realistic equilibrium EoS used herein is taken from Krakow parametrization of lattice data [22].

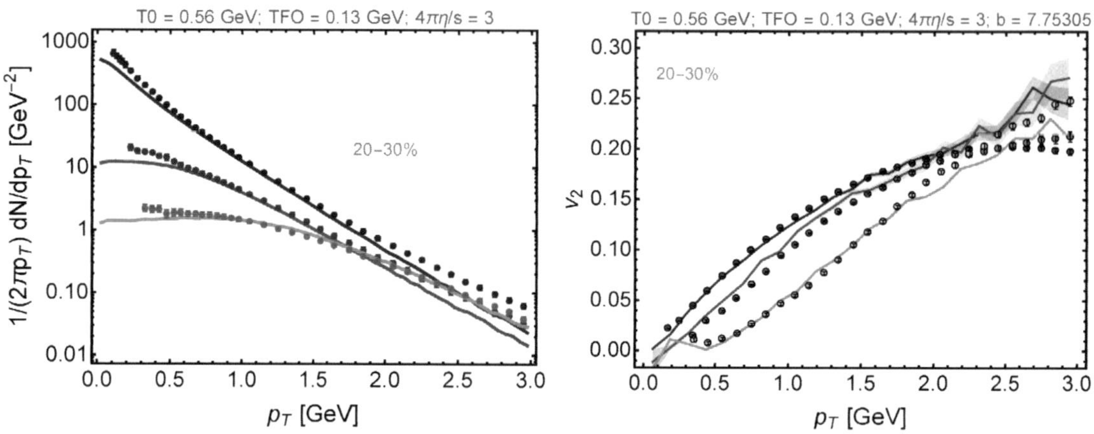

Figure 1. In this figure, we compare aHydro predictions to experimental data for π^+ (blue), K^+ (red), and p (green). The left panel shows the particle spectra and the right panel shows v_2, both as a function of transverse momentum p_T for the 20-30% centrality class. The best fit corresponded to $T_0 = 0.56$ GeV, $T_{\rm FO} = 0.13$ GeV, and $4\pi\eta/s = 3$. The system is initialized at $\tau_0 = 0.25$ fm/c using an smooth optical Glauber profile. Note that the experimental error bars shown are statistical only.

5. aHydro freezeout

Evolving to late times, the system undergoes a crossover from quarks and gluons to hadronic degree of freedom and later on the kinetic freezeout. In order to compare the result of our hydrodynamics model to experimental data, one needs to calculate the differential particle spectra at freezeout. This procedure is done by constructing a constant energy density hypersurface, defined through an effective temperature $T_{\rm FO} = \mathcal{E}^{-1}(\mathcal{E}_{\rm FO})$. Then, by computing for the particles which cross this hypersurface, one can determine the number of hadrons produced in heavy-ion collisions at the freezeout. For this purpose, we apply a generalized Cooper-Frye formula, modified for aHydro [20].

6. Results

In this section, we present our preliminary results and compare them to LHC experimental data for 2.76 TeV Pb+Pb collisions [23, 24]. The results are generated for many hadronic event samples at freezeout. In the plots shown herein we sampled 20,100 hadronic events using THERMINATOR 2 [25]. The system is initialized at $\tau_0 = 0.25$ fm/c using an smooth optical Glauber profile and assumed to be initially isotropic in momentum space. In our model, we have some free parameters including initial central temperature T_0, shear viscosity to entropy density ratio η/s, initial momentum-space anisotropies of the system, and the freezeout temperature $T_{\rm FO}$. By fitting to the experimental data and determining the best fit, we have estimated the appropriate values for these parameters. Based on our model, the best fit parameters values are: $T_0 = 0.56$ GeV, $T_{\rm FO} = 0.13$ GeV, and $\eta/s = 3/4\pi$.

In Fig. 1, we have plotted our results for LHC 2.76 TeV Pb+Pb collisions for 20-30% centrality class. In our plots, the results for charged pions (blue), kaons (red) and protons (green) are presented. The shaded bands are the statistical error associated with the THERMINATOR 2 Monte-Carlo sampling of the hypersurface. The left panel shows the p_T-differential particle spectra and the right panel shows v_2, both as a function of transverse momenta p_T. For $p_T < 2$ GeV the v_2 fit quality seems very good. Our results follow the experimental data not only in magnitude but also in curvature. For $p_T > 2$ GeV the fluctuations are significant because of the fewer number of events in this range of momentum. While the fit to v_2 seems

IOP Conf. Series: Journal of Physics: Conf. Series **832** (2017) 012054 doi:10.1088/1742-6596/832/1/012054

quite promising, for very small p_T our model underestimates the spectra, as can be seen from the left panel of Fig. 1. We are currently investigating the cause of this discrepancy.

7. Conclusions and outlooks

In conclusion, we have implemented a 3+1d aHydro code for studying the quark-gluon plasma created in ultra-relativistic heavy-ion collisions. This model can be used to study particle spectra, flow coefficients, and other quark-gluon plasma experimental observables. By fitting to experimental data, we have managed to extract a preliminary estimation for some unknown parameters including initial central temperature, freezeout temperature, and shear viscosity to entropy density ratio. Our estimation for $\eta/s \simeq 0.23$ is quite different from a recent estimation by viscous hydrodynamics, $\eta/s \simeq 0.095$ [26]. In order to improve the accuracy of the model for the future studies, one needs to consider fluctuating initial conditions for the aHydro evolution. Fluctuating initial conditions are already implemented in the 3+1d aHydro code, however, for this preliminary study, we restricted our attention to a smooth optical Glauber profile. Looking forward, one also can take into account off-diagonal components of anisotropy tensor and the effect of chemical potential(s) in the anisotropic distribution function.

8. Acknowledgements

M.N. and M.S. were supported by the U.S. Department of Energy, Office of Science, Office of Nuclear Physics under Award No. DE-SC0013470. R.R. was supported by Polish National Science Center Grant DEC-2012/07/D/ST2/02125.

References

[1] Jeon S and Heinz U 2016 Introduction to Hydrodynamics *Quark-Gluon Plasma 5* ed Wang X N pp 131–187
[2] Florkowski W and Ryblewski R 2011 *Phys. Rev.* **C83** 034907 (*Preprint* 1007.0130)
[3] Martinez M and Strickland M 2010 *Nucl. Phys.* **A848** 183–197 (*Preprint* 1007.0889)
[4] Bazow D, Heinz U W and Strickland M 2014 *Phys. Rev.* **C90** 054910 (*Preprint* 1311.6720)
[5] Tinti L and Florkowski W 2014 *Phys. Rev.* **C89** 034907 (*Preprint* 1312.6614)
[6] Nopoush M, Ryblewski R and Strickland M 2014 *Phys. Rev.* **C90** 014908 (*Preprint* 1405.1355)
[7] Tinti L 2016 *Phys. Rev.* **C94** 044902 (*Preprint* 1506.07164)
[8] Bazow D, Heinz U W and Martinez M 2015 *Phys. Rev.* **C91** 064903 (*Preprint* 1503.07443)
[9] Strickland M, Nopoush M and Ryblewski R 2016 *Nucl. Phys.* **A956** 268–271 (*Preprint* 1512.07334)
[10] Molnar E, Niemi H and Rischke D H 2016 *Phys. Rev.* **D93** 114025 (*Preprint* 1602.00573)
[11] Molnar E, Niemi H and Rischke D H 2016 (*Preprint* 1606.09019)
[12] Florkowski W, Ryblewski R and Strickland M 2013 *Nucl. Phys.* **A916** 249–259 (*Preprint* 1304.0665)
[13] Florkowski W, Ryblewski R and Strickland M 2013 *Phys. Rev.* **C88** 024903 (*Preprint* 1305.7234)
[14] Denicol G S, Heinz U W, Martinez M, Noronha J and Strickland M 2014 *Phys. Rev.* **D90** 125026 (*Preprint* 1408.7048)
[15] Denicol G S, Heinz U W, Martinez M, Noronha J and Strickland M 2014 *Phys. Rev. Lett.* **113** 202301 (*Preprint* 1408.5646)
[16] Nopoush M, Ryblewski R and Strickland M 2015 *Phys. Rev.* **D91** 045007 (*Preprint* 1410.6790)
[17] Heinz U, Bazow D, Denicol G S, Martinez M, Nopoush M, Noronha J, Ryblewski R and Strickland M 2016 (*Preprint* 1509.05818)
[18] Martinez M, Ryblewski R and Strickland M 2012 *Phys. Rev.* **C85** 064913 (*Preprint* 1204.1473)
[19] Alqahtani M, Nopoush M and Strickland M 2015 *Phys. Rev.* **C92** 054910 (*Preprint* 1509.02913)
[20] Alqahtani M, Nopoush M and Strickland M 2016 (*Preprint* 1605.02101)
[21] Nopoush M, Strickland M, Ryblewski R, Bazow D, Heinz U and Martinez M 2015 *Phys. Rev.* **C92** 044912 (*Preprint* 1506.05278)
[22] Chojnacki M and Florkowski W 2007 *Acta Phys. Polon.* **B38** 3249–3262 (*Preprint* nucl-th/0702030)
[23] Abelev B *et al.* (ALICE) 2013 *Phys. Rev.* **C88** 044910 (*Preprint* 1303.0737)
[24] Abelev B B *et al.* (ALICE) 2015 *JHEP* **06** 190 (*Preprint* 1405.4632)
[25] Chojnacki M, Kisiel A, Florkowski W and Broniowski W 2012 *Comput. Phys. Commun.* **183** 746–773 (*Preprint* 1102.0273)
[26] Ryu S, Paquet J F, Shen C, Denicol G S, Schenke B, Jeon S and Gale C 2015 *Phys. Rev. Lett.* **115** 132301 (*Preprint* 1502.01675)

IOP Conf. Series: Journal of Physics: Conf. Series **832** (2017) 012055 doi:10.1088/1742-6596/832/1/012055

Measurements of azimuthal anisotropy for high p_T charged hadrons at $\sqrt{s_{NN}} = 200$GeV in Au+Au at RHIC-PHENIX

Akari Takeda, for the PHENIX collaboration

Graduate School of Humanities and Sciences, Nara Women's University, Kitauoya-Nishimachi, Nara, Nara, 630-8506, Japan

E-mail: paa_takeda@cc.nara-wu.ac.jp

Abstract. The azimuthal anisotropy of particle emission is a good observable to study the property of the quark gluon plasma (QGP), from the point of view of not only its collective dynamics but also the energy loss of hard scattered partons in it. We measured charged hadron v_2 as a function of p_T in $0.5 < p_T < 10\,\mathrm{GeV}/c$ for minimum bias and several centrality Au+Au collisions at $\sqrt{s_{NN}}$=200 GeV using the RHIC Year-2014 data, which significantly improved the statistical precision compared to the previous results. It was found that the v_2's for charged hadrons are consistently above zero up to $p_T = 10\,\mathrm{GeV}/c$ over the centralities. It was also found that there is a difference in v_2 between charged hadrons and neutral pions for p_T<7 GeV/c.

1. Introduction

It has been said that in a few microseconds after the Big Bang, there was a state called quark gluon plasma (QGP) where quarks and gluons are not bound in nucleons. In order to create the QGP and investigate its property on the ground, high energy heavy ion collisions have been utilized for decades. The relativistic heavy ion collider (RHIC) at the Brookhaven National Laboratory was constructed specifically for this purpose. RHIC succeeded to accelerate and collide the gold ions up to $\sqrt{s_{NN}} = 200\,\mathrm{GeV}$.

During the first one and half decades of its experimental running, there have been many positive results that support the successful creation of the QGP at RHIC. The results also suggest that the QGP created at RHIC is strongly coupled. Among many observables, the transverse momentum distribution (p_T) and the emission angle anisotropy of the high p_T hadrons have been of strong interest. The partons scattered with a high Q^2 by the initial hard scattering lose their energies in the hot and dense medium created in the collisions, and result in a strongly suppressed yield compared to the one in the $p + p$ collisions scaled by the number of binary nucleon-nucleon collisions. The high p_T hadrons are considered as leading particles of hard scattered partons, i.e., jets.

The energy loss mechanism can further be investigated in detail with the azimuthal anisotropy of the high p_T hadrons. For non-central collisions, the initial collision geometry becomes an elliptic shape. In this case, the hard scattered partons will pass through the QGP by different path-lengths depending on the azimuthal angles with respect to the reaction plane. This gives different suppression strength and therefore results in a different shape of p_T spectra especially at high p_T depending on the azimuthal angle. The effect will be most clearly observed in the coefficient of the second term (v_2) of Fourier expansion of azimuthal distribution with respect to the reaction plane.

In this paper, we show the latest PHENIX results on the high p_T charged hadron v_2 in Au+Au collisions at $\sqrt{s_{NN}}$=200 GeV using the data taken in the RHIC Year-2014 run.

2. PHENIX detectors

The PHENIX experiment [1] at RHIC consists of four spectrometer arms (two central and two muon arms) and a set of global detectors as shown in the left and middle panels of Fig.1. Each

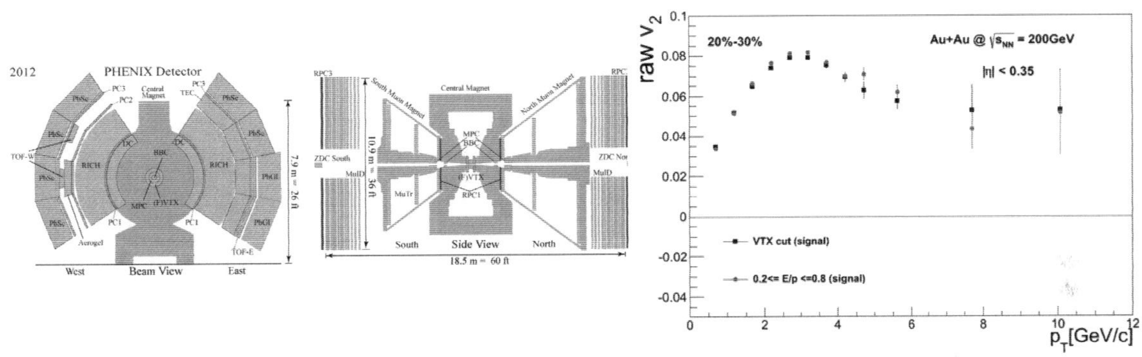

Figure 1. (left) Cross-sectional view of the PHENIX detector, (middle) top view of the PHENIX detector, and (right) raw v_2 for charged hadrons as a function of p_T in 20-30 % Au+Au collisions at $\sqrt{s_{NN}}$ = 200 GeV. Black points are the results with a DCA (VTX) cut and red points are with an E/p cut.

central arm covers the pseudorapidity range $\mid \eta \mid \leq 0.35$ and 90 degrees in azimuth. Charged particle tracks are reconstructed by the drift chamber (DC), and associated with the hits in the pad chamber (PC) and the electromagnetic calorimeter (EMCal). The silicon vertex tracker (VTX) installed in the RHIC Year-2011 run is used for further refining the track reconstruction and collision-vertex determination. The VTX consists of the two pixel detector layers followed by the two stripixel detector layers. The resolution of the distance of closest approach (DCA) of the VTX is 60μm at p_T >2 GeV/c [3]. The beam beam counters (BBC) in the global detector set are used to trigger events and determine the event centrality.

3. Analysis

The azimuthal angle distribution is expressed by the following equation:

$$\frac{dN}{d(\phi - \Psi)} \propto 1 + 2v_2 \cos\left[2(\phi - \Psi)\right], \tag{1}$$

where the ϕ is the azimuthal angle of particles, Ψ is the angle of the reaction plane, and the v_2 is second order anisotropy parameter. In this analysis, the eventplane was determined using the tower hits in the North and South BBC's located at 3.1< $|\eta|$ <3.9 [2]. We used the two-subevent method to calculate the resolution of the eventplane of the BBC as described below.

$$resolution = \sqrt{2\langle\cos\left[2(\Psi_{BBC\_North} - \Psi_{BBC\_South})\right]\rangle} \tag{2}$$

The true v_2 values are then obtained by the following equation:

$$v_2 = \frac{v_2^{measured}}{resolution} \tag{3}$$

One of the key analysis issues in the v_2 measurement at high p_T is to reduce background tracks and improve signal to background ratios. The main sources of the background (BG) are fake tracks produced in the DC, decay particles, and electrons from photon conversion. Of these source, the fake tracks contribute most under the high multiplicity in Au+Au collisions. The ratio of the fake tracks to genuine tracks increases as going to a higher p_T. We tried two different methods (so called E/p cut and DCA cut) to reduce the background. The E/p cut uses the energy measured by the EMCal (E) and the momentum (p) measured by the DC. The fake track will be measured with a certain momentum from the DC, but will be matched with an EMCal cluster that carries small or almost zero energy, resulting likely in a small E/p. Therefore, by applying E/p cut of $0.2 \leq E/p \leq 0.8$, we are able to remove most of the BG tracks. The DCA is defined as the closest distance between the collision vertex and the reconstructed track. Since the charged hadrons come from the collision vertex, the DCA should be small for a true track, while that for BG is large for a fake track. Therefore, BG can be removed by selecting the tracks with small DCAs (DCA cut). We compared the raw v_2 from the E/p cut and the DCA cut, and found that they give very consistent result, as shown in the right panel of Fig.1. The results presented in this paper employed the E/p cut because the statistical precision is better. The differences of v_2 from the two cuts are included in the systematic uncertainty of the results.

4. Results

The left panel of Fig.2 shows the charged hadron v_2 as a function of p_T in minimum bias Au+Au collisions at $\sqrt{s_{NN}} = 200$ GeV obtained from the data collected in the RHIC Year-2014 run. It can be seen that v_2 is consistently above zero up to p_T=10 GeV/c. The result implies that the charged hadrons are fragments of hard scattered partons that suffered the energy loss in QGP, and the path-length dependent energy loss is reflected to the yield difference between in-plane and out-of-plane.

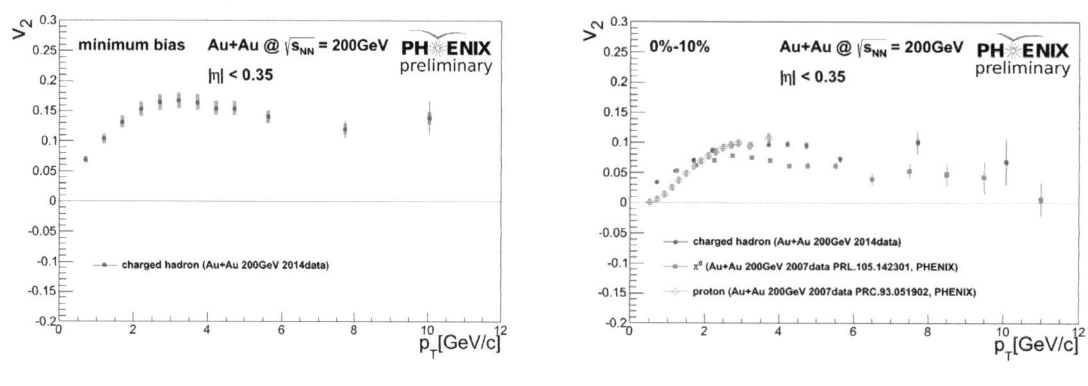

Figure 2. (left) v_2 of charged hadrons as a function of p_T in minimum bias Au+Au collisions at $\sqrt{s_{NN}} = 200$ GeV, and (right) that in 0-10 % centrality. Red points are from this work.

The right panel of Fig.2 shows the charged hadron v_2 in 0-10 % Au+Au collisions from the same dataset. The v_2 for 10-20 % and 20-30 % central collisions are shown in the Fig.3 in the same manner. In all of the four plots, red points are from this work, pink points are neutral

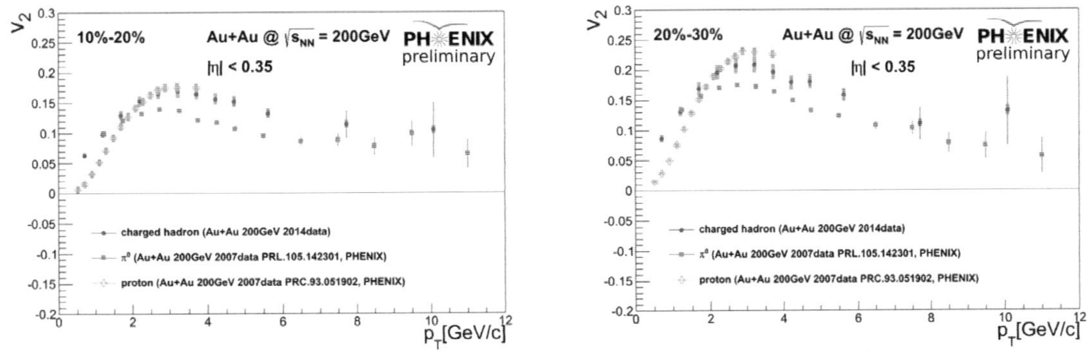

Figure 3. (left) v_2 of charged hadrons as a function of p_T in 10-20 % Au+Au collisions at $\sqrt{s_{NN}}$ = 200 GeV, and (right) that in 20-30 % centrality. Red points are from this work.

pions from the RHIC Year-2007 data [4], and light blue points are protons from the RHIC Year-2007 data [5]. There is a difference in v_2 between charged hadrons and neutral pions for $p_T <$ 7 GeV/c. The difference becomes smaller as going to more peripheral collisions. The trend is qualitatively consistent with the baryon anomaly phenomena which becomes prominent as going to central collisions. If the effect of the energy loss to pions and charged hadrons were the same and the energy loss were the only mechanism to create the v_2, their v_2 would also become same. However, the hypothesis is not supported by the measurement.

5. Summary

We measured charged hadron v_2 as a function of p_T in $0.5 < p_T < 10$ GeV/c for minimum bias and several centrality Au+Au collisions at $\sqrt{s_{NN}}$=200 GeV using the RHIC Year-2014 data, which significantly improved the statistical precision compared to previous analyses. It was found that the v_2's for charged hadrons are consistently above zero up to p_T = 10 GeV/c for the centralities shown in this paper. It was also found that there is a difference in v_2 between charged hadrons and neutral pions for p_T<7 GeV/c. Four times more data from the RHIC Year-2014 run are expected compared to the samples analyzed here. They are under the process of reconstruction and will become available in the very near future. We are planning to measure v_2 in more peripheral collisions and perform a detailed systematic study.

References

[1] K. Adcox et al. 2003 *Nucl. Inst. Meth.* **A499**, 469
[2] S. S. Adler et al. 2003 *Phys. Rev. Lett.* **91**, 182301
[3] A. Adare et al. 2016 *Phys. Rev.* **C93**, 034904
[4] A. Adare et al. 2010 *Phys. Rev. Lett.* **105**, 142301
[5] A. Adare et al. 2016 *Phys. Rev.* **C93**, 051902

A Blast Wave Model With Viscous Corrections

Z Yang and R J Fries

Cyclotron Institute and Department of Physics and Astronomy, Texas A&M University, College Station TX 77843, USA

E-mail: `zdyang@physics.tamu.edu`

Abstract. Hadronic observables in the final stage of heavy ion collision can be described well by fluid dynamics or blast wave parameterizations. We improve existing blast wave models by adding shear viscous corrections to the particle distributions in the Navier-Stokes approximation. The specific shear viscosity η/s of a hadron gas at the freeze-out temperature is a new parameter in this model. We extract the blast wave parameters with viscous corrections from experimental data which leads to constraints on the specific shear viscosity at kinetic freeze-out. Preliminary results show η/s is rather small.

1. Introduction

Heavy-ion collisions provide information about nuclear matter at high temperature and density. In the initial stage of heavy-ion collisions, a hot and dense fireball is created. The fireball cools rapidly and expands into the surrounding vacuum. This process continues until it reaches a final freeze-out stage when hadrons decouple from each other. Hadronic observables in the final stage of the heavy-ion collisions can be described well by fluid dynamics or blast wave parameterizations. In this contribution, we discuss an improved blast wave model by adding shear viscous corrections to the particle distributions in the Navier-Stokes approximation.

2. Viscous Corrections In Fluid Dynamics And Blast Wave

Fluid dynamics models the time evolution of the nuclear matter in heavy ion collisions, determined by initial conditions and the equation of state. When the system becomes dilute enough, kinetic freeze-out happens at a temperature $T = T_{\text{fo}}$. At the kinetic freeze-out fluid cells are translated into particles through the Cooper-Frye formula

$$E\frac{d^3N}{dp^3} = \frac{g}{(2\pi)^3}\int_\sigma f(p,T)p^\mu d\sigma_\mu\,. \tag{1}$$

Shear viscous corrections, and in particular the specific shear viscosity η/s will enter both the fluid dynamic equations of motion and the Cooper-Frye formula at freeze-out. Extractions of η/s from data using viscous fluid dynamics thus conflate the effects of shear viscosity on the dynamical evolution and on freeze-out. This is to some extent also true in calculations that switch to a hadronic transport model for freeze-out. Typical values extracted for η/s from fluid dynamics or fluid dynamics plus hadronic transport are $(1-2)/4\pi$ [1], averaged over all temperatures and including freeze-out in the averaging. Efforts are on the way to pin down the temperature evolution of η/s during the dynamical evolution [2]. In contrast, the blast wave model fits the flow field and shape of the fireball in the transverse plane. It becomes then sensitive to η/s at kinetic freeze-out only.

3. Viscous Blast Wave Parameterization

The hadron spectrum at freeze-out is given by Eq. (1) above where $f(p,T)$ is the distribution function of the given hadron species, g is its degeneracy and σ^μ is the normal vector on the freeze-out hypersurface. In equilibrium $f(p,T) = f_0(p,T)$ and with shear viscous corrections $f(p,T) = f_0(p,T) + \delta f$, where

$$f_0(p,T) = \frac{1}{e^{p\cdot u/T} \pm 1} \tag{2}$$

$$\delta f = \frac{1}{2}\frac{p_\mu p_\nu}{T^2}\frac{\pi^{\mu\nu}}{e+p} f_0(p,T) \tag{3}$$

and $\pi^{\mu\nu}$ is the shear stress tensor [3][4][5]. In the first-order (Navier-Stokes) approximation, $\pi^{\mu\nu}$ is given by

$$\pi^{\mu\nu} = 2\eta\langle\partial^\mu u^\nu\rangle \tag{4}$$

we plug Eq. (4) into Eq. (3) and find

$$\delta f = \frac{1}{T^3}\frac{\eta}{s} p_\mu p_\nu \langle\partial^\mu u^\nu\rangle f_0(p,T)\,. \tag{5}$$

We have neglected a Bose/Fermi enhancement/suppression factor since we will not work at very small momenta $p \lesssim T$. We see that the viscous correction at freeze out is directly proportional to η/s.

We choose constant $\tau = \sqrt{t^2 - z^2}$ as the freeze-out hypersurface. We follow [6] and parameterize the blast wave as follows. The coordinates are given by

$$x^\mu = (t,x,y,z) = (\tau\cosh\eta, \rho R_x\cos\phi, \rho R_y \sin\phi, \tau\sinh\eta) \tag{6}$$

where $\rho = \sqrt{\frac{x^2}{R_x^2} + \frac{y^2}{R_y^2}}$ is the normalized radial coordinate. The hadron momentum is

$$p^\mu = (m_T\cosh Y, p_T\cos\theta, p_T\sin\theta, m_T\sinh Y) \tag{7}$$

where $m_T = \sqrt{m^2 + p_T^2}$ is the transverse mass and Y is momentum rapidity. We will work with hadrons at $Y = 0$ here. The flow velocity can be written as [3][6]

$$u^\mu = (\cosh\eta_L\cosh\eta_T, \sinh\eta_T\cos\phi_b, \sinh\eta_T\sin\phi_b, \sinh\eta_L\cosh\eta_T) \tag{8}$$

where η_L is given by the longitudinal velocity $v_L = \tanh\eta_L$, η_T is given by the transverse velocity $v_T = \tanh\eta_T$. For longitudinal flow, we choose $v_L = \frac{z}{t}$, thus $\eta_L = \eta$, which enforces boost invariance. For transverse flow, we use the parameterization

$$v_T = \alpha\rho^n, \qquad \alpha = \alpha_0 + \alpha_2\cos 2\phi_b \tag{9}$$

where α is the surface velocity, α_0 is the average surface velocity, α_2 is an elliptic deformation of the flow field and n is a power term. Let us summarize the parameters in this blast wave: freeze-out temperature T, surface velocity α_0, velocity profile power n, velocity deformation α_2, ratio of event plane radii R_y/R_x and specific viscosity η/s.

We are left to calculate the Navier-Stokes shear stress tensor given our velocity field. In detail this tensor is $\pi^{\mu\nu} = 2\eta\langle\partial^\mu u^\nu\rangle$ where [4][5]

$$\langle\partial^\mu u^\nu\rangle \equiv [\frac{1}{2}(\Delta^\mu_\sigma\Delta^\tau_\nu) - \frac{1}{3}\Delta^{\mu\nu}\Delta_{\sigma\tau}]\partial^\sigma u^\tau, \qquad \Delta^{\mu\nu} = g^{\mu\nu} - u^\mu u^\nu\,. \tag{10}$$

The computation of the Navier-Stokes shear tensor is now straight forward but results in long expressions. We show one derivative as an example

$$\partial_2 u^1 - \frac{\partial u^1}{\partial y} = n \sinh \eta_{T'} \cosh^2 \eta_{T'} \frac{\sin\phi}{\rho R_y} \cos\phi_b - \sinh\eta_T \frac{\tan^2\phi_b}{(1+\tan^2\phi_b)^{3/2}} \frac{1}{\rho R_y \sin\phi}. \tag{11}$$

4. Extraction Of Freeze Out Shear Viscosity From Data

To extract the blast wave parameters from experimental data, we use a Bayesian package from the Models and Data Analysis Initiative (MADAI) project [2][7]. We first calculate training points in parameter space given a prior range for each parameter. The package then uses a Gaussian process emulator to estimate spectra at random parameter values. Finally it will do a likelihood analysis and give the maximum likelihood parameters.

The data used here is from the ALICE collaboration for Pb+Pb collisions at 2.76 TeV in the 40%-50% centrality bin [8][9]. Our fit ranges are 0.325-2.75 GeV/c, 0.225-1.95 GeV/c and 0.425-1.25 GeV/c for proton, kaon and pion respectively. We use both the transverse momentum spectra (without their absolute normalization) and elliptic flow v_2 to constrain the blast wave parameters. We set τ=9 fm/c, $R_x R_y$=12 fm^2 and assume $\partial_\tau u^\mu$=0.

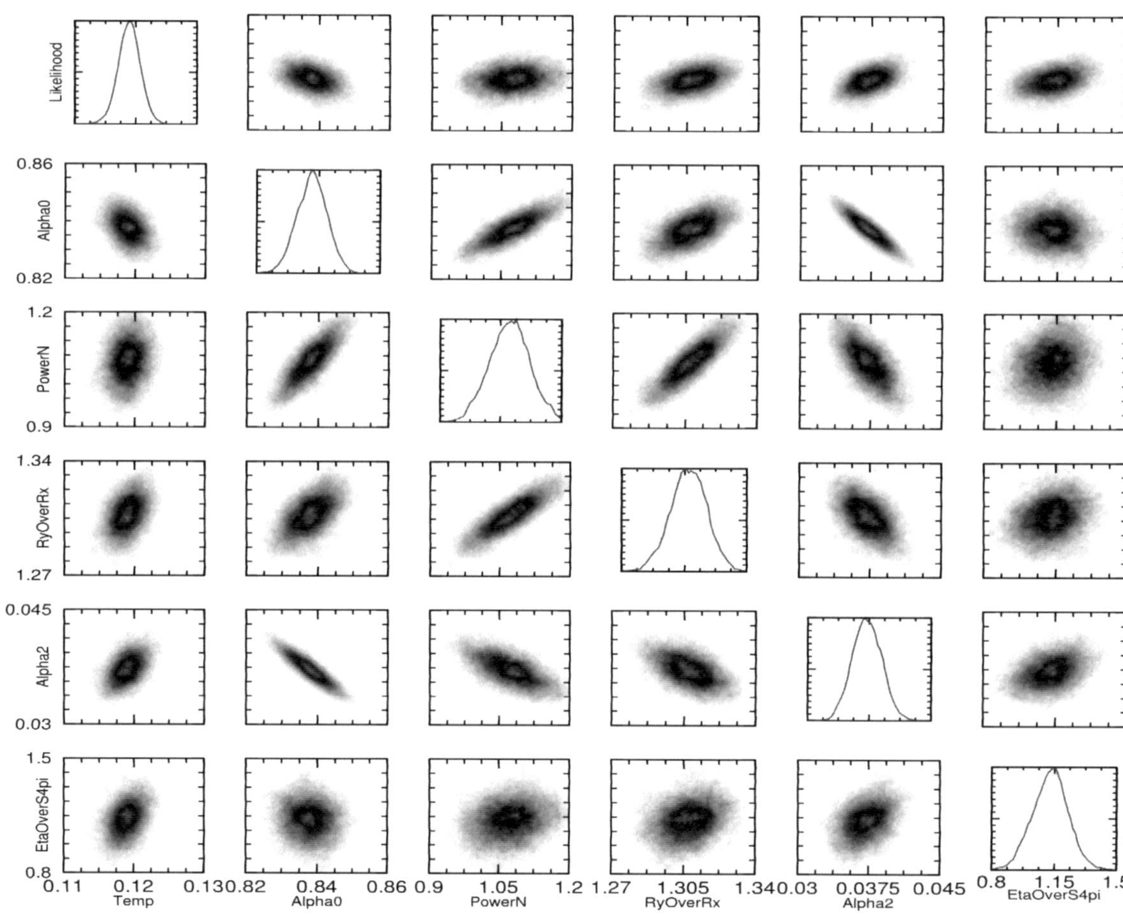

Figure 1. Likelihood analysis of our fit. The x and y axes show our chosen prior ranges for all parameters. The plots on the diagonal show likelihood distributions and the off diagonal plots show correlations.

IOP Conf. Series: Journal of Physics: Conf. Series **832** (2017) 012056 doi:10.1088/1742-6596/832/1/012056

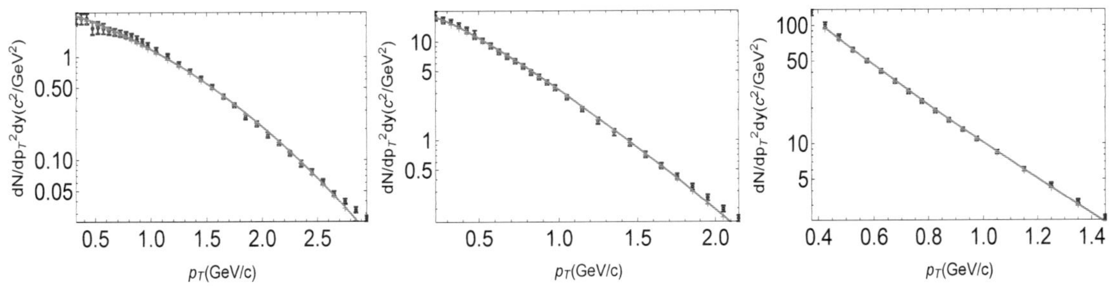

Figure 2. Transverse momentum spectra for protons (left), kaons (center) and pions (right). Solid line: viscous blast wave result with best fit parameters. Data: ALICE collaboration [8].

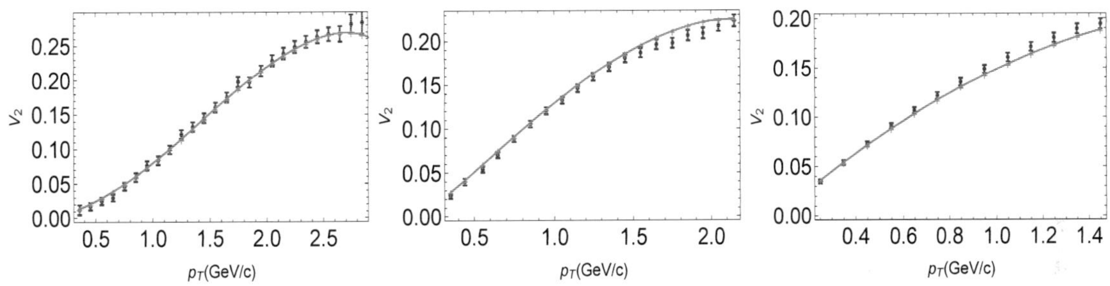

Figure 3. Same as Fig. 2 for elliptic flow v_2. Data: ALICE collaboration [9].

The likelihood analysis given by the MADAI package is shown in Fig. 1. The best parameter values are T=0.119 GeV, α_0=0.838c, n=1.07, R_y/R_x=1.31, α_2=0.0372c, η/s=1.13/4π. We can check the quality of our fit results by calculating the transverse momentum spectra, see Fig. 2 and elliptic flow v_2, see Fig. 3 for the best parameter values.

The p_T spectra with the best fit parameters describe the data rather well. Elliptic flow is generally described satisfactorily as well, but there is slight tension between the kaons and pions. This can be ameliorated by allowing pions to have a separate (lower) freeze-out temperature which improves the fits (not shown here) and leads to a very similar value of the specific shear viscosity.

5. Conclusions

We have developed a viscous blast wave model. We have used this model to extract η/s at kinetic freeze-out in a complimentary way. The preliminary result shows η/s is rather small. In the future, we will improve our statistical analysis and work on more centrality bins. We may also use the viscous blast wave for other projects like quark recombination.

References

[1] Gale C, Jeon S and Schenke B 2013 *Int. J. of Mod. Phys. A* **28** 1340011
[2] Bernhard J, Moreland J, Bass S, Liu J and Heinz U 2016 *Phys. Rev. C* **94** 024907
[3] Song H and Heinz U 2008 *Phys. Rev. C* **77** 064901
[4] Muronga A and Rischke D 2004 [arXiv:nucl-th/0407114]
[5] Dusling K, Moore G and Teaney D 2010 *Phys. Rev. C* **81** 034907
[6] Retiere F and Lise M 2004 *Phys. Rev. C* **70** 044907
[7] MADAI Collaboration 2013, https://madai-public.cs.unc.edu/. Accessed 20 June 2016.
[8] Adam J et al. ALICE Collaboration 2016 *Phys. Rev. C* **93** 034913
[9] Adam J et al. ALICE Collaboration 2015 *JHEP* **06** 190

IOP Conf. Series: Journal of Physics: Conf. Series **832** (2017) 012057 doi:10.1088/1742-6596/832/1/012057

The electric conductivity of a pion gas

J. Atchison and R. Rapp

Cyclotron Institute and Department of Physics & Astronomy, Texas A&M University, College Station, TX 77843-3366, USA

E-mail: jia502@physics.tamu.edu, rapp@comp.tamu.edu

Abstract. The determination of transport coefficients plays a central role in characterizing hot and dense nuclear matter. In the present work we calculate the electric conductivity of hot hadronic matter by extracting it from the ρ meson spectral function, as its zero-energy limit at vanishing momentum. Using hadronic many-body theory, we calculate the ρ meson self-energy in a pion gas. This requires the dressing of the pion propagators in the ρ self-energy with π-ρ loops, and the inclusion of vertex corrections to maintain gauge invariance. The resulting spectral function is used to calculate the electric conductivity of hot hadronic matter. In particular, we analyze the transport peak of the spectral function and extract its behavior with temperature and coupling strength. Our results suggest that, while obeying lower bounds proposed by conformal field theories in the strong-coupling limit, hot pion matter is a strongly-coupled medium.

1. Introduction

Transport coefficients are an important tool for characterizing hot nuclear matter. Recent calculations of the electric conductivity in hot hadronic matter have been performed, however, their results have varied considerably [1,2,3,4,5]. Furthermore, several of the calculations yield a result below a quantum lower bound proposed in Ref. [1]. Here, we perform a quantum many-body calculation, while attempting to maintain unitarity and gauge invariance.

Our calculation is rooted in dilepton emission calculations reviewed in Ref. [6]. The dilepton emission rate is proportional to the electromagnetic current correlator (Π_{EM}),

$$\frac{dR_{l+l-}}{d^4q} = \frac{-\alpha_{EM}^2}{\pi^3 M^2} f_B(q_0,T) Im[\Pi_{EM}(M,q,T,\mu_B)]. \tag{1}$$

Where f_B denotes the Bose distribution and $\alpha_{EM} = \frac{1}{137}$. One can readily extract the electric conductivity from Π_{EM} in the zero-momentum, low-energy limit using

$$\sigma(T) = -e^2 \lim_{q_0 \to 0} \frac{Im[\Pi_{EM}(q_0, \vec{q}=0,T)]}{q_0}. \tag{2}$$

Using weak-coupling techniques in a three-flavor quark-gluon plasma (QGP), it was found that $Im[\frac{\Pi_{EM}(q_0,\vec{q}=0,T)}{q_0}]$ displays a Lorentzian-like transport peak centered at zero energy [7]. For zero coupling the transport peak becomes a Dirac δ-function at zero energy, resulting in an infinite conductivity. As the coupling is increased the peak spreads out, allowing the extraction of a finite conductivity.

Within the vector meson dominance model, electromagnetic interactions in hadronic matter are mediated by the ρ, ω, and ϕ mesons, with the primary contribution coming from the ρ meson. Thus, one can relate Π_{EM} to the ρ propagator (D_ρ) via:

$$\Pi_{EM}(q,T) = \frac{m_\rho^4}{g_\rho^2} D_\rho(q,T) = \frac{m_\rho^4}{g_\rho^2} \frac{1}{q^2 - m_\rho^2 - \Sigma_\rho(q,T)}. \tag{3}$$

The ρ self-energy (Σ_ρ) characterizes the ρ's interaction with the medium. The contributions of thermal baryons to Σ_ρ were calculated for example, in Ref. [8]. However, in the low-T limit thermal baryon-antibaryon excitations are suppressed and the effects of pions will become dominant. Here, we calculate the effects of in-medium pions on Σ_ρ, requiring the determination of the relevant Feynman diagrams between thermal pions and the ρ meson.

In section 2 we calculate the contribution of in-medium pions to Σ_ρ. We calculate the vertex corrections required to maintain gauge invariance in section 3. In section 4 we discuss our preliminary results. We summarize and discuss future work in section 5.

2. Rho self-energy from in-medium pion cloud

In Ref. [9] it was found that there are two ρ self-energy diagrams required to maintain gauge invariance in vacuum, see the left and middle panels of Fig. 1. The pertinent self-energies are given by:

$$\Sigma_\rho^{\mu\nu}(q) = ig_\rho^2 \int \frac{d^4k}{(2\pi)^4} \frac{(2k+q)^\mu(2k+q)^\nu}{(k^2 - m_\pi^2 + i\epsilon)((k+q)^2 - m_\pi^2 + i\epsilon)} - 2ig_\rho^2 g^{\mu\nu} \int \frac{d^4k}{(2\pi)^4} \frac{1}{k^2 - m_\pi^2 + i\epsilon}, \tag{4}$$

where g_ρ is the $\rho\pi\pi$ coupling. We adopt the value of $g_\rho = 5.9$, as used in Ref. [8].

We now consider the effects of thermal pions on Σ_ρ. Resonant scattering of a cloud-pion with a thermal pion via an intermediate ρ modifies Σ_ρ (see the right panel of Fig. 1), requiring the dressing of the cloud pions' propagators with a self-energy given by:

$$\Sigma_\pi(k) = 2ig_\rho^2 \int \frac{d^4p}{(2\pi)^4} \frac{(k+p)_\mu(k+p)_\nu(-g^{\mu\nu} + \frac{(p-k)^\mu(p-k)^\nu}{m_\rho^2})}{(p^2 - m_\pi^2 + i\epsilon)((p-k)^2 - m_\rho^2 + i\epsilon)}. \tag{5}$$

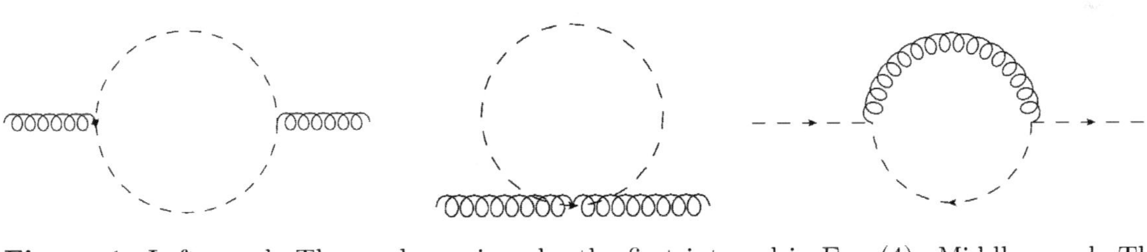

Figure 1. Left panel: The $\pi\pi$ loop given by the first integral in Eq. (4). Middle panel: The tadpole loop given by the second integral in Eq. (4). Right panel: The pion self-energy resulting from resonant scattering of a cloud-pion with a thermal pion through an intermediate ρ meson.

We resum the pion propagators in the ρ self-energy with Σ_π giving:

$$\begin{aligned}\Sigma_\rho^{\mu\nu}(q,T) = ig_\rho^2 \int \frac{d^4k}{(2\pi)^4} &\frac{(2k+q)^\mu(2k+q)^\nu}{(k^2 - m_\pi^2 - \Sigma_\pi(k,T))((k+q)^2 - m_\pi^2 - \Sigma_\pi(k+q,T))} \\ &-2ig_\rho^2 g^{\mu\nu} \int \frac{d^4k}{(2\pi)^4} \frac{1}{k^2 - m_\pi^2 - \Sigma_\pi(k,T)}.\end{aligned} \tag{6}$$

If the propagators are left undressed, an infinite conductivity will result from the calculation, due to the presence of non-interacting particles.

3. Vertex corrections

Dressing the pion propagators with Σ_π violates gauge invariance. This needs to be corrected by modifying the $\rho\pi\pi$ and $\rho\rho\pi\pi$ vertices, which amounts to considering the interactions of thermal pions with the vertices. We follow the lead of Refs. [8] and [10] to determine the required corrections. The calculation will be gauge invariant if the vertex corrections satisfy the following Ward-Takahashi identities:

$$q^\mu \Gamma^{(3)}_{\mu ab} = g_\rho \epsilon_{3ab}(-\Sigma_\pi(k+q) + \Sigma_\pi(k)) \tag{7}$$

$$q^\mu \Gamma^{(4)}_{\mu\nu ab} = i g_\rho(\epsilon_{3ca}\Gamma^{(3)}_{\nu bc}(k,-q) - \epsilon_{3bc}\Gamma^{(3)}_{\nu ca}(k+q,-q)), \tag{8}$$

where $\Gamma^{(3)}_{\mu ab}$ and $\Gamma^{(4)}_{\mu\nu ab}$ are the vertex corrections to the $\rho\pi\pi$ and $\rho\rho\pi\pi$ vertices, respectively.

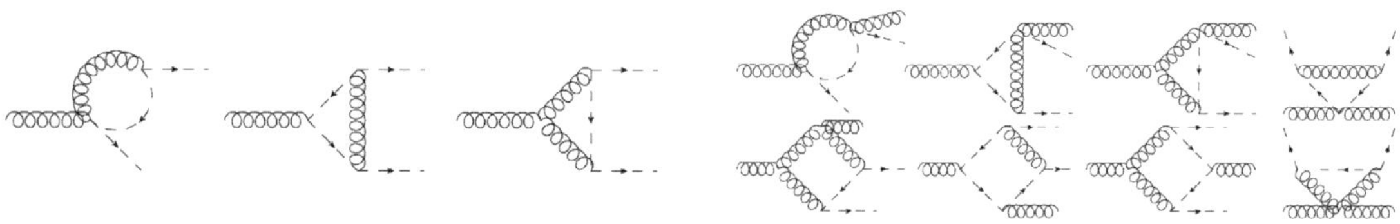

Figure 2. Left: Corrections to the $\rho\pi\pi$ vertex. Right: Corrections to the $\rho\rho\pi\pi$ vertex.

One must calculate all diagrams resulting from replacing a vacuum vertex with a corresponding vertex correction, see Fig. 2. The vertex corrections contain intermediate pion and ρ propagators, which must be dressed to extract a finite conductivity. We dress these propagators with widths, which are constant in energy and momentum. While introducing these widths also entails a violation of gauge invariance, it is controlled through a suppression by a thermal ρ mass.

4. Results

We have not completed the calculation of all the vertex corrections at this point. However, the majority of the corrections have been calculated, and we do not expect our final results to largely change upon completion.

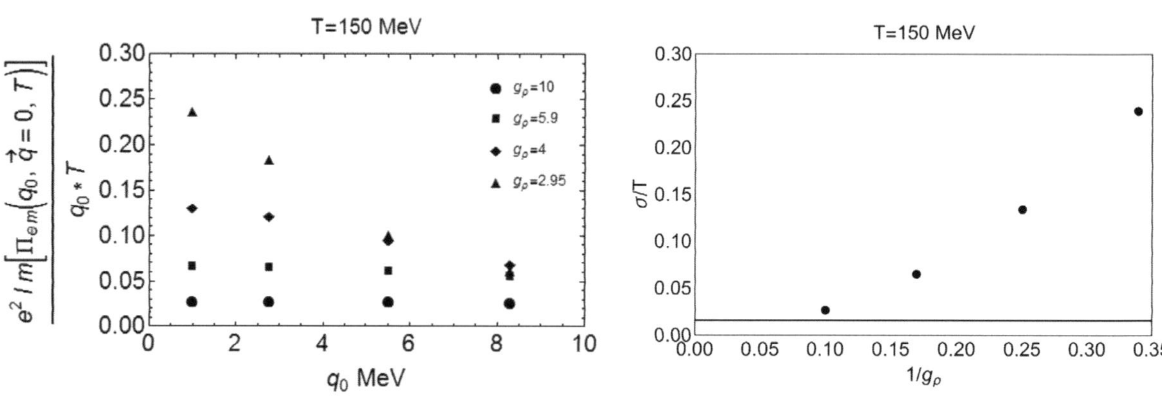

Figure 3. Left panel: The energy-scaled EM spectral function, σ/T can be extracted from the zero crossing. Values for several g_ρ's are displayed. Right panel: σ/T vs $1/g_\rho$, allowing one to examine the strong-coupling limit of the calculation. The solid line shows the proposed lower bound from Ref. [1].

Our results show that by $T = 150$ MeV the Lorentzian peak in $Im[\frac{\Pi_{EM}(q_0,\vec{q}=0,T)}{q_0}]$ has flattened out, see the left panel of Fig. 3. As we decrease g_ρ we see a re-emergence of the Lorentzian peak, suggestive of a weakly-coupled system with a well-defined conductivity peak, not unlike the one in the perturbative QCD calculation of Ref. [7]. Our findings thus suggest that the pion gas is a strongly coupled medium. The right panel of Fig. 3 shows that as the coupling increases our calculation may approach a lower limit compatible with the quantum bound.

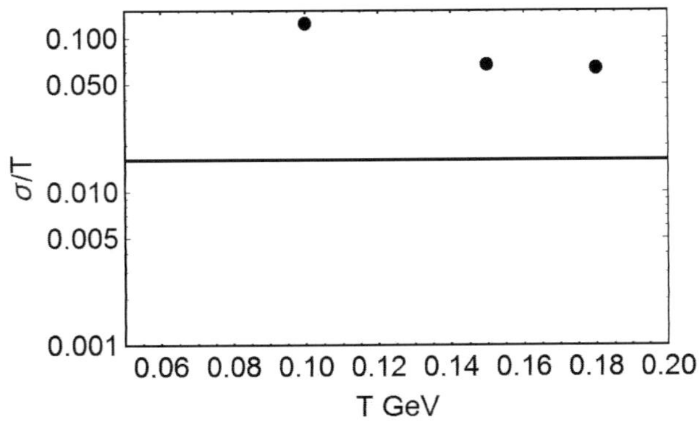

Figure 4. Results for the conductivity of a pion gas as a function of T. The solid line shows the proposed lower bound from Ref. [1].

Figure 4 shows our results for σ/T as a function of T for the nominal value of $g_\rho = 5.9$. Our results are significantly larger than previous calculations for a pion gas compiled in Ref. [11]. However, they respect the quantum bound proposed in Ref. [1].

5. Summary and future work

Our results suggest that a.) pion matter is a strongly coupled medium b.) the conductivity obeys a proposed quantum lower bound. Once we have completed the calculation of the spectral function at zero momentum, we will implement the baryonic effects calculated in Ref. [8] into our calculation. We will then extend the calculation to finite momentum and use the results to calculate corrections to dilepton emission rates of hot hadronic matter.

Acknowledgment

This work is supported by the US-NSF under grant no. PHY-1614484.

References

[1] S. C. Huot, P. Kovtun, G. D. Moore, A. Starinets, and L. G. Yaffe, JHEP 2006, 015 (2006).
[2] D. Fernández-Fraile and A. Gomez Nicola, Phys. Rev. D 73, 045025 (2006).
[3] S. I. Finazzo and J. Noronha, Phys. Rev. D 89, 106008 (2014).
[4] G. Aarts, C. Allton, A. Amato, P. Giudice, S. Hands, and J.-I. Skullerud, JHEP 2015, 186 (2015).
[5] A. Amato, G. Aarts, C. Allton, P. Giudice, S. Hands, and J.-I. Skullerud, Phys. Rev. Lett. 111, 172001 (2013).
[6] R. Rapp, J. Wambach and H. van Hees, Landolt-Börnstein 23, 134 (2010).
[7] G. Moore and J. Robert, arXiv:hep-ph/0607172v1 [nucl-th].
[8] M. Urban, M. Buballa, R. Rapp, and J. Wambach, Nucl. Phys. A 673, 357 (2000).
[9] M. Herrman, B. L. Friman and W. Nörenberg, Nucl. Phys. A 560, 411 (1993).
[10] C. Song, and V. Koch, Phys. Rev. C 54, 3218 (1996).
[11] M. Greif, C. Greiner and G. Denicol Phys. Rev. D 93, 096012 (2016).

IOP Conf. Series: Journal of Physics: Conf. Series **832** (2017) 012058 doi:10.1088/1742-6596/832/1/012058

Achieving high baryon densities in the fragmentation regions in heavy ion collisions at top RHIC energy

Ming Li and Joseph I. Kapusta

School of Physics and Astronomy, University of Minneosota, Minneapolis, MN 55455, USA

E-mail: `ml@physics.umn.edu`, `kapusta@physics.umn.edu`

Abstract. Heavy ion collisions at extremely high energy, such as the top energy at RHIC, exhibit the property of transparency where there is a clear separation between the almost net-baryon-free central rapidity region and the net-baryon-rich fragmentation region. We calculate the net-baryon rapidity loss and the nuclear excitation energy using the energy-momentum tensor obtained from the McLerran-Venugopalan model. Nuclear compression during the collision is further estimated using a simple space-time picture. The results show that extremely high baryon densities, about twenty times larger than the normal nuclear density, can be achieved in the fragmentation regions.

1. Introduction

In 1980 Anishetty, Koehler and McLerran [1] studied the fragmentation regions in heavy ion collisions at extremely high energy. They argued for the possible formation of quark-gluon plasma in these regions with an energy density of about 2 GeV/fm^3 enhanced by a nuclear compression factor of 3.5. However, after Bjorken's work in 1983 [2], people's interests have been mainly focused on the central rapidity region where even larger energy density could be achieved. This nearly net-baryon-free region is also easy to be probed experimentally. The fragmentation regions are only occasionally investaged since then [3, 4, 5, 6, 7]. We report our work on the high baryon densities in the fragmentation regions. To be specific, we use the glasma energy-momentum tensor obtained in the McLerran-Venugopalan model [8] to calculate the baryon rapidity loss and the nuclear excitation energy. We then use the same space-time picture of collision in [1] to estimate the nuclear compression factors and obtain the high baryon densities distributed in the fragmentation regions.

2. Baryon Rapidity Loss and Nuclear Excitation Energy

We consider central collisions of Au-Au nuclei at top RHIC energy $\sqrt{s_{NN}} = 200\,\mathrm{GeV}$. We study the simplified 1+1 dimensional motion with transverse motion ignored. Approximate the nucleus as a collection of slabs with unit surface area locating at different $\mathbf{r}_\perp$. Taking one pair of the colliding slabs as an example, the four-momentum of the projectile slab is $\mathcal{P}_P^\mu = (\mathcal{E}_P, 0, 0, \mathcal{P}_P)$. As the slabs recede, the loss of the energy and momentum from the slabs should be equal to the increase of the energy and momentum of the glasma in the regions swept by the slabs. The differential energy momentum conservation should be satisfied $d\mathcal{P}_P^\mu = -T^{\mu\nu}_{\text{glasma}} d\Sigma_\nu$ where $d\Sigma_\nu = (dz, 0, 0, -dt)$ is the differential hypersurface vector.

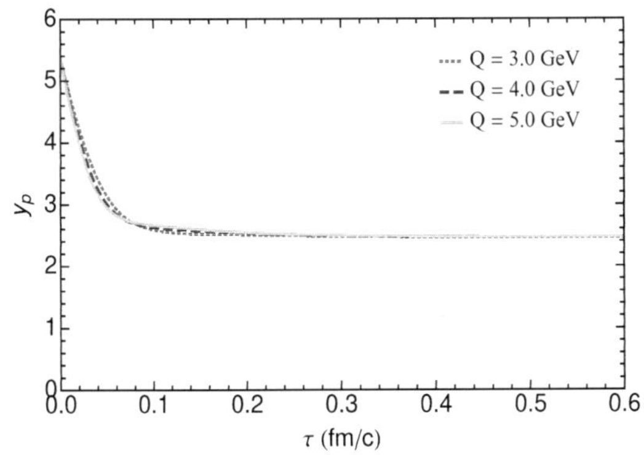

Figure 1. Projectile rapidity as a function of proper time for the central core slab of the nucleus after the collisions viewed in the center-of-mass frame.

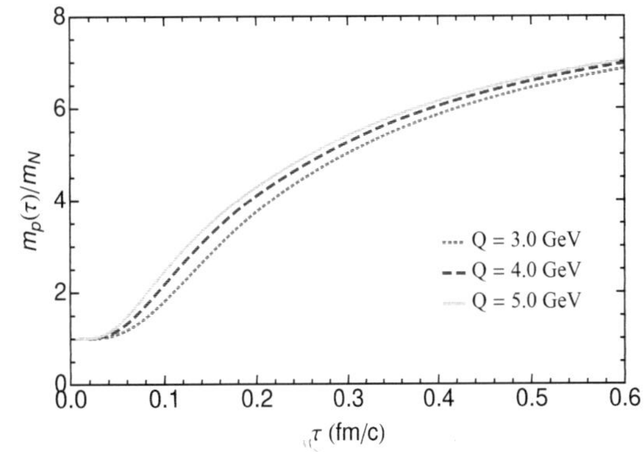

Figure 2. Time evolution of the nuclear excitation energy scaled by the nucleon rest mass for the central core of the nucleus.

The energy-momentum of the glasma has been calculated analytically using a small-τ power series expansion method in [9, 10]:

$$T^{\mu\nu}_{\text{glasma}} = \begin{pmatrix} \mathcal{A}+\mathcal{B}\cosh 2\eta & 0 & 0 & \mathcal{B}\sinh 2\eta \\ 0 & \mathcal{A} & 0 & 0 \\ 0 & 0 & \mathcal{A} & 0 \\ \mathcal{B}\sinh 2\eta & 0 & 0 & -\mathcal{A}+\mathcal{B}\cosh 2\eta \end{pmatrix}. \quad (1)$$

Notice that the glasma energy-momentum tensor is different from the energy-momentum tensor in the conventional string models [5] in the sense that off diagonal terms are nonvanishing. This is due to contributions from the transverse color electric and magnetic fields represented by $\mathcal{B}$. Using the relations $v_p = dz_p/dt = \tanh y_p$, $\mathcal{E}_P = \mathcal{M}_P \cosh y_p$ and $\mathcal{P}_P = \mathcal{M}_P \sinh y_P$, the two energy-momentun conservation equations are converted to the time evolution equations for $\mathcal{M}_P(\tau)$ and $y_P(\tau)$. To solve these equations, we need to specify the momentum space cut-off scale Q and the initial energy density distribution $\varepsilon(r_\perp, \tau = 0)$. The former is treated as a varying parameter ranging from 3 GeV to 5 GeV. The latter is determined by first matching the energy density at the central region to $\varepsilon(r_\perp = 0, \tau = 0.6 \text{ fm/c}) = 30 \text{ GeV/fm}^3$ [11] that is used as initial condition for the viscous hydrodynamics. Then we assume $\varepsilon(r_\perp)/\varepsilon(r_\perp = 0) \sim (T_A(r_\perp)/T_A(r_\perp = 0))^2$ at $\tau = 0$ with the nuclear thickness function $T_A(r_\perp) = \int_{-\infty}^{+\infty} dz\, \rho_A(r_\perp, z)$ and ρ_A the Woods-Saxon distribution for nuclear density.

From Figure 1, the rapidity for the central core of a gold nucleus decreases from the initial value 5.36 to about 2.5 within 0.1 to 0.2 fm/c. The final net rapidity loss does not depend on the momentum space cut-off scale Q. We can compare our calculations using the BRAHMS result [12] of average rapidity loss about $2.05 + 0.4/-0.6$ for 0-10% centrality class. Our calculation predicts an average rapidity loss at $\tau = 0.6$ fm/c to be about 2.4. We would expect the subsequent hydrodynamic evolutions and baryon diffusions only change the shape of net baryon distribution while leaving the average baryon rapdity loss unchanged. Therefore, our estimation of the baryon rapidity loss are consistent with the experimental results. For the nuclear excitation energy, Figure 2 exhibits a slow but monotonic increase. This is due to the dominant roles played by the transverse components of the color electric and magnetic fields after the longitudinal fields saturate the baryon rapidity loss.

IOP Conf. Series: Journal of Physics: Conf. Series **832** (2017) 012058 doi:10.1088/1742-6596/832/1/012058

3. High Baryon Densities

In the McLerran-Venugopalan model, the nucleus is assumed to propagate at the speed of light both before and after the collisions. Therefore the nucleus is infinitely contracted in the center-of-mass frame but not contracted in its own local rest frame. To estimate the nuclear compression in its own rest frame after the collisions, we resort to the simple space-time model of collisions used in [1]. In the local rest frame of the baryonic fireball after collisions, the baryon density is

$$n_B(\vec{r}_\perp, z') = e^{\Delta y(\vec{r}_\perp)} \rho_A(\vec{r}_\perp, e^{\Delta y(\vec{r}_\perp)} z'). \tag{2}$$

where $z' = z - z_P(r_\perp)$ is the longitudinal distance in the local rest frame and $\Delta y > 0$ is the rapidity change. The Δy is a Lorentz invariant quantity and can be calculated explicitly in the center-of-mass frame as shown in Figure 1. The $\rho_A(\vec{r}_\perp, z)$ is the Woods-Saxon distribution for nuclear density.

Figure 3 shows the proper baryon density distribution. The contours are drawn at $n_B = 3, 2, 1, 0.5$, and 0.15 baryons/fm^3. This plot is constructed in the local rest frames of nulcear fragments which depend on different transverse coordinate $\mathbf{r}_\perp$. Note that the maximum baryon density can be achieved is about 3 baryons/fm^3, which is about 20 times larger than the normal atomic nuclear density of 0.155 nucleons/fm^3. Moreover, the longitudinal extension of the high baryon density region is about 1 fm indicating that a finite volume region of high baryon densities can be formed. Figure 4 shows the proper baryon density against the space-time rapidity η. This is the shape of the net-baryon charge distribution at $\tau = 0.6$ fm/c, which should be the initial distribution for the subsequent dynamical evolution. The central core region experiences the largest rapidity loss, therefore the largest baryon density and the smallest space-time rapidity. On the other hand, the peripheral regions of the nucleus propagate with less rapidity loss, thus smaller baryon densities and larger space-time rapidities.

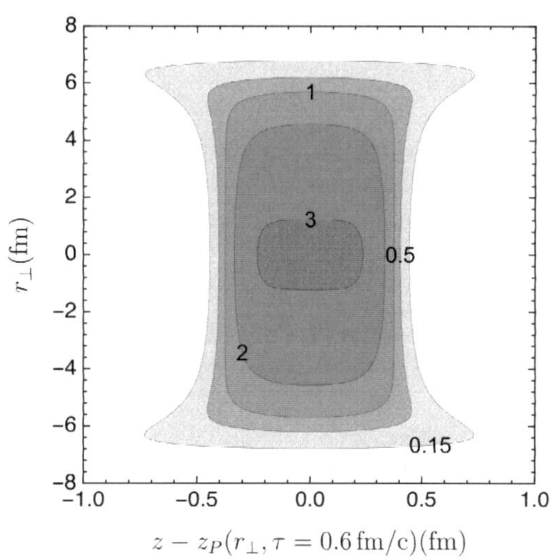

Figure 3. Contour plot of the proper baryon density in the local rest frames of nuclear fragments. The units are baryons per fm^3. The horizontal axis measures the distance in the local rest frame.

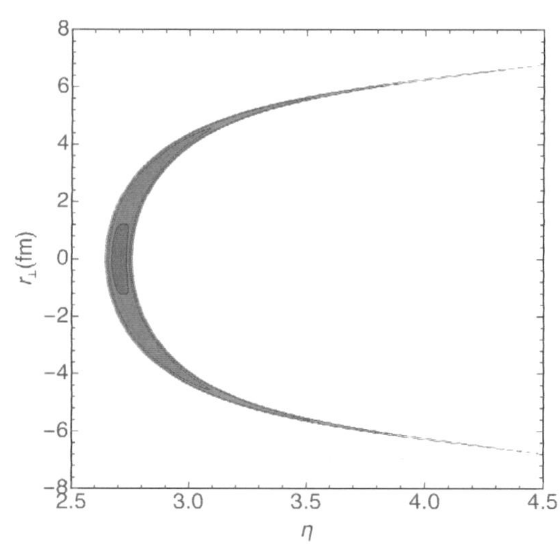

Figure 4. Contour plot of the proper baryon density against space-time rapidity η at $\tau = 0.6$ fm/c. The units are baryons per fm^3. The horizontal axis is the space-time rapidity.

The local energy density distribution is readily computed with the help of the nuclear excitation energy at $\tau = 0.6$ fm/c as exampled by Figure 2. However, one has to keep in mind that the energy density obtained could be overestimated due to the monotonic increasing feature of the nuclear excitation energy. For the central core region $r_\perp = 0$, the net baryon density is $n_B \simeq 3.0$ baryons/fm^3 and the corresponding energy density is $\varepsilon_B \simeq 20$ GeV/fm^3. For the peripheral region, let us pick $r_\perp = 5.25$ fm, the baryon density is $n_B = 1.5$ baryons/fm^3 and the energy density is $\varepsilon_B = 5.5$ GeV/fm^3. With the high baryon densities and large energy densities obtained, one is tempted to ask what the temperatures and baryon chemical potentials would be if the high baryon density matter in the fragmentation regions are thermalized. Let us approximate the high baryon density matter as a system of up, down and strange quarks and gluons. We can further assume that they are massless and noninteracting due to the large baryon density and energy density. Net strangeness is adjusted to be zero. The equation of state thus has the form $P(T, \mu_B) = \frac{19\pi^2}{36}T^4 + \frac{1}{9}T^2\mu_B^2 + \frac{1}{162\pi^2}\mu_B^4$. Typical values for the temperatures and baryon chemical potentials are $T = 299$ MeV, $\mu_{\rm B} = 1061$ MeV for $r_\perp = 0$ and $T = 205$ MeV, $\mu_{\rm B} = 1007$ MeV for $r_\perp = 5.25$ fm. The baryon chemical potentials are much larger than the temperatures indicating a very different region in the QCD phase diagram as compared to the matter created in the central rapidity region at top RHIC energy. In addition, we obtain the entropy per baryon ratio $s/n_B = 26.2$ and $s/n_B = 18.9$, respectively. If the following hydrodynamic expansions in the fragmentation regions are approximately adiabatic like what happened in the central rapidity region, these values would be relevant to the study of QCD phase transitions [13].

4. Conclusions

In conclusion, we used the glasma energy-momentum tensor from the McLerran-Venugopalan model to calculate both the baryon rapidity loss and the nuclear excitation energy in high energy heavy ion collisions. We obtained high baryon densities in the fragmentation regions with the largest value being 20 times larger than the normal atomic nuclear density. The initial distributions for the high baryon density and energy density in the fragmentation regions should be able to provide initial conditions for the subsequent hydrodynamic evolutions in these regions. It could also be relevant to the study of QCD phase diagram and the searching for the critical point of the phase transitions. Scanning through the large rapidity regions in high energy collisions would be equivalent to the conventional low energy beam energy scanning.

Acknowledgments

This work was supported by the U.S. Department of Energy grant DE-FG02-87ER40328.

References

[1] Anishetty R, Koehler P and McLerran L 1980 *Phys. Rev.* D **22** 2793
[2] Bjorken J D 1983 *Phys. Rev.* D **27** 140
[3] Csernai L P 1984 *Phys. Rev.* D **29** 1945
[4] Gyulassy M and Csernai L P 1986 *Nucl. Phys.* A **460** 723
[5] Mishustin I N and Kapusta J I 2002 *Phys. Rev. Lett.* **88** 112501
[6] Frankfurt L and Strikman M 2003 *Phys. Rev. Lett.* **91** 022301
Strikman M 2006 *AIP Conf. Proc.* **842** 92
[7] Mishustin I N and Lyakhov K A 2007 *Phys. Rev.* C **76** 011603.
[8] McLerran L D and Venugopalan R 1994 *Phys. Rev.* D **49** 3352
McLerran L D and Venugopalan R 1994 *Phys. Rev.* D **49** 2233
[9] Chen G, Fries R J, Kapusta J I and Li Y 2015 *Phys. Rev.* C **92** 064912
[10] Li M and Kapusta J I 2016 *Phys. Rev.* C **94** 024912.
[11] Song H, Bass S A and Heinz U W 2011 *Phys. Rev.* C **83** 024912
[12] Bearden I G *et al.* 2004 *Phys. Rev. Lett.* **93** 102301.
[13] Asakawa A, Bass S A, Muller B and Nonaka C 2008 *Phys. Rev. Lett.* **101** 122302

IOP Conf. Series: Journal of Physics: Conf. Series **832** (2017) 012059 doi:10.1088/1742-6596/832/1/012059

Kaon femtoscopy at the STAR experiment

Jindřich Lidrych for the STAR Collaboration
Department of Physics, FNSPE Czech Technical University in Prague, Břehová 7, Prague 115 19, Czech Republic

E-mail: jlidrych@gmail.com

Abstract. The properties of the quark-gluon plasma have been extensively studied in high-energy nuclear collisions at RHIC. Femtoscopic measurements of two-particle correlations at small relative momenta reveal the space-time characteristics of the system at the moment of particle emission. In comparison to analyses using the most abundant pions, like-sign kaons provide a cleaner probe of the emission source as they less frequently result from resonance weak decays. Additionally, kaons contain strange quarks so these measurements can be sensitive to different effects and earlier collision stages. Pairs of like-sign kaons exhibit correlations due to Coulomb interactions and Bose-Einstein quantum statistics. The system of unlike-sign kaons contains a narrow $\phi(1020)$ resonance in the final state. Femtoscopic measurements have been predicted to be particularly sensitive to the source size and momentum-space correlations in the region of this resonance.

In this proceedings, we present the STAR preliminary results on the like-sign and unlike-sign kaon femtoscopic correlation functions in 200 GeV Au+Au collisions at RHIC. Collision centrality and the transverse pair momentum k_T dependence of the radius parameters will be discussed. The results from unlike-sign kaon correlation functions will be compared with model predictions [1].

1. Introduction

Two-particle interferometry is a unique method which allows us to study spatial and temporal characteristics of a system created during the ultrarelativistic heavy-ion collisions at the moment of particle emission. This technique was for the first time used by G. Goldhaber [2] and his collaborations in study of angular distribution of identified pion pairs in proton-anti-proton annihilation in 1960. Since then the femtoscopic measurements of two-particle correlations at the low relative momenta became a standard tool for the extraction of the source sizes. Based on the typical extents which are of the order of tens of femtometers the two-particle interferometry method is called femtoscopy, as introduced by Richard Lednický [3]. Nowadays femtoscopic analyses span many particle species and include even non-identical interacting particle [4, 5, 6].

2. Two-particle correlation femtoscopy

The two-particle correlation function used in femtoscopy relates the space-time structure of the particle emitting source and the particle interaction via:

$$CF\left(p_1, p_2\right) = \int \mathrm{d}^3 r S\left(r, k^*\right) \left|\psi_{1,2}\left(r, k^*\right)\right|^2, \tag{1}$$

where $S\left(r, k^*\right)$ is the source function and $\psi_{1,2}\left(r, k^*\right)$ is the relative two particle wave function. The source function describes the emission of two particles at the relative distance r with the

IOP Conf. Series: Journal of Physics: Conf. Series **832** (2017) 012059 doi:10.1088/1742-6596/832/1/012059

relative momentum $(2k^*)^2 = -(p_1^\mu - p_2^\mu)^2 = q_{inv}^2$, where p_1^μ and p_2^μ are the four-momenta of the first and second particle, respectively.

Experimental three-dimensional correlation functions are constructed in the so-called "out-side-long" system, where the $\vec{q}$ is decomposed into three components: q_{out}, q_{side} and q_{long} [7]. In order to extract source radii R_{out}, R_{side} and R_{long}, the experimental correlation function for identical particles, where the correlation is caused by the quantum statistics and the Coulomb interaction, is fitted by the standard Bowler-Sinyukov form [8]:

$$CF\left(q_{out}, q_{side}, q_{long}\right) = N\left[(1-\lambda) + \lambda K\left(q_{inv}\right)\left(1 + e^{-R_{out}^2 q_{out}^2 - R_{side}^2 q_{side}^2 - R_{long}^2 q_{long}^2}\right)\right], \quad (2)$$

where $K(q_{inv})$ is the Coulomb function integrated over the source of size R_{inv}, N is the normalization and λ is the correlation strength. This approximation is based on the separation of the Coulomb interaction from the effect of the quantum statistics.

3. STAR Experiment

The results presented in this paper were obtained from Au+Au collisions at $\sqrt{s_{\rm NN}} = 200$ GeV measured by the Solenoidal Tracker at RHIC (STAR)[9] in 2011. The STAR is a multipurpose detector which excels in tracking and identification of charged particles at mid-rapidity with full azimuthal coverage which makes it ideal for multi-particle high statistics measurements. The most important subdetectors for kaon identification are the Time Projection Chamber (TPC)[9] and the Time of Flight detector (ToF)[10] which was fully installed in 2010.

4. Like-sign kaon femtoscopy

In comparison with the previous STAR analysis [11], thanks to higher statistics, it is now possible to construct experimental like-sign kaon correlation functions for more bins in centrality and transverse pair momentum $k_T = (p_1 + p_2)_T/2$. The presented analysis is performed for 4 centrality bins and 4 k_T bins. Measured correlation functions are corrected for kaon misidentification and momentum resolution. Figure 1 shows the STAR preliminary results on femtoscopic radii from like-sign kaon correlation functions from Au+Au collisions at $\sqrt{s_{\rm NN}} = 200$ GeV. The typical k_T and centrality dependence can be observed. While the transverse flow and expansion of the system are encoded in the falling of R_{out} and R_{side} with increasing k_T, the longitudinal flow exhibits a decrease of R_{long} with the increasing k_T. Such a behavior of source radii R_{out}, R_{side} and R_{long} confirms our expectation that the system under study undergoes space-time evolution and expands.

5. Unlike-sign kaon femtoscopy

In the previous section we have presented the results on measurements with identical charged kaons where the correlation is caused by Bose-Einstein quantum statistics and Coulomb interaction. The structure of the correlation function for unlike-sign kaons is more varied (see Figure 2). While at the low relative momenta the attractive Coulomb interaction and the strong interaction in s-wave can be observed, at higher $q_{inv} \sim 256$ MeV/c the strong interaction in p-wave through $\phi(1020)$ is presented. The measurement of space-time extents using unlike-sign kaon correlations can be a complementary study which should provide an additional information about the studied system. It has been predicted [1] that the correlation function will be sensitive to the space-time extents in the region of resonance in a system where the narrow, near-threshold, resonance is present. The height of peak should then scale as the inverse volume of the system i.e. as $\sim r^{-3}$ [1]. Figure 2 introduces the STAR preliminary results on unlike-sign kaon correlation functions from Au+Au collision at $\sqrt{s_{\rm NN}} = 200$ GeV. The left panel of Figure 2 shows the correlation function for 5 different centralities. The sensitivity on the transverse

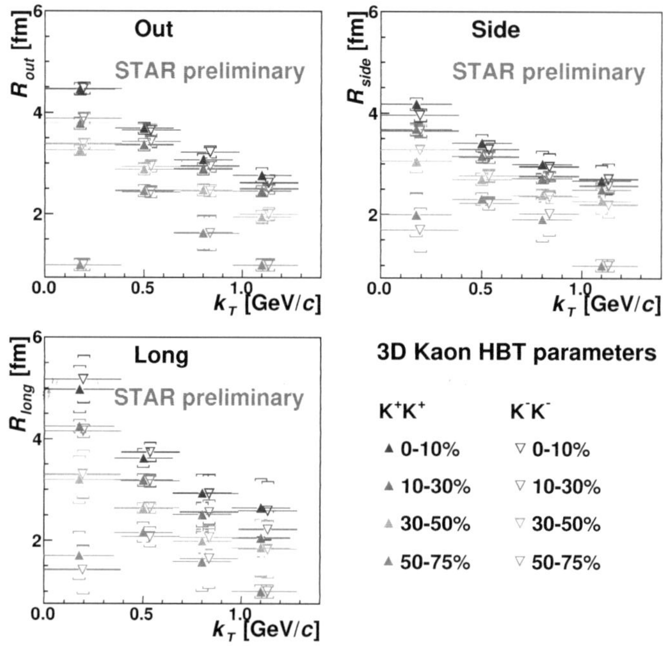

Figure 1. Fit results of three-dimensional K^+K^+ and K^-K^- correlation functions.

pair momentum k_T is presented in the right panel of Figure 2. In both cases it can be seen that the correlation function is sensitive to the source size. Especially in the region of the resonance a strong sensitivity can be observed. The experimentally measured and corrected correlation

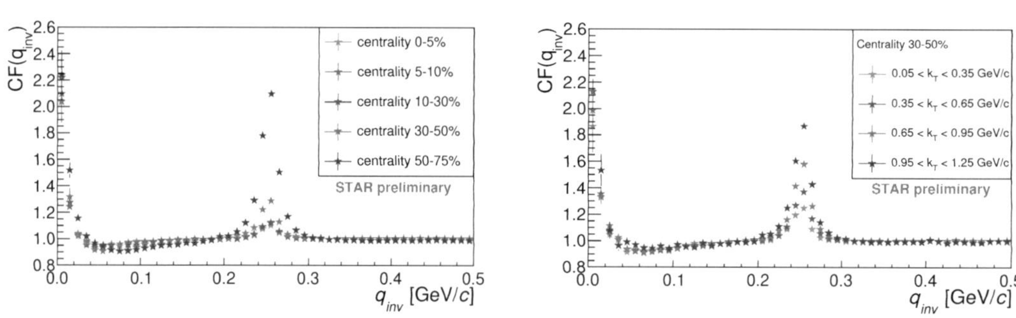

Figure 2. Left: Centrality dependence of one-dimensional K^+K^- correlation function from Au+Au collisions at $\sqrt{s_{\rm NN}} = 200$ GeV. Right: k_T dependence of one-dimensional K^+K^- correlation function for centrality 30-50% from Au+Au collisions at $\sqrt{s_{\rm NN}} = 200$ GeV.

functions of unlike-sign kaons can be compared to the theoretical prediction. In this analysis the Lednický model [1] of final state interaction is examined. Lednický model includes the treatment of the $\phi(1020)$ resonance in the final state. This model is used for the calculation of the two-particle wave function $\psi(r, k^*)$ which is included in the theoretical correlation function (see Eq. 1). The source function $S(r, k^*)$ in Eq. 1 is parameterized by the Gaussian and its size is equal to the source size obtained from the one-dimensional like-sign kaon correlation function. For the correct comparison which takes into the account the effects presented during the construction of the experimental correlation function, the theoretical function is scaled by the λ parameter via relation $CF = \left(CF^{theo} - 1\right)\lambda + 1$. Figure 3 shows the comparison of the experimental unlike-sign kaon correlation functions from Au+Au collisions at $\sqrt{s_{NN}}$=200 GeV to the theoretical calculations for the most central (0-5%) and peripheral (50-75%) collisions for $0.35 < k_T < 0.65$ GeV/c. The Lednický model is able to reproduce the overall shape of the correlation function especially for central collisions. However, the model underestimates the correlation strength in the region of the resonance for the peripheral collisions. The observed behavior can possibly be interpreted as a breakdown of the femtoscopic formalism which was developed for measurements at the very low relative momenta.

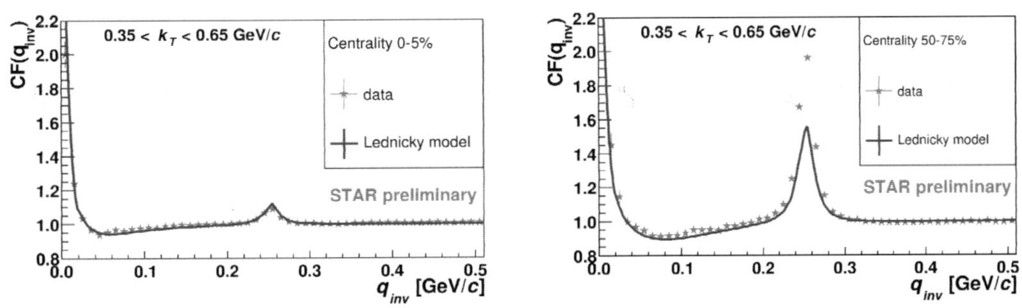

Figure 3. Comparison of experimental K^+K^- correlation functions from Au+Au collisions at $\sqrt{s_{NN}} = 200$ GeV to theoretical calculations for 2 different centralities for $0.35 < k_T < 0.65$ GeV/c.

6. Conclusions

In this paper, preliminary results of STAR analysis of like-sign as well as unlike-sign kaon femtoscopy in Au+Au at $\sqrt{s_{NN}} = 200$ GeV have been presented. The source radii R_{out}, R_{side} and R_{long} have been obtained from three-dimensional like-sign kaon correlation function. The measured unlike-sign kaon correlation function exhibits strong centrality and k_T dependence in the region of the $\phi(1020)$ resonance. The experimental data have been compared to the theoretical FSI model prediction. The model reproduces overall the shape of correlation function, however for peripheral collisions the model underpredicts the strength of the correlations in the region of the resonance.

Acknowledgments

This work was supported by the grant of the Grant Agency of Czech Republic n.13 - 20841S and by the Grant Agency of the Czech Technical University in Prague, grant No. SGS16/238/OHK4/3T/14.

References

[1] Lednicky R *et al.* 2012 *Prog. Theor. Phys. Suppl.* **193** 335–339
[2] Goldhaber G, Goldhaber S, Lee W Y and Pais A 1960 *Phys. Rev.* **120** 300–312
[3] Lednicky R and Lyuboshits V L 1982 *Sov. J. Nucl. Phys.* **35** 770 [Yad. Fiz.35,1316(1981)]
[4] Adamczyk L *et al.* (STAR) 2015 *Phys. Rev.* **C92** 014904 (*Preprint* 1403.4972)
[5] Adamczyk L *et al.* (STAR) 2015 *Phys. Rev. Lett.* **114** 022301 (*Preprint* 1408.4360)
[6] Adamczyk L *et al.* (STAR) 2015 *Nature* **527** 345–348 (*Preprint* 1507.07158)
[7] Lisa M A and Pratt S 2008 (*Preprint* 0811.1352)
[8] Bowler M G 1991 *Phys. Lett.* **B270** 69–74
[9] Ackermann K H *et al.* (STAR) 1999 *Nucl. Phys.* **A661** 681–685
[10] Llope W J (STAR) 2012 *Nucl. Instrum. Meth.* **A661** S110–S113
[11] Adamczyk L *et al.* (STAR) 2013 *Phys. Rev.* **C88** 034906 (*Preprint* 1302.3168)

IOP Conf. Series: Journal of Physics: Conf. Series **832** (2017) 012060 doi:10.1088/1742-6596/832/1/012060

Production of $\Sigma(1385)^{\pm}$ and $\Xi(1530)^{0}$ in p–Pb collisions at $\sqrt{s_{\mathrm{NN}}} = 5.02$ TeV measured by ALICE at the LHC

Jihye Song (for the ALICE Collaboration)

Department of Physics, Pusan National University, Pusan, South Korea

E-mail: Jihye.Song@cern.ch

Abstract. In order to study the hot hadronic matter created in heavy-ion collisions, it is important to compare particle production in large systems to that in smaller systems, such as proton-proton (pp) and proton-lead (p–Pb) collisions. In particular, resonances with different lifetimes are good candidates to probe the interplay of particle re-scattering and regeneration in the hadronic phase. The yields of the strange and double-strange hyperon resonances $\Sigma(1385)^{\pm}$ and $\Xi(1530)^{0}$ are measured in the rapidity range $-0.5 < y_{\mathrm{CMS}} < 0$ in p–Pb collisions at $\sqrt{s_{\mathrm{NN}}} = 5.02$ TeV with the ALICE detector at the LHC. We report on the transverse momentum distributions and mean transverse momentum as a function of the charged-particle multiplicity. These results complement the information derived from the measurements of other resonances such as $\mathrm{K}^*(892)^0$ and $\phi(1020)$. The multiplicity dependence of the integrated yield ratios of excited hyperons to longer-lived particles is discussed and compared to model predictions from pQCD-inspired models such as PYTHIA8 as well as statistical hadronization models.

1. Introduction

The measurement of resonances in ultra-relativistic heavy-ion collisions allows one to study the properties of the hadronic medium. Hyperon resonances with lifetimes comparable to the duration of the hadronic phase are good candidates to probe the interplay of particle re-scattering and regeneration which could result in a modification of their measured yield.

Mesonic resonances such as $\rho(770)^0$, $\mathrm{K}^*(892)^0$ and $\phi(1020)$ have been measured in pp, p–Pb and Pb–Pb collisions at different energies. The results show that the production of particles with short lifetime (Table 1), e.g. $\rho(770)^0$ [1] and $\mathrm{K}^*(892)^0$ [2], exhibits progressively larger suppression when going from pp to p–Pb and Pb–Pb collisions. Longer-lived particles such as $\phi(1020)$ do not appear to be significantly modified across these systems, suggesting that the short-lived particles are affected more by re-scattering in the larger collision systems. Measurements of baryonic resonances, $\Sigma(1385)^{\pm}$ and $\Xi(1530)^{0}$, complement the information obtained from mesonic resonances. Since the lifetime of $\Xi(1530)^{0}$ (Table 1) is between that of $\mathrm{K}^*(892)^0$ and $\phi(1020)$ and lifetime of $\Sigma(1385)^{\pm}$ is comparable with $\mathrm{K}^*(892)^0$, these results are expected to help to understand the properties of the hadronic medium.

The ratio of the p_{T}-integrated yields of resonances to ground-state hadrons are used along with model predictions to estimate the properties (temperature [3, 4] and lifetime [5, 6]) of the fireball produced in heavy-ion collisions. In addition, particle production mechanisms are studied by comparing the measurements of resonances with different mass and quark content. This paper present measurements from the ALICE experiment that are related to these topics.

	$c\tau$ (fm)	quark contents	Decay modes used
$\rho(770)^0$	1.3	$(u\bar{u}+d\bar{d})/\sqrt{2}$	$\pi^+ + \pi^-$
$K^*(892)^0$	4.2	$d\bar{s}$	$K^+ + \pi^-$
$\Sigma(1385)^+$	5.5	uus	$\Lambda\pi^+ \to (p\pi^-)\pi^+$
$\Sigma(1385)^-$	5.0	dds	$\Lambda\pi^- \to (p\pi^-)\pi^-$
$\Xi(1530)^0$	21.7	uss	$\Xi^-\pi^+ \to (\Lambda\pi^-)\pi^+ \to ((p\pi^-)\pi^-)\pi^+$
$\phi(1020)^0$	44	$s\bar{s}$	$K^+ + K^-$

Table 1. Lifetimes of the measured resonances with their quark contents and decay modes [9].

2. Analysis and Results

The $\Sigma(1385)^{\pm}$ and $\Xi(1530)^0$ baryons are reconstructed by the daughters of their hadronic decays. The intermediate decay products, Λ and Ξ^-, are identified through selections based on their decay topology. The signals were reconstructed by invariant mass analysis of candidates for the decay products pairs from same events for various p_T and multiplicity intervals. Combinatorial backgrounds, constructed using event mixing technique, are subtracted from the invariant mass distributions. The invariant mass distributions are fitted using Breit-Wigner functions for $\Sigma(1385)^{\pm}$ and Voigtian functions for $\Xi(1530)^0$. First- or second-order polynomials are used to describe the residual background; see [14] for further details. The p_T spectra, which are corrected for efficiency, acceptance, and branching ratios, are fitted using Lévy-Tsallis functions to extract the total integrated yields and mean transverse momenta $\langle p_T \rangle$.

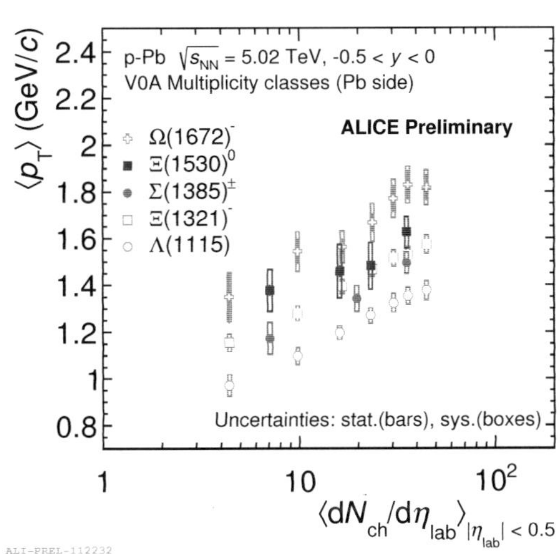

Figure 1. Mean transverse momenta $\langle p_T \rangle$ of Λ, Ξ^-, $\Sigma(1385)^{\pm}$, $\Xi(1530)^0$ and Ω^- in p–Pb collisions at $\sqrt{s_{NN}} = 5.02$ TeV as a function of the mean charged-particle multiplicity density $\langle dN_{ch}/d\eta_{lab} \rangle$, measured in the pseudo-rapidity range $|\eta_{lab}| < 0.5$.

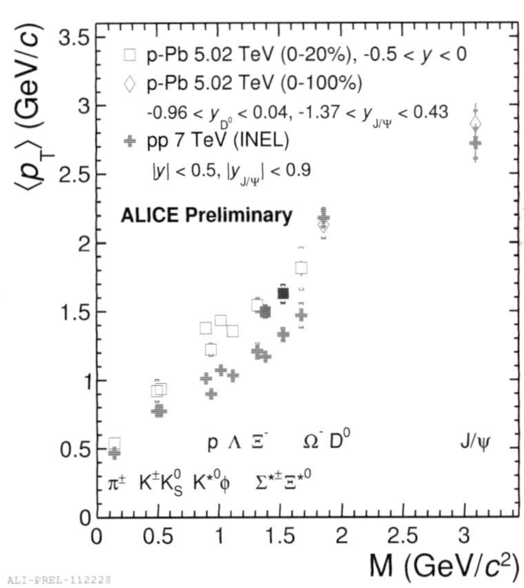

Figure 2. Mass dependence of the mean transverse momenta of identified particles for the $0-20\%$, 0-100% for D^0 and J/ψ, multiplicity class defined by scintillator detector, V0A, and with $-0.5 < y < 0$ in p–Pb collisions at $\sqrt{s_{NN}} = 5.02$ TeV [10, 11] ([12], [13]) and minimum-bias pp collisions at $\sqrt{s} = 7$ TeV [14],[15] with $|y| < 0.5$.

Figure 1 shows the mean transverse momentum as a function of the mean charged-particle multiplicity density at midrapidity $\langle dN_{ch}/d\eta_{lab}\rangle$. The results for $\Sigma(1385)^{\pm}$ and $\Xi(1530)^0$ are compared with those for other hyperons measured by the ALICE collaboration in p–Pb collisions at $\sqrt{s_{NN}} = 5.02$ TeV [10, 11]. For both $\Sigma(1385)^{\pm}$ and $\Xi(1530)^0$ the mean transverse momenta increase as the collision system size grows, similar to the other hyperons.

In all multiplicity classes, the $\langle p_T\rangle$ follows mass ordering: $\langle p_T\rangle_{\Lambda} < \langle p_T\rangle_{\Xi^-} \simeq \langle p_T\rangle_{\Sigma(1385)^{\pm}} < \langle p_T\rangle_{\Xi(1530)^0} < \langle p_T\rangle_{\Omega^-}$. This hierarchy of mass-ordering, including D^0 and J/ψ, is presented in Figure 2. A similar mass dependence is observed in p–Pb and pp collisions.

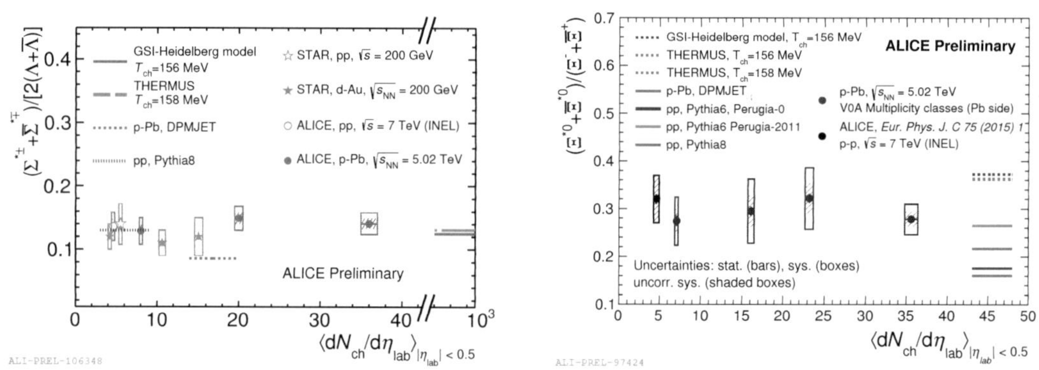

Figure 3. Ratio of $\Sigma(1385)^{\pm}$ to Λ (Left) and ratio of $\Xi(1530)^0$ to Ξ (Right) measured in pp [14, 16], d–Au [17] and p–Pb [10, 11] collisions, as a function of $\langle dN_{ch}/d\eta_{lab}\rangle$ at midrapidity. A few model predictions are also shown as lines at the appropriate abscissa.

The integrated yield ratios of excited to ground-state hyperons [10, 11, 14, 16] with the same strangeness content are shown in Figure 3 as a function of the mean charged particle multiplicity density for different collision systems and energies. The integrated yield ratios of both particles do not vary with multiplicity within the experimental uncertainties.

The results are compared with model predictions, PYTHIA8 [8] for pp and DPMJET [7] for p–Pb collisions. The $\Sigma^{*\pm}/\Lambda$ ratios (left panel in Figure 3) are consistent with the values predicted by PYTHIA8 in pp, whereas the DPMJET prediction for p–Pb collisions is lower than the experimental data. The measured $\Xi(1530)^0/\Xi$ ratios (right panel in Figure 3) appear higher than the corresponding event generator predictions for both systems.

The integrated yield ratios of excited hyperons to pions are shown in Figure 4 to study the evolution of relative strangeness production with the increase of the collision system size. Both $\Sigma(1385)^{\pm}/\pi^{\pm}$ (left panel in Figure 4) and $\Xi(1530)^0/\pi^{\pm}$ (right panel in Figure 4) show a gradual increase with multiplicity and approach (grand canonical) thermal model prediction [3, 4] in the highest multiplicity event class. The measured particle ratios indicate that the enhancement of hyperons, both resonances and ground-state, is due to their strangeness content and not their masses.

3. Summary

The baryonic resonances $\Sigma(1385)^{\pm}$ and $\Xi(1530)^0$ have been measured in p–Pb collisions at $\sqrt{s_{NN}} = 5.02$ TeV as a function of multiplicity. The $\langle p_T\rangle$ of those resonances increases with multiplicity and exhibits mass ordering. The $\Sigma(1385)^{\pm}/\Lambda$ and $\Xi(1530)^0/\Xi$ ratios do not vary with system in pp and p–Pb collisions. Integrated particle ratios of excited hyperon to pion yields ($\Sigma(1385)^{\pm}/\pi^{\pm}$ and $\Xi(1530)^0/\pi^{\pm}$) increase with the multiplicity density and strangeness content.

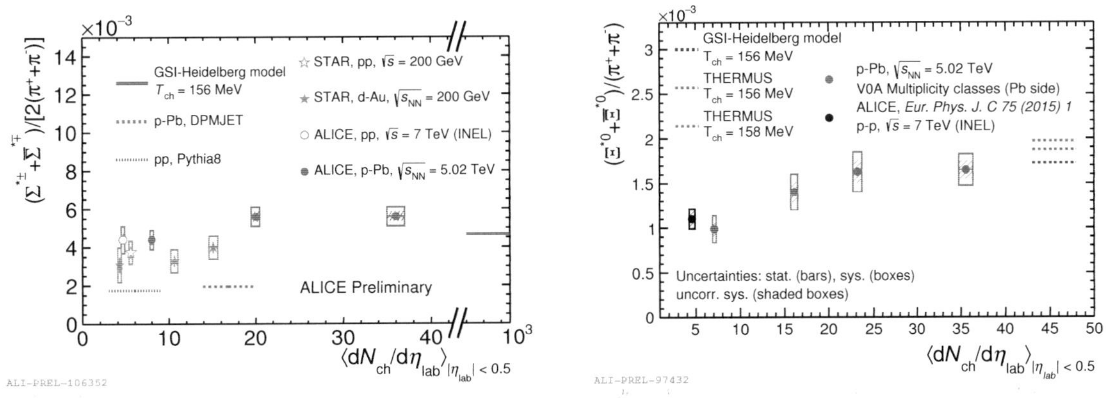

Figure 4. Ratio of $\Sigma(1385)^{\pm}$ to $\pi^{\pm}$ (Left) and ratio of $\Xi(1530)^0$ to $\pi^{\pm}$ (Right) measured in pp [19], d–Au [17] and p–Pb [10, 14] collisions, as a function of the average charged particle density ($\langle dN_{ch}/d\eta_{lab} \rangle$) measured at midrapidity. A few model predictions are also shown as lines at the appropriate abscissa.

The ALICE Collaboration is currently extending these measurements to higher energies, other resonances and other collision systems. Especially, the measurements of $\Sigma(1385)^{\pm}$, $\Xi(1530)^0$ and $\Lambda(1520)$ in Pb–Pb collisions are being carried out and these will provide important information to further constrain the properties of the hadronic phase.

References

[1] A. G. Knospe (for the ALICE Collaboration) 2016 *Proceedings of the Strangeness in Quark Matter 2016 conference, to appear in JPCS (Journal of Physics : Conference Series)* arXiv:1610.09529
[2] Phys. Rev. C 91 (2015) 024609 B. Abelev *et al.* (ALICE Collaboration) 2015 *Phys. Rev.* C **91** 024609.
[3] A. Andronic, P. Braun-Munzinger, and J. Stachel 2009 *Phys. Lett.* B **673** 1420145.
[4] S. Wheaton, J. Cleymans, and M. Hauer 2009 *Comp. Phys. Comm.* **180** 84-106.
[5] G. Torrieri and J. Rafelski 2002 *J. Phys.* G **28** 1911.
[6] J. Rafelski, J. Letessier, and G. Torrieri 2001 *Phys. Rev.* C **64** 054907.
[7] S. Roesler, R. Engel, and J. Ranft 2000 *ConferenceProceedings, MC2000, Lisbon, Portugal, October23 – 26* 1033. arXiv:hep-ph/0012252.
[8] Sjöstrand, T. and Mrenna, S. and Skands, P. 2008 *Comput.Phys.Commun.* **178** 852-867.
[9] K.A. Olive *et al.* (Particle Data Group) 2014 *Chin. Phys.* **C38** 090001.
[10] J. Adam *et al.* (ALICE Collaboration) 2014 *Phys. Lett.* B **728** 25-38.
[11] J. Adam *et al.* (ALICE Collaboration) 2016 *Phys. Lett.* B **758** 25-38.
[12] J. Adam *et al.* (ALICE Collaboration) 2016 *submitted to JHEP* arXiv:1605.07569.
[13] J. Adam *et al.* (ALICE Collaboration) 2015 *JHEP* **06** 055.
[14] B. Abelev *et al.* (ALICE Collaboration) 2015 *Eur. Phys. J.* C **75** 1.
[15] B. Abelev *et al.* (ALICE Collaboration) 2012 *Phys. Lett.* B **718** 295.
[16] B. Abelev *et al.* (ALICE Collaboration) 2012 *Phys. Lett.* B **712** 309.
[17] B. I. Abelev *et al.* (STAR Collaboration) 2008 *Phys. Rev.* C **78** 044906.
[18] T. Sjostrand, S. Mrenna, and P. Z. Skands 2006 *JHEP* **05** 026.
[19] J. Adam *et al.* (ALICE Collaboration) 2015 *Eur. Phys. J.* C **75** 226.

(Hyper-)nuclei and exotica production measured with ALICE at the LHC

Stefano Trogolo

Dipartimento di Fisica, Università degli Studi di Torino, Via P. Giuria 1, 10125, Torino, Italy

E-mail: stefano.trogolo@to.infn.it

Abstract. The high collision energies reached at the LHC open the possibility to study extensively the production of light (anti-)(hyper-)nuclei and exotic bound states in proton-proton (pp), proton-lead (p-Pb) and lead-lead (Pb-Pb) collisions.

ALICE has excellent particle identification (PID) capability which allows the detection of these rarely produced particles. PID is perfomed using several techniques, namely by exploiting the measurement of the specific ionization energy loss in the Time Projection Chamber (TPC) and the information of the Time-Of-Flight (TOF) detector.

The transverse momentum spectra and the yields ($\mathrm{d}N/\mathrm{d}y$) of light nuclei and anti-nuclei were obtained and results are discussed here, and the first measurement of deuteron elliptic flow is presented. Furthermore, the study of the production of (anti-)hypertriton and the status of the searches for exotic bound states, i.e. Λ-Λ and Λ-n, are discussed.

Results are compared to predictions from thermal and coalescence models in order to enquire into the production mechanisms of nuclei, hypernuclei and exotic bound states.

The study of light (anti-)(hyper-)nuclei and exotic bound states in proton-proton and heavy-ion collisions is important because it allows to test production models. On one side these measurements can be useful for the comparison with QCD-inspired models which predict the existence of multi-baryon states with strangeness content. On the other hand, an open point in heavy-ion physics is whether the production of nuclei and multi-baryon states, such as hypernuclei and exotica, can be described by thermal models [1] or by coalescence [2, 3] of hadrons. The thermal model predicts that hadrons are produced in the interaction region, which is assumed to be in thermal equilibrium, when it reaches the chemical freeze-out temperature (T_{chem}). The coalescence model assumes that (anti-)baryons which are close enough in phase-space at kinetic freeze-out, can form a multi-baryon state.

1. Nuclei

The ALICE experiment has measured the p_{T}-differential production spectra for (anti-)deuterons in pp collisions at $\sqrt{s} = 7$ TeV, p-Pb at $\sqrt{s_{\mathrm{NN}}} = 5.02$ TeV and Pb-Pb at $\sqrt{s_{\mathrm{NN}}} = 2.76$ TeV and for (anti-)^{3}He in p-Pb and Pb-Pb [4]. The deuteron and ^{3}He spectra exhibit a hardening with increasing centrality in Pb-Pb collisions, which can be explained by hydrodynamic models as an effect of the radial flow. A similar behaviour with increasing multiplicity is also observed for deuteron in p-Pb. Blast-Wave (p-Pb and Pb-Pb) and Levy-Tsallis (pp) fits were used to extrapolate the spectra to the unmeasured p_{T} region in order to compute the integrated yields.

The integrated yields of deuteron and proton were used to compute the d/p ratio as a function of charged particle multiplicity, as shown in Fig.1 (*left*). The rise of the ratio from pp to Pb-Pb

IOP Conf. Series: Journal of Physics: Conf. Series **832** (2017) 012061 doi:10.1088/1742-6596/832/1/012061

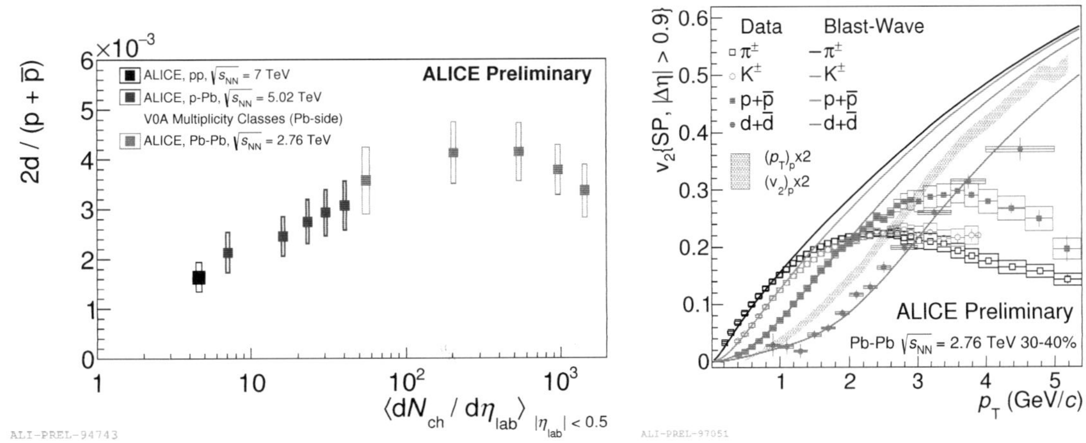

Figure 1: (*left*) d/p ratio as a function of charged particle multiplicity for pp, p-Pb and Pb-Pb collisions; (*right*) deuteron v_2 in Pb-Pb collisions for 30-40% centrality class with Blast-Wave (red line) and coalescence model (magenta band) predictions.

collisions is consistent with the coalescence model [5] which predicts an increase of deuteron production for higher nucleon densities. The flattening reached in Pb-Pb can be described in the coalescence picture assuming an increase of the size of the emitting source. This result is also consistent with thermal model expectations in which chemical freezeout is attained at a temperature of 156 MeV for all centralities.

The p_{T} spectra of deuteron and ^{3}He, were also used to compute the ratio between the production of anti-nuclei and nuclei [4] in the central rapidity region. These ratios are in agreement with unity within uncertainties and it is possible to conclude that, as expected, there is an equal production of matter and anti-matter in Pb-Pb collisions at the LHC energies.

Deuteron elliptic flow was measured in Pb-Pb collisions, as shown in Fig.1 (*right*), and follows the mass ordering up to $p_{\mathrm{T}} = 3$ GeV/c as expected from hydrodynamics models. A Blast-Wave parameterization in Fig.1 (*right*) (red line) obtained from lower mass species can describe deuteron v_2 while the coalescence model (magenta band) is not able to reproduce it.

ALICE also studied the mass dependence of production yields in Pb-Pb and p-Pb collisions, using the measured abundances of proton, deuteron, ^{3}He and ^{4}He. From the distribution of dN/dy as a function of nucleus mass m_A, the penalty factors for adding one nucleon were extracted: they are ~300 in Pb-Pb and ~600 in p-Pb collisions.

2. Hypernuclei

Hypernuclei are nuclei where a nucleon has been replaced with a hyperon and the hypertriton, a bound state of a proton, a neutron and a Λ, is the lightest known hypernucleus. Its mass is 2.991 ± 0.002 GeV/c^2 [7] and its lifetime is expected to be compatible with the free Λ lifetime.

The (anti-)hypertriton production yields were measured in Pb-Pb collisions exploiting the mesonic weak decay into two charged particles [6] ($^3_\Lambda\mathrm{H} \rightarrow {}^3\mathrm{He}+\pi^-$ and $^3_{\bar{\Lambda}}\overline{\mathrm{H}} \rightarrow {}^3\,\overline{\mathrm{He}}+\pi^+$). The decay products were identified via specific energy loss in the TPC and topological cuts were applied for the identification of secondary vertices. The signal extraction was performed by means of a fit to the invariant mass spectra with a function that includes signal and background contributions. The measured production yields dN/dy $\times$ B.R. have been compared to different models as a function of the branching ratio ($B.R.$), as shown in Fig.2 (*left*). The result is in

IOP Conf. Series: Journal of Physics: Conf. Series **832** (2017) 012061 doi:10.1088/1742-6596/832/1/012061

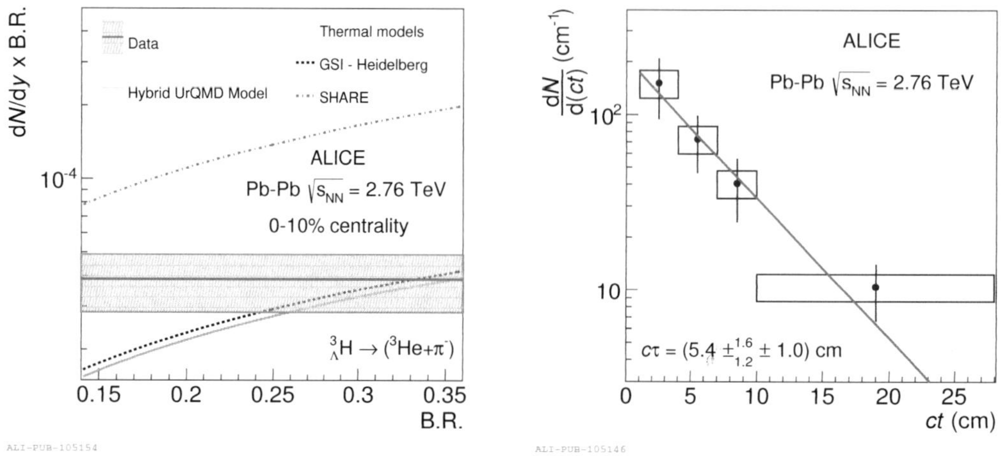

Figure 2: (*left*) Measured ${}^{3}_{\Lambda}$H d$N/$d$y\times$ B.R. compared to three different theoretical models; (*right*) exponential fit to the ct differential spectrum to evaluate the $c\tau$ of the hypertriton.

agreement with the equilibrium thermal model (T$_{chem}$ = 156 MeV) [1] and with the Hybrid UrQMD model, if a theoretical value of $B.R.$=25% [8] is assumed.

From the ct-differential production spectrum, shown in Fig. 2 (*right*), it has been possible to compute the lifetime of the hypertriton with an exponential fit (red line). The value reported by the ALICE experiment is $\tau = 181^{+54}_{-39}(stat.)\pm 33(syst.)$ ps and it is in agreement with the lifetime world average of 216^{+19}_{-16} ps [9]. However, being a loosely bound object, the theory predicts a value of the lifetime compatible with the free Λ lifetime. The fact that the ALICE value, as well as previous heavy-ion experiments results, is lower than the free Λ lifetime is still an open question. An improvement on this result might come from the study of the mesonic weak decay into three charged particles, ${}^{3}_{\Lambda}\mathrm{H} \rightarrow \mathrm{d}+\mathrm{p}+\pi^{-}$ and ${}^{3}_{\Lambda}\overline{\mathrm{H}} \rightarrow \overline{\mathrm{d}}+\overline{\mathrm{p}}+\pi^{+}$.

3. Exotica searches

The ALICE experiment is also focused on the search for exotica, such as hyperon-hyperon and hyperon-nucleon bound states. In particular the analysis looked for two particular exotic bound states in Pb-Pb collisions at $\sqrt{s_{NN}}$ = 2.76 TeV: the H-dibaryon and $\overline{\Lambda n}$ [10].

The H-dibaryon is a hypothetical $uuddss$ bound state ($\Lambda\Lambda$) first predicted by Jaffe [11] using a bag model approach. In this analysis the $\Lambda p\pi^{-}$ decay channel of the H-dibaryon was investigated, under the assumption of a weakly-bound state. Since no evidence of a signal for the $\Lambda\Lambda$ was found in the invariant mass distribution, an upper limit (99% C.L.) on d$N/$dy was obtained, assuming a B.R. = 64% [12] and the free Λ lifetime. The limit d$N/$d$y \sim 2.0\times10^{-4}$ is a factor 30 below the values predicted by thermal models (d$N/$d$y \approx 6\times10^{-3}$) [10].

The $\overline{\Lambda n}$ bound state was investigated in the decay channel $\overline{\mathrm{d}}\pi^{+}$, but no signal was found in the invariant mass distribution. Assuming a B.R. = 54% and the free Λ lifetime, this led to an upper limit (99% C.L.) on d$N/$d$y \sim 9.0\times10^{-4}$ that is a factor 45 below values predicted by thermal models (d$N/$d$y \approx 4\times10^{-2}$) [10]. The upper limits on $\Lambda\Lambda$ and $\overline{\Lambda n}$ d$N/$dy were compared to different thermal and coalescence (only for H-dibaryon) model predictions as a function of the branching ratio and they are below the predictions which assume the same B.R.

IOP Conf. Series: Journal of Physics: Conf. Series **832** (2017) 012061 doi:10.1088/1742-6596/832/1/012061

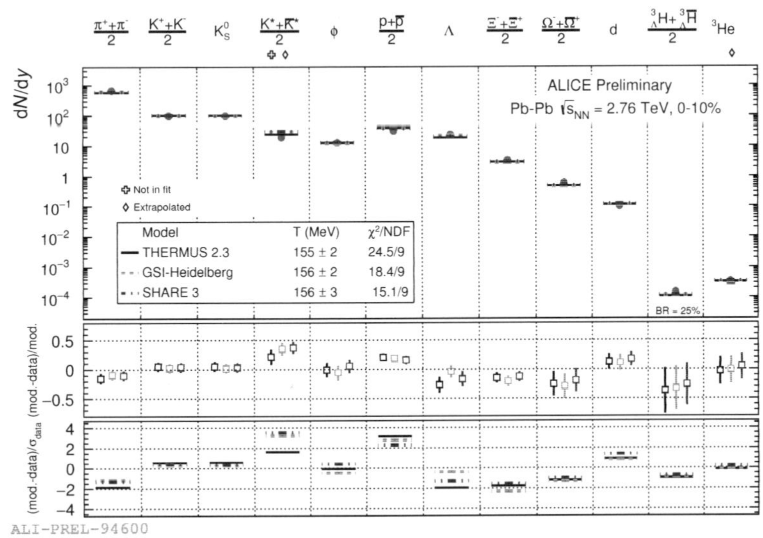

Figure 3: Grand-canonical fits to the particle yields measured in central Pb-Pb collisions at $\sqrt{s_{\rm NN}} = 2.76$ TeV.

4. Conclusion and perspective

ALICE's excellent performance in tracking and particle identification allows the detection of light (anti-)(hyper-)nuclei and the search for exotic bound states.

The results on nuclei and hypernuclei production in heavy-ion collisions can be rather well described by thermal and coalescence models. In this direction it is crucial to measure with better precision the d/p ratio. The production spectra show the effect of the radial flow due to the presence of a hydrodynamically expanding source, as already observed for lighter particles in Pb-Pb and p-Pb collisions. We also confirmed the dN/dy mass hierarchy predicted by the grand-canonical statistical model and the expected symmetry between matter and anti-matter production at the LHC. For the first time ALICE measured the elliptic flow in the nuclei sector, finding an agreement with hydrodynamic models. The thermal fits to the particle yields, shown in Fig.3, including nuclei and hypertriton, show that equilibrium thermal models with $T_{chem} =$ 156 MeV are able to describe almost all the particle abundances.

Two open points are still left: the lifetime of the hypertriton, which shows a deviation from theoretical expectations, and the existence of exotic bound states. We expect to give hints on these two open points using the increased statistics (factor ~10 in Pb-Pb) of the Run 2 data taking period of LHC and improving the existing measurements.

References

[1] Andronic A., Braun-Munzinger P., Stachel J. and Stocker H., *Phys. Lett.* B **697** (2011) 203-207
[2] Kapusta J.I., *Phys. Rev.* C **21** (1980) 1301-1310
[3] Steinheimer J. et al, *Phys. Lett.* B **714** (2012) 85-91
[4] ALICE Collaboration, *Phys. Rev.* C **93** (2016) 024917
[5] Kapusta J.I., Csernai L.P., *Phys. Rep.* **131** (1986) 223-318
[6] ALICE Collaboration, *Phys. Lett.* B **754** (2016) 360-372
[7] STAR Collaboration, *Science* **328** (2010) 58
[8] W. Gloeckle et al., *Nucl. Phys.* A **639** (1998) 297
[9] C.Rappold et al., *Phys. Lett.* B **728** (2014) 543-548
[10] ALICE Collaboration, *Phys. Lett.* B **752** (2016) 267-277
[11] Jaffe R.L., *Phys. Rev. Lett.* **38** (1977) 195, Erratum ibid. **38** (1977) 617
[12] J.Schaffner-Bielich et al., *Phys. Rev. Lett.* **84** (2000) 4305

IOP Conf. Series: Journal of Physics: Conf. Series **832** (2017) 012062 doi:10.1088/1742-6596/832/1/012062

Testing the hadronic spectrum in the strange sector

P. Parotto

Department of Physics, University of Houston, Houston, TX 77204, USA

E-mail: pparotto@uh.edu

Abstract. Heavier resonances are continually being added to the hadronic spectrum from the Particle Data Group that follow an exponentially increasing mass spectrum. However, it has been suggested that even further states predicted from Quark Models are needed in the hadronic spectrum in order to improve the agreement between the hadron resonance gas model predictions and lattice QCD data. We find that the inclusion of such states with extrapolated branching ratios slightly decreases the freezeout temperature. To eliminate ambiguities, we introduce a first principle method to extract the freeze-out temperature for charged kaons from experimental data, which yields a lower bound of $T_{\rm fo} \gtrsim$145 MeV for the highest collision energy at RHIC.

1. Introduction

The Hadron Resonance Gas model has been extremely successful in reproducing Lattice Quantum Chromodynamics results in the hadron gas phase [1, 2]. The underlying assumption is that it is possible to describe an interacting gas of hadrons in the ground states through a non-interacting gas of hadrons and resonances. The expression for thermodynamic quantities in the HRG model is then simply a sum over the contributions from each resonance. In this sense, the only "variable" in the model is the hadronic spectrum one feeds into it.

The original idea by Rolf Hagedorn was that a limiting temperature T_H for matter exists, called Hagedorn temperature, which cannot be exceeded, and therefore a system of hadrons responds to an arbitrary increase in the energy density by creating more and more resonances rather than by increasing the temperature [3]. Following these arguments, he showed that these resonances would need to populate a spectrum which is exponentially increasing in mass, when counting all states with the corresponding degeneracy:

$$N(m) = \sum_i d_i \, \Theta_i(m - m_i)$$

summing over the resonances with degeneracy d_i and mass m_i. It was later shown [4] that T_H can be interpreted as the limiting temperature for hadronic matter, if a phase transition occurs to a deconfined state of quarks and gluons. Presently, we know from Lattice QCD calculations that the deconfinement transition is a smooth crossover for vanishing baryonic chemical potential [5]. In this sense, the Hagedorn temperature can be analogous to the critical temperature of the deconfinement transition, and, therefore, it is reasonable to expect the effects related to the increased states population to appear and become relevant close to such transition.

The list of experimentally measured resonances is continuously updated by the Particle Data Group [6], and confirms the original idea from Hagedorn [9, 10]. Moreover, there exist Quark

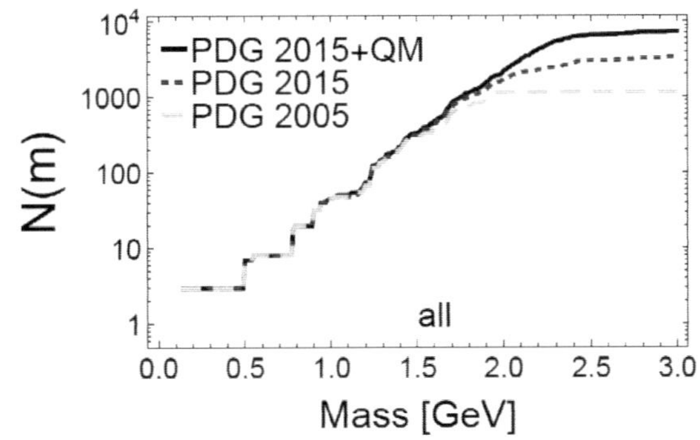

Figure 1. Exponentially increasing mass spectrum for different PDG lists, and with the inclusion of further Quark Model states

Models [11, 12] that predict the presence of further, not yet discovered, hadronic states. The latter states also show an exponential behavior in the mass spectrum, as shown in Fig.1. Including further states in an exponentially increasing mass spectrum can have far reaching effects in the field of heavy-ions. For instance, they play a role in the transport coefficients close to the critical temperature region [13, 14, 15, 16], influence elliptical flow [17, 18], can improve thermal fits [19], and affect the chemical equilibration time in the hadron gas phase [20, 21, 22, 23]. Thus, it is important to determine the precise number of resonances that exist otherwise a systematic bias affects numerous comparisons between theory and experiment.

2. Strange states and chemical freeze-out

The Hadron Resonance Gas model has been employed to extract information about chemical freeze-out in heavy ion collisions through comparison of particle yields [24], particle ratios [25] and ratios of fluctuations of conserved charges (susceptibilities) [26], with experimental data. In the past few years, such comparisons of yields and particle ratios with data from both RHIC and the LHC seemed to indicate the existence of different freeze-out temperatures for light and strange matter [27]. When comparing to the experiment, effects due to resonance decays, acceptance cuts and isospin randomization need to be taken into account [26], which can also be affected by the inclusion of more states in the hadronic population [8, 7].

For strangeness, the agreement of the Hadron Resonance Gas model with Lattice results is not always satisfactory, so that it was suggested [29] that the inclusion of Quark model states is necessary. It was also pointed out that this inclusion would reduce the difference in temperature between light and strange freeze-out. However, in [29] the effects of resonance decays were not taken into account. Thus, we used the known branching ratios from the PDG list [6] to extrapolate the branching ratios for the Quark Model states, carefully enforcing quantum numbers and mass conservation. By comparing net-charge and net-proton fluctuations (χ_1/χ_2) with STAR data [30, 31] from the Beam Energy Scan, it was possible to extract the values of T and μ_B at chemical freeze-out. The addition of Quark model states only has a small decrease of the freeze-out temperature, as seen in Fig.2.

To extract freeze-out parameters from first principles, one can use ratios of susceptibilities as discussed in [32, 33]. If Lattice QCD showed a separation between light and strange freeze-out [34], this would have an heavy impact for example on the hadronization schemes used in hydrodynamical simulations. In principle, one could compare Lattice QCD to experiment directly from susceptibilities of strangeness. However, lattice only has global contributions whereas the only strangeness-related fluctuations presently available from experiment are those

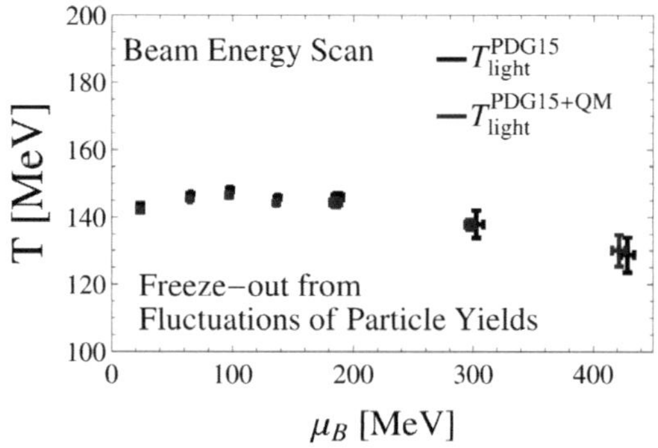

Figure 2. Freeze-out points from particle yields fluctuations comparison with HRG with and without Quark Models states

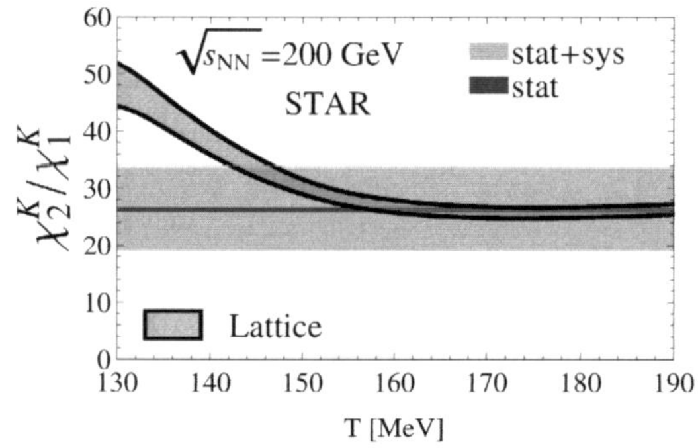

Figure 3. Comparison of charged kaons χ_2^K/χ_1^K from Lattice QCD (Wuppertal Budapest) to preliminary STAR data for $\sqrt{s_{NN}}$ =200 GeV.

of charged kaons. In order to isolate charged kaons on the lattice, partial pressures for charged strange mesons can be used [28], where ratios of susceptibilities are shown to be of the form

$$\frac{\chi_2^K}{\chi_1^K} = \frac{\cosh(\hat{\mu}_S + \hat{\mu}_Q)}{\sinh(\hat{\mu}_S + \hat{\mu}_Q)} , \tag{1}$$

where $\hat{\mu}_S = \mu_S/T$ and $\hat{\mu}_Q = \mu_Q/T$ are supplied by the Wuppertal Budapest collaboration. In Fig.3 $\frac{\chi_2^K}{\chi_1^K}$ is compared from Lattice QCD with preliminary STAR data. Due to large uncertainties in the experimental data, only a lower bound at $T_{fo}^K \gtrsim 145\,\text{MeV}$ can be extracted at the moment.

3. Conclusions

In this proceedings two different approaches for studying light and strange freeze-out temperatures are shown. Adding Quark model states with their decays to the hadronic spectrum, lowers the light freeze-out temperature by a small amount. Second, a new method for extracting strangeness freeze-out parameters from first principles using strange charged sucpetibilities is presented to resolve the tension between the light and strange sector. In both cases, smaller error bars are needed to better check the strange sector.

Acknowledgments

This material is based upon work supported by the National Science Foundation under grant no. PHY-1513864 and by the U.S. Department of Energy, Office of Science, Office of Nuclear Physics, within the framework of the Beam Energy Scan Theory (BEST) Topical Collaboration. This work contains lattice QCD data provided by the Wuppertal-Budapest Collaboration. An award of computer time was provided by the INCITE program. This research used resources of the Argonne Leadership Computing Facility, which is a DOE Office of Science User Facility supported under Contract DE-AC02-06CH11357. The work of J. G. and A. P. was supported by the DFG grant SFB/TR55. The authors gratefully acknowledge the Gauss Centre for Supercomputing (GCS) for providing computing time for a GCS Large-Scale Project on the GCS share of the supercomputer JUQUEEN [30] at Jülich Supercomputing Centre (JSC).

IOP Conf. Series: Journal of Physics: Conf. Series **832** (2017) 012062 doi:10.1088/1742-6596/832/1/012062

References

[1] C. Ratti *et al.* [Wuppertal-Budapest Collaboration], Nucl. Phys. A **855**, 253 (2011) doi:10.1016/j.nuclphysa.2011.02.052
[2] M. Bluhm, P. Alba, W. Alberico, A. Beraudo and C. Ratti, Nucl. Phys. A **929**, 157 (2014) doi:10.1016/j.nuclphysa.2014.06.013
[3] R. Hagedorn, Nuovo Cim. Suppl. **3**, 147 (1965).
[4] N. Cabibbo and G. Parisi, Phys. Lett. B **59**, 67 (1975). doi:10.1016/0370-2693(75)90158-6
[5] Y. Aoki, G. Endrodi, Z. Fodor, S. D. Katz and K. K. Szabo, Nature **443**, 675 (2006) doi:10.1038/nature05120
[6] K. A. Olive *et al.* [Particle Data Group Collaboration], Chin. Phys. C **38**, 090001 (2014). doi:10.1088/1674-1137/38/9/090001
[7] J. Noronha-Hostler and C. Greiner, Nucl. Phys. A **931**, 1108 (2014) doi:10.1016/j.nuclphysa.2014.08.101
[8] J. Noronha-Hostler and C. Greiner, arXiv:1405.7298 [nucl-th].
[9] W. Broniowski, W. Florkowski and L. Y. Glozman, Phys. Rev. D **70**, 117503 (2004) doi:10.1103/PhysRevD.70.117503
[10] P. M. Lo, M. Marczenko, K. Redlich and C. Sasaki, Phys. Rev. C **92**, no. 5, 055206 (2015) doi:10.1103/PhysRevC.92.055206
[11] S. Capstick and N. Isgur, Phys. Rev. D **34**, 2809 (1986) [AIP Conf. Proc. **132**, 267 (1985)]. doi:10.1103/PhysRevD.34.2809, 10.1063/1.35361
[12] S. Godfrey and N. Isgur, Phys. Rev. D **32**, 189 (1985). doi:10.1103/PhysRevD.32.189
[13] J. Noronha-Hostler, J. Noronha and C. Greiner, Phys. Rev. Lett. **103**, 172302 (2009) doi:10.1103/PhysRevLett.103.172302
[14] J. Noronha-Hostler, J. Noronha and C. Greiner, Phys. Rev. C **86**, 024913 (2012) doi:10.1103/PhysRevC.86.024913
[15] G. P. Kadam and H. Mishra, Nucl. Phys. A **934**, 133 (2014) doi:10.1016/j.nuclphysa.2014.12.004
[16] S. Pal, Phys. Lett. B **684**, 211 (2010) doi:10.1016/j.physletb.2010.01.017
[17] J. Noronha-Hostler, J. Noronha, G. S. Denicol, R. P. G. Andrade, F. Grassi and C. Greiner, Phys. Rev. C **89**, no. 5, 054904 (2014) doi:10.1103/PhysRevC.89.054904
[18] J. F. Paquet, C. Shen, G. S. Denicol, M. Luzum, B. Schenke, S. Jeon and C. Gale, Phys. Rev. C **93**, no. 4, 044906 (2016) doi:10.1103/PhysRevC.93.044906
[19] J. Noronha-Hostler, H. Ahmad, J. Noronha and C. Greiner, Phys. Rev. C **82**, 024913 (2010) doi:10.1103/PhysRevC.82.024913
[20] J. Noronha-Hostler, C. Greiner and I. A. Shovkovy, Phys. Rev. Lett. **100**, 252301 (2008) doi:10.1103/PhysRevLett.100.252301
[21] J. Noronha-Hostler, M. Beitel, C. Greiner and I. Shovkovy, Phys. Rev. C **81**, 054909 (2010) doi:10.1103/PhysRevC.81.054909
[22] M. Beitel, K. Gallmeister and C. Greiner, Phys. Rev. C **90**, no. 4, 045203 (2014) doi:10.1103/PhysRevC.90.045203
[23] M. Beitel, C. Greiner and H. Stoecker, Phys. Rev. C **94**, no. 2, 021902 (2016) doi:10.1103/PhysRevC.94.021902
[24] J. Stachel, A. Andronic, P. Braun-Munzinger and K. Redlich, J. Phys. Conf. Ser. **509**, 012019 (2014) doi:10.1088/1742-6596/509/1/012019
[25] A. Andronic, P. Braun-Munzinger and J. Stachel, Phys. Lett. B **673**, 142 (2009) Erratum: [Phys. Lett. B **678**, 516 (2009)] doi:10.1016/j.physletb.2009.02.014, 10.1016/j.physletb.2009.06.021
[26] P. Alba, W. Alberico, R. Bellwied, M. Bluhm, V. Mantovani Sarti, M. Nahrgang and C. Ratti, Phys. Lett. B **738**, 305 (2014) doi:10.1016/j.physletb.2014.09.052
[27] M. Floris, Nucl. Phys. A **931**, 103 (2014) doi:10.1016/j.nuclphysa.2014.09.002
[28] J. Noronha-Hostler, R. Bellwied, J. Gunther, P. Parotto, A. Pasztor, I. P. Vazquez and C. Ratti, arXiv:1607.02527 [hep-ph].
[29] A. Bazavov *et al.*, Phys. Rev. Lett. **113**, no. 7, 072001 (2014) doi:10.1103/PhysRevLett.113.072001
[30] X. Luo [STAR Collaboration], PoS CPOD **2014**, 019 (2015)
[31] L. Adamczyk *et al.* [STAR Collaboration], Phys. Rev. Lett. **113**, 092301 (2014) doi:10.1103/PhysRevLett.113.092301
[32] F. Karsch, Central Eur. J. Phys. **10**, 1234 (2012) doi:10.2478/s11534-012-0074-3
[33] S. Borsanyi, Z. Fodor, S. D. Katz, S. Krieg, C. Ratti and K. K. Szabo, Phys. Rev. Lett. **113**, 052301 (2014) doi:10.1103/PhysRevLett.113.052301
[34] R. Bellwied, S. Borsanyi, Z. Fodor, S. D. Katz and C. Ratti, Phys. Rev. Lett. **111**, 202302 (2013) doi:10.1103/PhysRevLett.111.202302